Hibernation and Torpor
in Mammals and Birds

PHYSIOLOGICAL ECOLOGY

A Series of Monographs, Texts, and Treatises

A complete list of titles in this series appears at the end of this volume.

Hibernation and Torpor in Mammals and Birds

CHARLES P. LYMAN
Department of Anatomy
Harvard Medical School
Boston, Massachusetts

JOHN S. WILLIS
Department of Physiology and Biophysics
University of Illinois
Urbana, Illinois

ANDRÉ MALAN
Laboratoire de Physiologie Respiratoire du C. N. R. S.
Strasbourg, France

LAWRENCE C. H. WANG
Department of Zoology
University of Alberta, Edmonton
Alberta, Canada

ACADEMIC PRESS, INC.
Harcourt Brace Jovanovich, Publishers
San Diego   New York   Berkeley   Boston
London   Sydney   Tokyo   Toronto

ACADEMIC PRESS, INC.
San Diego, Caliifornia 92101

United Kingdom Edition published by
ACADEMIC PRESS LIMITED
24-28 Oval Road, London NW1 7DX

Library of Congress Cataloging in Publication Data
Main entry under title:

Hibernation and torpor in mammals and birds.

   (Physiological ecology)
   Includes bibliographies and index.
   1. Hibernation. 2. Mammals--Physiology. 3. Birds--
Physiology. I. Lyman, Charles Peirson, Date.
QL755.H525   599.05'43   82-1779
ISBN 0-12-460420-X   AACR2

PRINTED IN THE UNITED STATES OF AMERICA
 89 90 91   9 8 7 6 5 4 3 2

Contents

Preface

Since Raphael Dubois' incomparable "Physiologie Comparée de la Marmotte" (1896), a few short books have been published with the subject of mammalian hibernation as their central theme. The latest of these are C. Kayser's "The Physiology of Natural Hibernation" published in 1961, and P. Raths' charming book for lay readers "Tiere im Winterschlaf" published in 1975. In addition, the proceedings of five international symposia have been reported, the latest being "Strategies in Cold: Natural Torpidity and Thermogenesis" edited by L. C. H. Wang and J. W. Hudson (1978). Although these proceedings are invaluable sources of information on sharply delineated subjects, they do not give an overall view on the phenomena of torpor and hibernation in birds and mammals. Kayser's book was the last to do this, and it was published twenty-one years ago. Thus, there is a gap which we hope to fill.

As research unfolds it becomes increasingly clear that hibernation in mammals, and probably in birds, is under precise physiological control. It is this control and its ramifications that are the subject of this book. In our approach we have been critical, and we have indicated what we observe to be weaknesses and strengths in experimental designs and in conclusions. We have not attempted to cover the whole field in equal depth. In fact, we have purposely omitted references to some articles because they appeared to us to add nothing to the general knowledge. A few overall concepts have been treated very briefly. For example, yearly or "circannian" cycles have been mentioned only incidentally because the onset and termination of hibernation have proved to be such variable end points that valid conclusions concerning controlling mechanisms are elusive. At the cellular level, there is no chapter on the effect of cold and hibernation on the function of membranes because a review paper on this subject is presently being published by one of the co-authors [J. S. Willis *et al., In* "Effects of Low Temperature on Biological Membranes" (G. J. Morris and A. Clarke, eds.). Academic Press, New York (in press)].

We have attempted to avoid two pitfalls which are prevalent in the literature on hibernation. First, although hibernation occurs in at least six Orders of mammals, most of the research has been carried out on a very few species, and conclusions are often made which may, in fact, be applicable only to the single species that was used as the experimental animal. For this reason the species is identified, and generalizations are made only if the accumulated evidence indicates that the observation applies to the majority of hibernators. Second, confusion has occurred concerning the total "season" of hibernation and the "bouts" of hibernation which occur during this season. As far as is known, hibernation in the wild occurs during a specific period or season which may be as long as nine months of the year. During this season, hibernation is not continuous, for the animal spontaneously arouses from time to time. Clearly, the period of arousal, the period of euthermia, and the period of re-entrance into hibernation are not the same physiologically as the periods of deep hibernation, yet many studies do not take this into consideration. We have treated with skepticism reports which assume that hibernation is continuous throughout the season.

In reading this book, the sophisticated experimentalist must be patient with the methods employed in the various investigations. In spite of all the research that has been carried out, the physiological factors which control the onset of hibernation are poorly understood, and it is not yet possible to cause, or to force, an animal to enter a state of natural hibernation. Thus, the investigator must await the capricious whim of the animal before the experiment can be started. By the same token, experiments which "prevent" hibernation must be viewed with caution, not because they are unimportant, but because so many factors can prevent hibernation that it is difficult to separate the primary from the secondary cause. Manipulations which appear to increase the incidence of hibernation are few, but they are the more revealing because they may give clues to the primary factor or factors which bring on the hibernating state.

During the preparation of this book I have constantly turned to Regina C. O'Brien for assistance, and she has never failed me. I am deeply grateful to her for everything she has done.

Charles P. Lyman

Hibernation and Torpor in Mammals and Birds

1

Why Bother to Hibernate?

The phenomenon of hibernation in mammals and birds has excited the interests of biologists for many years. One cannot fail to be impressed with the changes that take place during one cycle of hibernation. An animal that was active and alert on the previous day is now cold to the touch, moribund, and either motionless or capable only of slow, uncoordinated movements. When disturbed, the hibernator starts the orchestrated process of arousal, and in a few hours is once again in its original active state. There is no doubt that this is a type of hibernation, but it is so specialized that it is in a class by itself and must be dignified by a more precise rubric.

In its common usage the verb "to hibernate" means to pass the winter in a torpid or lethargic state. In this broad sense the study of hibernation may encompass reptiles, amphibians, fish, invertebrates, and even many plants. The subject of hibernation as it occurs in mammals and birds differs from hibernation in the other vertebrates, for it is inextricably bound up with the phenomenon of warm-bloodedness and the control of body temperature (T_b). Most mammals and birds maintain a high, steady body temperature throughout their adult life. However,

1

under rather specific circumstances, some mammals and birds experience a profound reduction in T_b and metabolic rate with T_b dropping to a few degrees above the freezing point of water and metabolic rate experiencing a concurrent decline. The T_b of other specific groups of mammals or birds may decline to a lesser extent, and, in either case, the animal is capable of rewarming using only self-generated heat.

Quite obviously, these are special physiological states, and before one enters into the subject of this volume, it is necessary to advance some clear-cut definitions, while admitting that one condition may shade imperceptibly into another. Some years ago, when the variability in temperature control in mammals had not yet been explored in detail, we suggested that the term "deep hibernation" be applied to the condition in which the animal remained at a T_b of about 5°C for a period of days or weeks. It was believed that this term was preferable to the phrase "true hibernation" which implied that any other type of hibernation was false (Lyman, 1948). As studies of hibernation continued, it became clear that many mammals underwent a sort of intermediate type of hibernation in which T_b declined markedly, but not usually below 15°C. The word "torpor" has often been used to refer to this condition, and it will be used here while recognizing that it covers a very broad field. Examples of torpor will be described in Chapter 2, from the long winter sleep of the bear to the brief period of chilling of the tiny mouse *Baiomys*.

The words "euthermia" and "euthermic" will be used to describe the warm-blooded or active state, whether the animal is asleep or awake. Euthermic means "promoting warmth" and in this sense is preferable to "normothermic," which suggests that this is the normal condition and hence torpor and deep hibernation are abnormal. The words "homeothermia" and homeothermic" are often used to describe this condition, but these imply a steady T_b, and animals in torpor or deep hibernation may maintain a steady, controlled T_b, even though it is much lower than during the euthermic state; therefore, these words are not sufficiently descriptive. The word "hypothermia" will be used to refer to a depressed T_b which does not occur under normal conditions and from which the animal cannot rewarm without heat from external sources. Finally, "poikilothermia" will be used to denote a condition in which the body temperature is dependent on environmental temperature.

These definitions all involve control of body temperature and this control is of paramount importance in mammals and birds for the maintenance of a steady internal environment. In the evolutionary development from fish to higher vertebrates, there is an uneven progression toward a warm and stable body temperature, yet the hibernators appear to fly in the face of this progression and to revert to their poikilothermic ancestors. The concept that hibernation is an atavistic condition is not generally accepted, but, in order to compare it to other states, it is necessary to give a brief background of the present knowledge of temperature control and warm-bloodedness in the various classes of living vertebrates.

Starting, then, with aquatic vertebrates, it is well known that most fish maintain a T_b which is slightly above that of the water in which they live, and the geographic distribution of many fish is limited by water temperature. Fish are essentially "water cooled," in that the temperature of the blood approaches that of the surrounding water as the blood is passed through the gills. Thus, even a violently exercised fish shows little or no rise in body temperature (Lyman, 1968). Exceptions to this general rule are found in the tunas and the poor-beagle and mako sharks. These fast-swimming fish are only remotely related, yet they have a similar countercurrent arrangement in the vascular supply of the powerful muscles which are used in swimming. Because of the countercurrents, heat is trapped in the muscles and their temperature may rise as much as 21°C above the surrounding water. Carey et al. (1971) were able to record the abdominal temperature of a bluefin tuna while the fish remained at an ambient temperature (T_a) of 5°C. At the start of the observation, T_b was 21°C and declined only 2°C over a period of 4 hr.

In further work it has been established that the bluefin tuna actually maintains a fairly steady T_b irrespective of the temperature of the water in which it is swimming, whereas the skipjack and yellowfin tuna tend to simply hold body temperature several degrees above that of the water (Carey and Lawson, 1973). Due to another countercurrent system, the brain and the eye of the bluefin are also kept warmer than the surrounding water. This appears to be true temperature regulation and is the more remarkable because fish cannot raised their T_b, as can terrestrial vertebrates, by increasing their metabolic rate. The higher the metabolic rate, the more oxygen is needed and hence the more blood is pumped through the gills and chilled to the temperature of the water. Thus, fish must regulate their temperature by controlling the efficiency of the vascular heat exchangers, but how this is accomplished is not known.

Tuna can tolerate water temperatures ranging from 30°C to 5°C and have been known to travel 4200 miles in 50 days. At least part of this mobility can be attributed to the lack of dependence on water temperature.

Of course, it is also true that the rates of physiological processes increase, within certain upper limits, with increasing temperatures. Carey et al. (1971) point out that the contraction–relaxation cycle of frog muscle triples in speed with a 10°C increase in temperature. Since there is no loss in contractile force, there is a threefold increase in power. Yellowfin tuna and wahoo are among the fastest of fish, attaining a speed of 70 km/hr (43 mph) for short bursts. The correlation between their speed and high body temperature is clear, but it must be realized that temperature is not the only factor involved in the speed of the contraction and relaxation cycle of muscle. The cycle is many times more rapid in a mouse than in a horse, yet their body temperature is virtually the same. Perhaps, as Linthicum and Carey (1972) suggest, the steadiness of body temperature is at least as important as the level at which T_b is maintained.

These "fish with warm bodies" (Carey, 1973) are the exception, for many fish, particularly marine fish, migrate each year and exhibit a preference for a certain range of water and body temperatures. This might be regarded as a type of behavioral temperature regulation, but migration often results in a better supply of food, so that body temperature cannot be regarded as the only factor involved in the movement of fish.

Whether fish travel far or remain in one place, they are surrounded by water and water gives them a relatively stable environment. In contrast, a terrestrial vertebrate is faced with an environment which can change very rapidly, and the seasonal changes can be extreme. For example, the extremes of hot and cold often vary as much as 50°C in a temperate climate and a high of 31°C (88°F) in summer and a low of minus 71°C (-96°F) in winter have been reported west of the Lena river in Russia. In order to cope with daily and seasonal changes, terrestrial vertebrates have developed means to hold their bodies at a steady state in spite of environmental change.

Amphibians are comparatively unsuccessful at controlling their internal environment, and none of them are known to seek to maintain their T_b at a high level. Much of their respiration occurs through the skin, which is permeable to gases and to water. Hence, exposure to warm environments means an increase in water loss and eventual desiccation (Pearson and Bradford, 1976). Actually, the evaporative water loss of an amphibian on land may reduce the T_b to a lower level than that of the environment. Amphibians are able to sense the level of their body temperature, seeking a cooler place or taking refuge in the water if T_b becomes too high. The Manitoba toad, *Bufo hemiophrys,* presents an example of an amphibian that can sense temperature change even at an extremely low body temperature. When winter approaches it burrows into the soft soil and enters into a torpid state. It apparently can detect the approach of frost, digging deeper as winter progresses, and reversing the process with the coming of spring. Presumably the T_b of this animal is very close to 0°C as it responds to the onset of cold (Tester and Breckenridge, 1964).

Unlike amphibians, reptiles can remain in a warm environment and allow their body temperature to rise without undergoing the risk of rapid desiccation. Since the pioneer work of Colbert *et al.* (1946) on temperature tolerance of alligators, it has become increasingly apparent that reptiles maintain a chosen or "eccritic" T_b during the active part of their diurnal cycle. This temperature is held within narrow limits and is higher than the T_b of the animal during its nocturnal, inactive period. The eccritic temperature is regulated by behavioral means such as sunning or seeking warm or cool retreats. Although this is a type of temperature regulation, still reptiles undergo a decline in body temperature when exposed to a cold environment, and will enter into a state of hypothermia from which they cannot arouse until there is an increase in ambient temperature. Their main protection against lethal freezing in winter depends on choosing a suitably protected spot before being overcome by hypothermia, and the animal that makes the wrong choice will probably pay with its life.

It has been shown that the python undergoes muscular contractions while brooding its eggs and that these contractions generate a measurable amount of heat, much as shivering does in mammals (Hutchinson *et al.*, 1966). This is a step toward the true warm-bloodedness that is unique to mammals and birds, for the heat is endothermic or generated within the animal.

True warm-bloodedness occurs only in mammals and birds and involves a combination of attributes. It is characteristic of birds and mammals that they can maintain a gradient between their body temperature and that of the environment. When T_a is low, endothermic heat is vital in maintaining this gradient. In this regard, the metabolic rate of most mammals is much higher than the rate of a reptile of the same size at the same T_b (Table 1-1). Birds have even higher metabolic rates than mammals and Bligh (1973) has proposed that the term *"tachymetabolism"* (fast metabolism) be substituted for *"warm-bloodedness."* This sensible suggestion emphasizes the great difference in the level of metabolism in mammals and birds compared with other vertebrates, and overcomes the difficulty of referring to a lizard basking in the hot sun as a "cold-blooded" animal.

It is also characteristic of mammals and birds that their normal body temperature is set at a very high level, and this level changes little during their adult life. It has been said that mammals live internally in the tropics, and this is even more true of birds. Monotremes, the most primitive of living mammals, maintain a T_b of not much more than 30°C (Table 1-1); marsupials average five or more degrees higher; 37°–38°C approximates the T_b of eutherian mammals with exceptions such as some of the edentates, whereas birds hold to three or four degrees

TABLE 1-1

Basal Metabolism of Mammals and Reptiles[a]

Metabolism	Reptiles	Mammals		
	Lizards	Monotremes	Marsupials	Placentals
Approximate T_b (°C)	30	30	35.5	38.0
Basal metabolic rate (kcal/kg$^{3/4}$ per day)	7.5	34	49	69
Corrected (38°C) BMR (kcal/kg$^{3/4}$ per day)	19.5	62	62	69

[a] After Dawson (1972).

above this figure. Because of the low T_b in monotremes, with progressively higher temperatures in marsupials and eutherian mammals, it has been argued that primitive mammals maintained relatively low T_b's. If this is the case, it is difficult to explain why birds, with their direct ancestry from the saurapsid reptiles, have even higher T_b's than eutherian mammals.

The reasons for the high body temperatures are still debated. It has been pointed out that the only way an animal can remain cooler than the environment is by evaporation of body water, and that the rate of evaporation is limited and is dependent on the atmospheric water vapor pressure. Because of this, it would be difficult if not impossible for a mammal to invade and live in the tropics if its T_b were set at a steady level much below the average environmental temperature. Thus, it is argued that a steady high body temperature can be maintained under almost any earthly conditions, whereas a T_b set at a lower level would inevitably rise in warmer environments (Burton and Edholm, 1955). Increases in T_b above "normal" can be very dangerous to mammals, for it is a peculiarity of this "normal" T_b that it is very close to lethality. Proteins start to denature at 45°C, so this would seem to be the upper limit, but fevers lower than this cause severe disruptions of bodily functions which can result in death. The balance of rates of enzyme activity, neural function, and all the intricate relationships that maintain a steady state appear to be set within a very narrow temperature range, and any perturbation of that range, particularly an increase in T_b, can lead to disastrous results.

In this regard, Crompton et al. (1978) have advanced an intriguing theory concerning the development of warm-bloodedness in mammals. They postulate that a temporally wide nocturnal ecological niche was exploited by primitive insectivores. These animals maintained a T_b of about 30°C and their metabolic rate was reptilian—perhaps one-fourth of the rate of a modern mammal. They also had to maintain their T_b within narrow limits, and thus were confined to a nocturnal existence, for they could not endure a higher T_b, and their small size rendered evaporative water loss impossible as a means of holding in check the increase in temperature which would result if the animal was active during the day. In order to test this concept, Crompton et al. trained several primitive species of mammals to run on a treadmill, and measured their oxygen consumption and body temperature at various treadmill speeds. From previous data they were able to compare the energetics of the experimental animals with that predicted for a lizard and a modern mammal of the same weight and T_b, and running at the same speed. In these experiments, they used a monotreme—the spiny anteater or echidna; three primitive insectivores—the European hedgehog and two species from Madagascar, *Tenrec ecaudatus* and *Setifer setosis;* and one species of marsupial—the American opossum. Although all five species are now nocturnal, the ancestors of the opossum and echidna were diurnal, whereas the insectivores used in the test are believed to have remained nocturnal throughout their evolutionary history. The echidna and opossum had high metabolic rates, comparable with those of modern mammals, whereas the rates of the three insectivores were at the low reptilian level. Thus, the experimental results support the concept that tachymetabolism was achieved in two steps, first by small nocturnal insectivores with low body temperatures and reptilian metabolic rates, and later by mammals that were able to escape from the nocturnal niche by increasing their basal metabolic rate and body temperature.

An important factor in the maintenance of a euthermic condition is the relationship between body size and metabolic rate. It is universally recognized that the basal metabolic rate of mammals varies more in proportion to body surface than to body weight. This would seem to make sense because the rate of heat flow from a body to an environment is roughly proportional to the surface area. If an animal's body was a perfect sphere, the exponent of body weight with which to relate metabolic rate would be 2/3 ($W^{2/3}$). In actual measurements of metabolic rates, it has been found that the correct exponent is nearer 3/4 ($W^{3/4}$). The evidence for this has been reviewed by Kleiber (1975) and need not be considered here, except to point out that effective surface of an animal is extremely difficult to determine. Rarely is the whole surface at the same temperature, and the lungs and respiratory passages play an important role in heat exchange which is hard to measure. Furthermore, the proportions of small and large animals are not the same, and this exerts an influence. Finally, it is difficult to measure metabolic rate in wild animals under basal conditions of rest and in the postabsorbtive state, and Crompton et al. (1978) maintain that the treadmill technique yields more accurate results when comparisons of metabolic rate are needed.

Whatever the precise exponent, the fact remains that a small mammal has proportionately a larger body surface compared to its body weight than does a large mammal, and the former's metabolic rate per unit weight of tissue is greater. Benedict's famous "mouse to elephant curve" is an early attempt to quantitate this relationship and shows, for example, that the metabolic rate of a mouse per unit weight of tissue is about twenty times that of a sheep (Kleiber, 1975). With small mammals particularly, the relationship of $W^{3/4}$ to metabolic rate must be taken as a general principle rather than a fixed rule. For example, Pearson (1947) has measured the metabolic rate of small mammals weighing 16–26 g and has found that each genus had a "metabolic personality" of its own and that this was strong enough to mask the weight–metabolic rate relationship. However, the long-tailed shrew (Sorex cinereus), which weighed less than 4 g and is one of the smallest mammals in the world, showed a higher metabolic rate per unit weight than any other mammal used in this study and served to emphasize the energy problems of a minute mammal with a proportionately large body surface. In a later paper, Pearson (1948) reported metabolic measurements on other small shrews and suggested that no active adult mammal could be smaller than 2.5 g because it would be unable to gather enough food to support its extremely high metabolic rate. Neal and Lustick's (1973) metabolic measurements of the American short-tailed shrew confirm Pearson's earlier work, but British shrews and the mole appear to have metabolic rates near the predicted value (Hawkins and Jewell, 1962).

The high metabolic rate and high surface-to-mass ratio impose a strain on small mammals and birds, particularly those which live in cold climates. This can be compensated for in part by insulation. Insulation can be a very important factor in maintaining a high T_b with the minimum of metabolic cost, and the observation that a husky dog can sleep at a T_a of $-30°C$ (Scholander *et al.*, 1950) without increasing its metabolic rate serves to emphasize the importance of this factor. The smaller animals cannot fully compensate for the heat loss through their proportionately large body surface by increasing their surface insulation. There must be a critical point when insulation becomes so bulky that its survival value as protection against cold is outweighed by the inevitable increase in clumsiness. The amount of insulation an animal can carry depends on its size and its way of life, but it is obvious that a wild animal the size of a rat could not survive with the insulation of a husky dog. Furthermore, as Poczopko (1971) points out, insulation is most effective when it covers a flat surface and the smaller the animal the less flat surface there is.

In actual measurements it has been found that small mammals acclimatize to the cold climate of winter by increasing their basal metabolic rate rather than their insulation, whereas the opposite is true of mammals the size of an arctic fox or larger. Birds also have little change in insulation as winter approaches, though this may be due to the importance of plumage for flight (Hart, 1964). Kleiber (1947) has dramatized the dependence of small homeotherms on high metabolic rate, rather than insulation. He has calculated that a 60-g mouse with the same metabolic rate per unit weight as a steer would need the equivalent of a steer's surface covering in a layer 20 cm thick to maintain a normal T_b at an environmental temperature of $3°C$.

Small mammals compensate to some degree for lack of their own insulation by using insulation provided by the environment. In the colder areas these animals live most of the winter beneath the snow, and burrows and other shelters give them added protection. With an increased metabolic rate to compensate for the low T_a, the small mammal can live through the winter provided it can obtain enough food to stoke its "internal furnace." Sufficient food at frequent intervals is absolutely essential, for a small animal can only store enough spare energy in the form of fat to carry it over a very short period. Winter and early spring are the times when food is most scarce and the demand for food is greatest. Many species of birds avoid this problem by migrating in the autumn to a warmer climate where food is more abundant. For an animal that can fly, migration is a possibility, but for a small terrestrial mammal it is out of the question, and it must stay in the unfavorable environment and find enough food to carry it through or die in the attempt.

Reduction of metabolic rate with a concurrent decrease in T_b would provide surcease for the constant need for food, and some mammals and a few birds have adopted this strategy to escape from an unfavorable environment. The physiological states that accompany these reductions are the subject of this volume, but it is necessary to mention here that most mammals have not adopted this strategy and it is interesting to speculate why this is the case. As was emphasized previously, most mammals and birds maintain their body temperature within very narrow limits throughout their adult, healthy life. The physiological mechanisms that are involved in this precise balancing of heat generated and heat lost have been the object of much study. Though these mechanisms are not yet completely understood, it is clear that the whole process is a very complicated one, and it is reasonable to assume that such a complex system would not have developed if it were not essential for the successful life of the animal. It is also clear that a steady T_b is necessary for the efficient function of the central nervous system. When the brain and body temperature are lowered only a few degrees centigrade, most mammals become ataxic and an increase in brain temperature above "normal" produces a similar result, often followed by death. When exposed to high T_a's, some mammals use extraordinary means, such as spreading saliva over the body, to keep T_b from rising. Present evidence emphasizes that temperature of the brain is the critical factor and various ways have been developed that serve to shield the central nervous system from wide changes in temperature. For example, it has been shown that the deep body temperature of antelopes can rise to as high as 44°C, whereas the brain, protected by a circulatory heat exchange mechanism, reaches only 40.5°C (Taylor and Lyman, 1972). On the low side of the temperature scale, mammals and birds may permit their extremities to chill, but T_b and brain temperature do not suffer a similar decline (Burton and Edholm, 1955; Chatfield et al., 1953). Unlike the situation with high temperatures, there appears to be no case in which the brain alone is maintained at a different, and in this case higher, temperature than the rest of the body, and, in fact, it is difficult to imagine how this could be brought about.

Though brain temperatures a few degrees above normal are fatal, there is a greater margin of safety with temperatures below the normal setting. Most mammals can tolerate hypothermia down to about 15°C T_b, though lack of muscular coordination can be noted when the T_b is only about 5°C below normal. This lack of coordination may be of vital importance to the survival of the animal, and the T_b of the normal, active mammal does not usually slip to this level. There are a few exceptions to this generalization, including the nocturnal insectivores with reptilian metabolic rates discussed earlier, and some of the edentates. The T_b of the three-toed sloth may be as low as 30°C under conditions which it encounters during its life in the tropics and can sink to 23°C when exposed to a T_a of 12°C (Wislocki, 1933). However, in its pendulous, slow motion life, coordination is not at a premium and even the sloth is said to maintain its T_b within narrower limits during pregnancy (Morrison, 1945).

As a general rule, mammals that experience low T_b's are inactive and secrete themselves in sheltered places where coordination is not necessary for survival. Some sacrifice is necessary to avoid the expense of a high metabolic rate and that sacrifice is the fine tuning of the tachymetabolic state. Whether T_b drops only a few degrees or whether it approaches 0°C, these animals are temporarily out of the mainstream of competition. However, hibernation is not simply a matter of hiding and abandoning the control of body temperature. The homeostatic heritage is too ingrained for that. Rather, as research on this subject continues, it becomes increasingly clear that hibernation is under precise physiological control. The mammalian or avian hibernator must temporarily sacrifice portions of its homeostatic armamentarium, but it must also maintain its physiological balance as it sinks into its lethargy, remains in this state for hours, days, or even weeks, and then warms itself spontaneously to regain its euthermic state. The goals in the study of hibernation are to discover how these profound changes can take place in mammals and birds in which almost all the evolutionary emphasis has been toward a maintenance of a high body temperature and metabolic rate. How can a mammal regulate its T_b with precision one day, and enter hibernation the next? How can cells and tissues from hibernators function at temperatures approaching 0°C when similar cells and tissues of nonhibernators lose their effectiveness at much higher temperatures?

To reemphasize the uniqueness of hibernation, it should be mentioned that mammals and birds may undergo an enforced hypothermia, brought about by an overpowering of the physiological defenses against extreme cold. The condition is really a pathological one, from which the animal will not recover unless warmed artificially. It undoubtedly occurs quite often in small mammals in the wild. Wetting of the insulating pelage, forced exposure, or lack of food may each result in hypothermia and ultimately in death. Hypothermia does not usually occur in healthy, large mammals because their defenses against cold are less precarious. Hypothermia is sometimes used for surgical operations in man, and various drugs, including curare to block muscular shivering, are often used in conjunction with the chilling process. Unfortunately, this type of hypothermia is sometimes referred to as "artificial hibernation" though, except for the low T_b, it bears little resemblance to the state of natural hibernation.

REFERENCES

Bligh, J. (1973). "Temperature Regulation in Mammals and Other Vertebrates." North-Holland Publ., Amsterdam.

Burton, A. C., and Edholm, O. G. (1955). "Man in a Cold Environment." Arnold, London.

Carey, F. G. (1973). Fishes with warm bodies. *Sci. Am.* **228,** 36–44.

Carey, F. G., and Lawson, K. D. (1973). Temperature regulation in free-swimming bluefin tuna. *Comp. Biochem. Physiol. A* **44A,** 375–392.

Carey, F. G., Teal, J. M., Kanwisher, J. W., Lawson, K. D., and Beckett, J. S. (1971). Warm-bodied fish. *Am. Zool.* **11,** 135-143.

Chatfield, P. O., Lyman, C. P., and Irving, L. (1953). Physiological adaptation to cold of peripheral nerve in the leg of the herring gull (*Larus argentatus*). *Am. J. Physiol.* **172,** 639-644.

Colbert, E. H., Cowles, R. B., and Bogert, C. M. (1946). Temperature tolerances in the American alligator and their bearing on the habits, evolution, and extinction of the dinosaurs. *Bull. Am. Mus. Nat. Hist.* **86,** 331-373.

Crompton, A. W., Taylor, C. R., and Jagger, J. A. (1978). Evolution of homeothermy in mammals. *Nature (London)* **272,** 333-336.

Dawson, T. J. (1972). Primitive mammals and patterns in the evolution of thermoregulation. *In* "Essays on Temperature Regulation" (J. Bligh and R. E. Moore, eds.), pp. 1-18. North-Holland Publ., Amsterdam.

Hart, J. S. (1964). Geography and season: mammals and birds. *Handb. Physiol., Sect. 4: Adapt. Environ.* pp. 295-321.

Hawkins, A. E., and Jewell, P. A. (1962). Food consumption and energy requirements of captive British shrews and the mole. *Proc. Zool. Soc. London* **138,** 137-155.

Hutchinson, V. H., Dowling, H. G., and Vinegar, A. (1966). Thermoregulation in a brooding female Indian python, *Python molurus bivittatus. Science* **151,** 694-696.

Kleiber, M. (1947). Body size and metabolic rate. *Physiol. Rev.* **27,** 511-541.

Kleiber, M. (1975). "The Fire of Life." Krieger, Huntington, New York.

Linthicum, D. S., and Carey, F. G. (1972). Regulation of brain and eye temperatures by the bluefin tuna. *Comp. Biochem. Physiol. A* **43A,** 425-433.

Lyman, C. P. (1948). The oxygen consumption and temperature regulation of hibernating hamsters. *J. Exp. Zool.* **109,** 55-78.

Lyman, C. P. (1968). Body temperature of exhausted salmon. *Copeia* No. 3, 631-633.

Morrison, P. R. (1945). Acquired homeothermism in the pregnant sloth. *J. Mammal.* **26,** 272-275.

Neal, C. M., and Lustick, S. I. (1973). Energetics and evaporative water loss in the short-tailed shrew, *Blarina brevicauda. Physiol. Zool.* **46,** 180-185.

Pearson, O. P. (1947). The rate of metabolism of some small mammals. *Ecology* **28,** 127-145.

Pearson, O. P. (1948). Metabolism of small mammals, with remarks on the lower limit of mammalian size. *Science* **108,** 44.

Pearson, O. P., and Bradford, D. F. (1976). Thermoregulation of lizards and toads at high altitudes in Peru. *Copeia* No. 1, 155-170.

Poczopko, P. (1971). Metabolic levels in adult homeotherms. *Acta Theriol.* **16,** 1-21.

Scholander, P. F., Hock, R., Walters, V., Johnson, F., and Irving, L. (1950). Heat regulation in some arctic and tropical mammals and birds. *Biol. Bull. (Woods Hole, Mass.)* **99,** 237-258.

Taylor, C. R., and Lyman, C. P. (1972). Heat storage in running antelopes: independence of brain and body temperatures. *Am. J. Physiol.* **222,** 114-117.

Tester, J. R., and Breckenridge, W. J. (1964). Winter behavior patterns of the Manitoba toad, *Bufo hemiophrys,* in northwestern Minnesota. *Ann. Acad. Sci. Fenn., Ser. A4* **71,** 421-431.

Wislocki, G. B. (1933). Location of the testes and body temperature in mammals. *Q. Rev. Biol.* **8,** 385-396.

2

Who Is Who among the Hibernators

The progression in mammals and birds from species in which T_b and metabolic rate drop slightly to animals that undergo the profound changes of the deep hibernators is not an orderly one and it seems unwise in our present state of knowledge to be too categorical concerning the various types of dormancy, torpidity, estivation, and hibernation that may occur in different species. This chapter will attempt to describe the variations from euthermia that have been observed among birds and mammals. To give some structure to this listing, the approach will be phylogenetic, starting with the birds and proceeding through the primitive to the more advanced mammals. It must be kept in mind that the total picture is far from complete. Geographic and economic stringencies usually play an important part in the choice of species for research and the result is an uneven overall view. Thus, the number of research facilities in Canada and the United States and the variety of environments and species in the northern hemisphere of the New World have resulted in a relatively complete study of the various forms of hibernation and torpor found throughout this area. In contrast, our knowledge of the bioenergetics and occurrence of hibernation is lacking for many groups of animals that occur in large areas on the earth. For example, there is virtually no information on the occurrence of hibernation in the hystricomorph rodents or the edentates in South America, though some live in environments in which hibernation might be advantageous or expected (Pearson, 1951).

The inaccessability of some animals is often compounded by the problems encountered under the artificial conditions of the laboratory. It is difficult to maintain some species in healthy condition in captivity, either because they are highly excitable or because an adequate diet has not been developed for them. A normal, healthy physical state is obviously essential in any study concerned with energetics and temperature regulation, and even apparently healthy animals in the laboratory may differ from animals in the field in their tendency to hibernate, as Herreid (1963) has shown for bats.

By process of elimination, certain species have been chosen for physiological experimentation because they have proved to be more compliant to laboratory conditions. However, these species may be only remotely related phylogenetically and have little in common but the fact that they hibernate. After an exhaustive series of experiments on a single species, it is tempting to generalize and assume that the observed reactions occur in every animal that hibernates. This sort of assumption occurs repeatedly and confuses rather than clarifies, particularly if the observation or conclusion is an appealing one, so that it becomes ingrained in the literature. There are many characteristics and reactions that are common to all hibernators, and well-founded generalizations must form the basis of knowledge of the subject, but many species have specializations of their own and the limitations of our knowledge must be kept in mind.

BIRDS

The birds are the most glaring example of difficult experimental subjects and, probably because of this, our knowledge of torpor or hibernation in birds has lagged in comparison to our information on mammals. Samuel Johnson, after an argument about whether scorpions commit suicide, is quoted by Boswell (1927) as saying, ''Swallows certainly sleep all the winter. A number of them conglubulate together, by flying round and round, and then all in a heap throw themselves under water and lye in the bed of a river.'' To bring this closer to home, McAtee (1947) quotes one Samuel Williams who, in about 1760, observed a swift that had been dug out of 2 ft of mud on the bank of the Charles river in Cambridge, Massachusetts. The bird was torpid, but revived in half an hour. The location was said to be covered with water to a depth of 4–5 ft at high tide. In spite of these observations, torpor in birds was not generally accepted as a fact until the

1950s. Dawson and Hudson (1970) have thoroughly reviewed the subject and present a list of various species in which torpor is known to occur, and Calder and King (1974) made the list current.

Because knowledge of the physiology of torpor in birds is so limited, the following chapters will deal almost exclusively with mammals. For this reason, observations on the various degrees of torpor in birds are given emphasis here.

Torpor has been observed principally in the orders Apodiformes (hummingbirds and swifts) and Caprimulgiformes (nightjars, nighthawks, goatsuckers, poorwills, and the whippoorwill). In all the hummingbirds that have been studied to date, the torpor takes place during the period in the 24-hr day when the animal does not actively search for food. In many species the onset of torpor is associated with lack of food, but it can also occur in well-nourished birds. Torpor usually lasts for less than 12 hr and we find no reports of healthy hummingbirds remaining in deep hibernation for days at a time. At the end of a torpid period the bird rewarms itself using endogenously generated heat.

Field studies of the broad-tailed hummingbird (*Selasphorus platycercus*) show that the temperature of incubating females and their eggs may drop to about 6.5°C during the night when food has been in short supply and rise again in the early morning before T_a rises (Calder and Booser, 1973). In contrast, the Calliope hummingbird (*Stellula calliope*), which is the smallest bird in North America north of Mexico, maintains its T_b while incubating its eggs, but may undergo torpor at other times of the year (Calder, 1971). The Andean hillstar hummingbird (*Oreotrochilus estella*) has been studied in its native habitat in Peru at altitudes of 3800–4300 m (Carpenter, 1974). Diurnal torpor occurs more frequently and lasts longer in the winter than in the summer in this species and it is believed that this is governed by an innate annual cycle which has developed in response to the harsh conditions of winter.

Some hummingbirds that experience torpor have little control of their T_b during the torporous period. If T_a drops, T_b follows until a temperature is reached which is either fatal or from which the bird cannot rewarm unless T_a increases. Other species possess the ability to regulate their T_b during torpor. Hainsworth and Wolf (1970) showed that the oxygen consumption of a West Indian hummingbird (*Eulampis jugularis*) decreased linearly with a decrease of T_a from 30° to 18°C, but below T_a of 18°C the relationship was reversed and the oxygen consumption increased linearly with the decrease in T_a. Body temperature declined with oxygen consumption to 18°C, and then actually increased slightly as T_a dropped to 5°C. They noted a similar response in two species of hummingbirds that live in the high mountains of Costa Rica where the temperatures are much lower, but with these birds the oxygen consumption of the torpid birds increased when T_a reached 10°–12°C. Carpenter (1974) found evidence of the same sort of regulation in her field studies of the Andean hillstar hummingbird, for below a T_a of 7°C, periods of apnea ceased and T_b remained about

7°C, though T_a was as low as 3°C. There appears to be a correlation between the average minimum environmental temperature that the species encounters and the temperature at which it starts to increase its oxygen consumption while in torpor, and it has been reasonably suggested that the temperature at which regulation occurs has been set as a metabolic adaptation to different climatic origins (Wolf and Hainsworth, 1972). If this is true, it is one of the few cases where logical orderliness has been shown to occur among a group of hibernators.

It has been tacitly assumed that the occurrence of daily torpor in hummingbirds is simply a means of conserving energy, but Hainsworth *et al.* (1977) present evidence to show that torpor occurs in the black-chinned hummingbird, *Archilochus alexandri,* and in Rivoli's hummingbird, *Eugenes fulgens,* only when a minimum threshold of daily energy reserves is reached. This implies that torpor involves some disadvantage so that the hummingbirds enter into this state only during energy emergencies.

It is clear that there are differences in the way various species of hummingbirds react to environmental stress. Beuchat *et al.* (1979) have shown that the Anna hummingbird, *Calypte anna,* and the rufous hummingbird, *Selasphorus rufus,* undergo a daily gain in body mass, which is then lost during the night. When exposed to cold, *C. anna* increased its gain of body mass by reducing its energy expended in flight and by increasing its intake of food. This compensated for an increase in loss of mass during the night. Under the same conditions, the rufous hummingbird decreased its overnight mass loss by spending longer periods of the night in torpor. A single Anna hummingbird, when denied an unlimited food supply, also resorted to torpor as a metabolic saving. This again brings up the concept that some species of hummingbirds avoid torpor until it is forced upon them.

There are no hummingbirds in Africa and their niche is filled by the African sun birds. In the montane regions these birds are exposed to warm days and cold nights. Cheke (1971) reports that T_b drops at night and begins to rise again, often while T_a is still decreasing, as daylight approaches. The lowest recorded cloacal temperature was about 24°C in *Nectarina mediocris,* which weighs about 7.5 g.

Among the Caprimulgiformes, the poorwill (*Phalaenoptilus nuttalli*) has been thoroughly studied. Jaeger (1949) repeatedly found a torpid poorwill in its roosting place in a hole in a rock, though there was no evidence that torpor was continuous. Captive torpid birds in the laboratory aroused spontaneously approximately every 4 days, and one bird was able to warm itself from a T_b of 6°C at a T_a of about 1°C (Lignon, 1970). In the laboratory at least, lack of food usually results in torpor, but it has not been established that this is a prerequisite in the wild, and torpor is not seasonal but may occur at any time of year. The ecological significance of this type of torpor has not been clearly established. For example, specimens of the common nighthawk (*Chordeiles minor*) only became torpid after a loss of about 30% of their body weight and only one out of four was able to arouse in a cold environment. Lasiewski and Dawson (1964) conclude that torpor rarely occurs under natural conditions in this highly migratory species. On the other hand, poorwills return to their breeding grounds in Idaho in spring when it is apt to be cold and wet, with little or no insects for food. At this time, torpor may serve a useful purpose in conserving energy. Their migration in the fall takes place when insects are still abundant and the weather is warm. The locations of torpid poorwills found in the field are confined to the southwestern portion of the United States where the cold periods are short and flying insects are available sporadically throughout the winter. Presumably, resident birds resort to torpor during cold periods when insects are in short supply (Lignon, 1970).

Calder and King (1974) emphasize that the variations in the depth and duration of torpor can be correlated with the size of the birds. In the minute hummingbirds, the metabolic rate per gram of body weight is very high, cooling and warming are rapid, and, therefore, the short periods of torpor result in a substantial metabolic saving. In the caprimulgids, body mass is greater, cooling and rewarming are slower, and duration of torpor is longer. Birds larger than this are limited to relatively small fluctuations in T_b. The data as they are known to date fit well with this concept when considering birds, but, as is described in Chapter 6 the length of a bout of hibernation in mammals is not correlated with size, so that the relationship in birds may be in part fortuitous in spite of the limitations that increase in size imposes.

The regulation of T_b described by Hainsworth and Wolf (1970) suggests that these hummingbirds are protecting themselves against a decline in T_b which might be fatal. Dawson and Hudson (1970) list hummingbirds and Caprimulgiformes that can survive low T_b's from which they are incapable of rewarming without artificial heat. In contrast to the relatively voluminous information on mammals, there appears to be little information on tolerance to hypothermia in birds. It has been shown that peripheral nerves of hens and herring gulls cease to conduct *in vivo* at about 10°C (Chatfield *et al.*, 1953), so that the lethal temperature for these two species must be above that which is tolerated by some hummingbirds and Caprimulgiformes, but it is not known whether the birds which enter torpor are the only ones which are able to tolerate low temperatures. Baldwin and Kendeigh (1932) tested the effect of hypothermia on various species of birds and concluded that they could survive T_b's of 32°C. An eastern house wren recovered from a T_b of 21.7°C, but this was close to lethality. There is a mixed bag of other species of birds that undergo various depths of torpor. Bartholomew *et al.* (1957) reported that the white-throated swift (*Aeronautes saxatalis*) chilled to a T_b of about 20°C and was able to arouse from that temperature at a T_a of 4°–5°C using endothermic heat. However, birds died at T_b's below 20°C. The speckled mouse bird, a tropical bird of the monogeneric order Coliiformes, tolerated about the same T_b and aroused spontaneously in the morning (Bartholomew and Trost, 1970). In the Inca dove (*Scardapella inca*), which lives in a warm climate, T_b dropped to about 30°C at night in birds that had lost 15% of their body weight. The birds were able to warm themselves from this temperature at a T_a of 20°C. Two birds survived T_b's of 25.5°C, but only recovered if T_a was raised (MacMillen and Trost, 1967). The smooth-billed ani (*Crotophaga ani*), which is a tropical bird of the cuckoo family, may achieve a T_b of 32.6°C without obvious signs of torpidity (Warren, 1960).

Steen (1958) measured T_b's of six species of small birds that wintered around Oslo in Norway: the titmouse, green finch, brambling, house sparrow, tree sparrow, and the red poll. He measured metabolic rate at T_a's of −25° to +25°C and found that the birds increased their insulation during the night hours by tucking their heads under their wings and fluffing their feathers. This resulted in a lower metabolic rate than was found in the inactive birds during the daylight. Furthermore, body temperature was reported to drop as low as 30°C during the night, with further metabolic saving, but the birds warmed again spontaneously at dawn. Chaplin (1974) found that T_b of black-capped chickadees dropped to 30°C at night when birds were exposed to low T_a's. Both Steen and Chaplin conclude that moderate declines of T_b at night may occur in many small birds which overwinter in cold climates.

MAMMALS

Prototheria

Information concerning hibernation in the primitive Prototheria is confined to the echidna or spiny anteater (*Tachyglossus aculeatus*). Recent detailed studies of temperature regulation in the duck-billed platypus (*Ornithorynchus anatinus*) make no mention of hibernation in this species, and it is a reasonable assumption that it does not occur (Grant and Dawson, 1978a,b; Griffiths, 1978). Griffiths (1978) reports that T. J. Dawson has measured the standard metabolic rate of two *Zaglossus bruijnii* and found it much lower than in *Tachyglossus aculeatus,* but he does not mention hibernation.

When *Tachyglossus* was exposed to a T_a of 25°C, it maintained a T_b of 32.2°C with little variation. At a T_a of 5°C, T_b averaged 2°C cooler in the morning than in the evening, with the lowest morning T_b registering 12.4°C. Prolonged periods of torpor did not occur until the echidnas were denied food, but then T_b's eventually declined to about 6°C, with the smaller animals entering hibernation before the larger ones. The echidnas were able to arouse from a T_a of 5°C if physically disturbed, but the time involved for a complete arousal was about 20 hr, which is much longer than arousal times for eutherian mammals of comparable size (Augee and Ealey, 1968). The increase in oxygen consumption paralleled the increase in T_b and the overshoot in oxygen consumption, which is typical of rodents arousing from hibernation, did not occur. [According to Griffiths (1978), the oxygen consumptions depicted in the figures of Augee's 1968 paper are too high by a factor of 100.] Augee *et al.* (1970) later reported that an echidna aroused from a T_a of 10°C in about 9 hr, again much slower than a comparable arousal in a rodent. Echidnas weighing over 4 kg would not enter torpor. They also found that echidnas periodically entered torpor during the winter months at an average T_a of 10°C and spontaneously aroused from time to time. The longest continuous torpor was 9.5 days, the shortest was 5 days, and the euthermic periods between bouts of torpor varied from 30 hr to 11 days. The T_b of a free ranging echidna with an implanted temperature transmitter remained between 10° and 15°C for about 5 days and then returned to the active level, indicating that hibernation may be a natural and perhaps seasonal occurrence. Remarkably, the arousing echidna is able to make coordinated digging movements at a T_b of 10.5°C (Augee and Ealey, 1968).

Thus, the echidna appears to fulfill most of the criteria of a deep hibernator, but the slow arousal from torpor, which has been confirmed by Allison and Van Twyver (1972), is not typical of the classical model. According to Griffiths (1978), Augee has theorized that the initial thermogenesis of arousal differs from that found in other hibernators, possibly due to the absence of brown fat. However, this theory has not been tested.

Metatheria

Information on torpidity and hibernation in the marsupials is also scanty. Using a single specimen of South American murine opossum (probably *Marmosa microtarsus*), Morrison and McNab (1962) reported periods of dormancy lasting about 6 hr. Body temperature dropped to as low as 16°C, accompanied by a reduced metabolic rate, and the animal was able to rewarm from this temperature using only self-generated heat. Godfrey (1968) found that two small species of insectivorous marsupial mice (Dasyuridae) from Australia reacted in a similar manner. If food was withheld from *Sminthopsis crassicaudata* (wt. 12–22 g) or *S. larapinta froggatti* (wt. 14–30 g), the animals underwent periods of dormancy lasting as long as 9–12 hr. Temperatures were measured by thermocouples in the nests, which recorded as low as 17°C, and the animals were able to rewarm from these temperatures.

Antechinus stuartii, another dasyurid of about the same size, exhibited a circadian fluctuation in T_b of about 4°C and entered into torpor if denied food in the winter. However, it was unable to rewarm itself from a T_b of 18°–21°C (Wallis, 1976). Dawson and Wolfers (1978) studied three species of the genus *Planigale* in order to compare their metabolic rate with eutherian shrews. As with other marsupials, their metabolic rate was about 30% lower than that reported for shrews of comparable size. All three of these dasyurids entered a torporous state, but they were not exposed to T_a's below 14°C, so their ability to recover from the

near-zero temperatures of deep hibernation is unknown. The resting metabolic rate of *Planigale maculata* is more than 20% lower than is predicted for marsupials and the oxygen consumption is unusually labile. It undergoes short periods of torpor even when adequately fed, but the duration of these periods under natural conditions and the lowest tolerated T_b have not been reported (Morton and Lee, 1978).

To date, deep hibernation has only been reported in two species of Tasmanian pigmy opossums or dormouse phalangers, *Cercaertus nanus* and *Eudromicia lepida*. They have been found curled up in hollow logs in a dormant state in the field, and when kept in captivity, they underwent periods of hibernation as long as 6 days in *E. lepida* and 12 days in *C. nanus*. There appears to be no particular season for hibernation with these animals, though they were more active from September to March (Hickman and Hickman, 1960). The mountain pigmy opposum, which was only discovered about 15 years ago, is reported not to hibernate under conditions that produced hibernation in *Cercaertus nanus*. However, one fat female entered a state of torpor which lasted a maximum of 7 days and appears to be a deep hibernator though no T_b's are given (Dimple and Calaby, 1972).

Bartholomew and Hudson (1962) were able to make physiological measurements on *Cercaertus nanus* in the laboratory and found that their torpor differed in some ways from typical deep hibernation. Oxygen consumption per unit weight was higher than in hibernation in rodents, and T_b always remained $2°-3°C$ above T_a in steady-state conditions. The slowest heart rate was 28 beats per minute at T_a's of $3.2°-4.6°C$, and rates of 60–80 beats per minute were often recorded. Unlike the typical hibernator, no skipped beats occurred in the heart rate, though respiration had typical periods of apnea. During arousal, the rectal temperature did not lag behind temperatures of the anterior part of the body, and this may indicate that the circulation is not under the precise control exhibited by the typical hibernator. However, the measurements on arousal were made at T_a's of $22°-26°C$, and the differential between anterior and posterior temperatures is not as great under these conditions as it is at lower T_a's in any deep hibernator. The dramatic rise and overshoot of oxygen consumption seen in the typical hibernator did not occur in *C. nanus*.

Here, then, is a marsupial which is virtually a "deep" hibernator. It lacks some of the refinements of the "perfect" model, but it has incorporated the basic principles. The observation that this species can hibernate continuously for 12 days shows that this type of hibernation can be successful.

Insectivora

The order Insectivora contains the most primitive of eutherian mammals, and one would hope that the development of hibernation in the Eutheria could be traced in this group. Unfortunately, every family of living insectivores, in spite of their primitive characteristics, is highly specialized in some particular way, so that a sequence of development can not be clearly established.

The tenrecs (family Tenrecidae) are probably the most primitive of the insectivores. They have been confined to the island of Madagascar for more than 20 million years and have undergone adaptive radiation there. Specimens for laboratory study are difficult to obtain, but the limited data indicate that many species maintain a low T_b and regulate this T_b imprecisely. For example, an active tenrec (*Tenrec ecaudatus*) maintained an average T_b of 28.4° with a range of 24.1°–34.8°C at a T_a of 12.0°–25.4°C (Eisentraut, 1955). The tenrec and the Madagascan "hedgehog" (*Setifer setosus*) may enter torpor during the dry austral winter, and, when moved to the northern hemisphere, they persist in their annual cycle and become lethargic during the boreal summer (Kayser, 1961; Hildwein, 1964, 1970; Eisenberg and Gould, 1970). They apparently cannot tolerate the low T_b's that are usual in deep hibernators, for Kayser quotes an experiment by Lachiver in which *Setifer* became flaccid and ceased to remain in a curled position when T_b reached 10°C, and the tenrec in torpor protects itself by increasing its oxygen consumption when T_a drops from 15°C to 12°C (Kayser, 1961). Reports from the field suggest that the tenrec is overwhelmed by lethargy during the season for torpor, as animals dug from their burrows returned to the torpid state when placed unprotected in a box (Rand, 1935). It may be argued that this type of lethargy is simply a manifestation of seasonal breakdown of thermoregulation in animals whose regulation was inferior at the outset. However, the increase in oxygen consumption when T_b approaches potentially lethal levels and the ability to warm from a T_b of 15°C using only endogenous heat (Kayser, 1961, p. 31) indicates that these animals have retained at least part of their thermoregulatory ability.

If *Echinops telfairi* is exposed to T_a's between about 28° and 33°C, it maintains its T_b between approximately 30° and 32°C. When T_a is held between 13° and 24°C, this animal undergoes a diurnal cycle, maintaining an elevated T_b during the night and dropping its T_b to a low of 13°C during the day. When in torpor, *Echinops* is capable of sluggish, but coordinated, movements and it warms itself during arousal from torpor (Herter, 1962). Scholl (1974) found that the diurnal cycle of temperature continued even when this species was maintained in continuous light or total darkness, and that long term periods of torpor also occurred. During warming, heat is generated from muscular shivering. A histological examination of three specimens revealed only a small amount of brown fat in one animal, so that this putative source of heat is apparently not involved in the arousal process.

Torpor occurs in both *Hemicentes semispinosus* and *H. nigriceps*, but it has not been studied in detail. *H. semispinosus* is also capable of uncoordinated locomotion and will eat and drink while in the torporous state. The lowest reported T_b was 22°C, but the animals were not exposed to T_a's below 18°C. A torpid *Geogale surita* was found in the wild, but there is no information on its T_b (Gould and Eisenberg, 1966).

Looking at the tenrecs alone, one could reasonably conclude that primitive eutherian mammals maintain a low and uneven T_b and that they may seasonally or daily enter a period of torpor. However, in the rest of the order Insectivora there are remarkably few species that are known to hibernate or enter a torporous state. The European hedgehog, *Erinaceus europaeus,* appears to have all the characteristics of a typical deep hibernator and many of the European studies on mammalian hibernation have used this animal as the experimental subject. The family Erinacidae of which *Erinaceus* is a member contains other genera that may hibernate, but little is known of their natural history. Shortridge (1934) reported that the African hedghog, *Aethechinus,* hibernates as well as some species of golden moles (Chrysochloridae), but no details are given. The diminutive shrew, *Suncus etruscus,* is said to be the smallest mammal in the world (2 g or less), and its T_b was reported to drop to 18°-20°C for as long as 7.5 hr when available food was reduced, and to rewarm from these periods of torpor using only internally generated heat. The animals were not curled up when in the chilled state but were stretched out and stiff. When well fed, these shrews were able to maintain their euthermic body temperature if exposed to T_a's as low as 0°C (Vogel, 1974).

Except for the above, there appear to be no detailed reports of hibernation or torpor in other groups of the Insectivora. This lack of hibernators is puzzling because the order encompasses many species that appear to be ideal candidates for hibernation. None of the members of this order are as large as the largest hibernators, and the ubiquitous family Soricidae, the shrews, are all small in size. If hibernation is an escape from the maintenance of high temperature in a small body during periods of scarcity of food, one would expect that the shrews would hibernate, but, with the exception of *Suncus etruscus,* they apparently show no tendency to enter the hibernating state. Of course, all of the shrews have not been subjected to scientific scrutiny, but a fair sampling has been made and it is reasonable to assume that the subfamily Soricinae or red-toothed shrews, which contains the New World shrews, probably are not hiding a hibernator in their midst. For example, *Microsorex hoyi* rivals *Suncus etruscus* for smallness of size and its weight of 2.3-4 g is comparable to that of a large hummingbird, but its brief and very active life is spent in a continuous search for food punctuated by short periods of sleep. Shrews have a higher metabolic rate per gram than do mice of the same size (Pearson, 1948) and this might be related to the absence of torpor and hibernation, but hummingbirds have a still higher metabolic rate per gram and torpor is usual with them. Gebczynski (1971) exposed *Sorex minutus* (wt. 3.7 g), *Sorex araneus* (wt. 7.0 g), and *Neomys fodiens* (wt. 17.1 g), all of the subfamily Soricinae, to various regimes of starvation at T_a's between 3° and 25°C, and was unable to unearth any tendency to hibernate.

The shrews were recently caught and Hudson (1978), who has had long experience with torpor in small mammals, suggests that such a tendency might have been found if the animals had been accustomed to captivity before being exposed to the experimental conditions. However, Vogel (1974) recognized that torpor in shrews was unusual and pointed out that *Suncus etruscus* was a member of the subfamily Crocidurinae, or white-tooth shrews, whereas the lack of torpor had been reported for red-toothed shrews of the subfamily Soricinae. He suggested that the Crocidurinae had a comparably lower basal metabolic rate, and that this was related to their ability to become torpid.

We find no references to experimental studies of hibernation in other families of the Insectivora, though passing references in field studies could easily be overlooked. Because of their specializations, many of the Insectivora would pose problems as experimental animals, and precise data on their temperature regulation and possible hibernation would be difficult to obtain. For example, the family Talpidae, or moles, are fossorial and telemetry would appear to be the only satisfactory way to study them.

Thus, among the order Insectivora there appear to be two extremes, the tenrecs on the one hand, and the red-toothed shrews or Soricinae on the other. For those who maintain that hibernation and torpor are primitive traits, the tenrecs offer some comfort, for their ability to regulate their body temperature is limited, their metabolic rate is low for their size, and, under certain conditions, they appear unable to resist entering into torpor. Further study of this interesting group should be rewarding and possible, for *Echinops* can be raised in captivity (Gould and Eisenberg, 1966).

In contrast, the red-toothed shrews, in spite of their small size, defend their euthermic virginity with a high metabolic rate and caloric intake and only yield when overcome by hypothermia and death. From our present knowledge of temperature regulation in the Insectivora, the tenrecs and shrews are so different that it is difficult to believe that they are closely related phylogenetically. The order Insectivora is often referred to as a "scrap basket" order to which most primitive eutherian mammals have been relegated (Simpson, 1945). From a metabolic point of view, the tenrecs and the red-toothed shrews should be in two different scrap baskets.

Chiroptera

Bats are the only small mammals that can migrate long distances, yet various types of torpor and hibernation occur within this group. Henshaw (1970) and Lyman (1970) have made tables which, when combined, summarize the information up to 1968, though the generic names in the latter work were unfortunately omitted to economize in space. McNab (1969) and Dwyer (1971) add further data on this subject. In general, bats that inhabit temperate zones are capable of deep hibernation during the winter months. In the summer, body temperature may decline while the bats roost during the day, and rise again before feeding time at dusk.

Some species of bats that are confined to the tropical or subtropical areas are capable of a partial torpor. For example, Kulzer (1965) reports that several species of *Tadarida* and *Rhinopoma*, when at rest, permit their T_b to drop to about 17°C, but increase their oxygen consumption and attempt to maintain their T_b at this level when challenged by a lower T_a. They are able to rewarm from this temperature using only endogenous heat, but cannot recover if their T_b drops below 17°C. Pagels (1975) reports that *Tadarida brasiliensis cynocephala* can tolerate T_b's as low as 10°C, but it is not clear whether they can rewarm themselves from this temperature.

It was long believed that all "fruit-eating" bats of the suborder Megachiroptera maintained a constant temperature throughout their adult life. Bats of this suborder are confined to tropical and subtropical areas, and many are large, attaining a wing span of over three feet. Bartholomew *et al.* (1970) have demonstrated that the smaller fruit bats, *Nyctimene albiventer* (wt. 28 g) and *Paranyctimene raptor* (wt. 21 g), follow a diurnal cycle and permit their T_b to drop to about 25°C when exposed to a T_a of 25°C. It is not known whether these bats would tolerate lower T_b's as they were not exposed to lower temperatures. Both species could warm from a T_b of 25°C using only endogenously generated heat. These animals are remarkably coordinated at low temperatures and are capable of eating at a T_b of 25°C. The larger *Nyctimene major* (wt. 79.5 g) maintains its euthermic temperature under the same experimental conditions, and Bartholomew *et al.* (1970) suggest that size is a critical factor in the dormancy of both mega- and microchiropteran bats, with the smaller bats in each suborder having very labile T_b's.

Although there are many species of bats that hibernate, and although some of these species are available from certain areas in large numbers, still bats are not good subjects for experimental work. Their specialized diets make them difficult to maintain in the laboratory, and there is evidence that they may change their "metabolic personality" during captivity. Thus, Studier and Wilson (1979) report that *Artibeus jamaicensis* develops a much more rigid control of body temperature when caught and held in the laboratory.

Primates

The occurrence of torpor in primates appears to be limited to the dwarf lemurs of Madagascar and information about this group is scanty. These animals are small, arboreal, and nocturnal, so that field studies are difficult, but Petter (1962a) reports that the native Madagascans believe that the dwarf lemurs become torpid during the colder periods of the austral winter. In the laboratory, the fat-tailed dwarf lemur (*Cheirogaleus medius*) periodically increased the amount of fat in the body and tail and entered a torpid state for over a week at a time.

During this period the T_b dropped to 17.5°C at a T_a of 16°C, but the animals did not appear to be in deep hibernation, for they were capable of slow, though uncoordinated, movements. The period of torpor did not take place at the same time each year with laboratory animals, but it is not known how regular this is in the wild. Daily fluctuations in T_b may also occur in this species, with the T_b sometimes dropping to 20°C at a T_a of 15°C. Periods of torpor last only 2–3 days in the larger *C. major* and T_b does not decline to as low levels (Bourlière *et al.*, 1956; Petter, 1962b).

The mouse lemurs, *Microcebus murinus,* are among the smallest living primates. Russell (1975) found that they gained weight in August and September from a low in January through March under unchanging laboratory conditions of light, dark, and T_a in North Carolina. He reported that similar changes in weight under the same laboratory conditions occurred in *Cheirogaleus medius.* During September through December, *M. murinus* became torpid in the light part of the day with a mean T_b of 22.5° ± 3.6°C. They rewarmed spontaneously from this lethargy without visible shivering before the onset of darkness, and were nocturnally active. No periods of torpor lasting more than a few hours were observed. Bourlière *et al.* (1956) reported a lower T_b in December and January with animals kept in the laboratory in France and it seems probable that daily torpor occurs also in the wild, though we find no record of it. Andriantsiferana and Rahandraha (1974) found that *Microcebus* entered a state of hypothermia in the laboratory with T_b as low as 7°C, and that the lemurs could remain alive in this condition for several days. They were unable to rewarm without heat from external sources, and it was concluded that the hypothermia was due to exhaustion. However, it is interesting that the animals shivered violently when rewarming at a T_a of 18°–20°C, and this would not be expected to occur if the lemurs were totally exhausted.

Periods of torpor have not been recorded for other primates and probably do not occur. The tropical and semitropical distribution of this group is not conducive to this type of metabolic saving. Indeed, the slow loris, *Nycticebus coucang,* presents an example of a mammal with variable T_b and low metabolic rate that does not indulge in torpor. Its metabolic rate is only 40% of that expected for an animal of its weight, it is well insulated, and its T_b varies from 33° to 37°C, but it remains euthermic (Müller, 1979).

Carnivora

The winter sleep of bears is traditional, but the bear is definitely not a deep hibernator. There is no question that bears may "den up" for the winter and spend long periods of time in a somnolent state. For 100 days or more they do not eat, and a fecal plug called a tappen forms in the rectum. Many Indian tribes believed that the denned bears obtained nourishment by sucking their feet. The keratinous foot pads are shed in the winter and the bears lick their tender feet, no doubt giving rise to this ancient tale (Rogers, 1974). The underdeveloped young are born and suckled during this period. Understandably, there are few physiological measurements from bears of any species. Hock (1960) in his studies of black bears (*Ursus americanus*) reported the lowest recorded rectal temperature as 31.2°C in comparison to 33°C measured by Rausch (1961) and oral temperatures of 35°C found by Svihla and Bowman (1954). Hock further reported that the metabolic rate of a dormant black bear was about 50–60% of the euthermic state, and Folk (1967) found that the heart rate of a dormant bear in winter was as low as 10 beats per minute compared to 40 beats per minute for an animal sleeping in summer. In contrast to the immobility of an animal in deep hibernation, the bear is capable of coordinated movement and will usually shuffle off in retreat if disturbed. As far as we know, no simultaneous recordings of oral or brain temperature and rectal temperature have been made. It would be interesting to know if the brain remains warmer than the rest of the body during the winter sleep.

Clearly, dormancy in the bear is not the deep hibernation found in small mammals. The drop in metabolism results in a sparing of energy so that some bears are quite fat when the winter is over. The physiological adjustments that must occur in deep hibernation at a T_b of 3°–5°C are not necessary when the T_b is less than 6°C below the euthermic level. Some desert-living ungulates experience nightly declines in T_b which are almost as low as those reported for the bear. The lowest recorded temperature for the camel appears to be 34.2°C (Schmidt-Nielsen *et al.*, 1957) and 33.9°C for the eland (Taylor and Lyman, 1967). In point of fact, it has been questioned whether deep hibernation could be an advantage to an animal as large as a bear (Morrison, 1960).

In the past few years dormant bears have been studied in an attempt to clarify problems that occur with kidney disease in man, and some interesting observations have been reported, though the number of experimental animals is necessarily quite small (Brown *et al.*, 1971; Nelson *et al.*, 1973, 1975). Given the fact that bears do not eat during dormancy, it is reasonable to suppose that their metabolism during this time is the same as that of other mammals during starvation. Under conditions of starvation, fat is the chief fuel of euthermic mammals, but protein is also utilized and the nitrogenous waste products are excreted in the urine, and to a lesser extent, in the feces. During observations lasting as long as 100 days, the dormant bears in the winter neither urinated nor defecated. The formation of urine was greatly reduced and it was postulated that this urine was reabsorbed through the wall of the bladder. The dormant bears metabolized fat almost exclusively as a source of energy so that the excretion of nitrogenous waste products in the urine was not necessary. Bears that were starved in the summer months did not become dormant even when kept at 5°–10°C and protein was used as well as fat as a source of energy, with urine being formed and excreted.

The intermediary metabolism of dormant bears will be discussed further in

Chapter 8, but the observations on these animals during the study of kidney function, coupled with previous knowledge of the biology of bears, suggest that a yearly endogenous cycle may be involved in this lethargy. The bears used in these studies were denied food in the autumn, and subsequently entered a state of dormancy. In the wild state, bears fatten in the fall and then become dormant. It is apparent that some change must take place in the autumn so that hunger or the desire to obtain food is suppressed. It might be expected that animals in deep hibernation, with a T_b a few degrees above 0°C, would lose all sense of hunger, but the T_b in these experiments was only 3°C below the T_b of euthermic bears, and this deviation is so slight that one would expect little change in the hunger drive. Dormancy, accompanied by lack of food, is in marked contrast to the effect of hunger on some small mammals. For example, it has long been known that hunger in the laboratory rat results in purposeless activity (Wald and Jackson, 1944; Richter and Rice, 1954), and it has been suggested that the mass emigrations of lemmings are motivated by hunger.

Although lethargy in bears does not involve the profound declines in T_b of the deep hibernators, the seasonal changes in intermediary metabolism may be just as fundamental as those which occur during deep hibernation in smaller mammals. This suggests that the use of the decline in T_b as the basic criterion for hibernation is perhaps too arbitrary, for a dormant bear in winter is certainly not simply a sleepy, hungry bear.

Several carnivores, such as the skunk (*Mephitis*), the raccoon (*Procyon*), the American badger (*Taxidea*), and the European badger (*Meles*) disappear for long periods during the winter and presumably experience some sort of lethargy. Remarkably little is known about this condition. Lethargic or dormant animals in their dens have not been noted by field naturalists (Twichell and Dill, 1949). Because negative results are not generally accepted for publication, failure to induce dormancy in the laboratory by exposure to cold and reduction of food is not generally reported. Slonim and Shcherbakova (1949), in a rather complete study, concluded that the European badger (*Meles meles*) did not hibernate. In our laboratory, a skunk (*Mephitis m. nigra*) did not hibernate when kept at 5°C all winter in a suitable nesting box. Hock (1960) recognized the difference between winter dormancy of the bear and deep hibernation and proposed the phrase "carnivorean lethargy" for the former condition. Whether other carnivores undergo this same change must await further investigation.

Rodents

The order Rodentia includes many deep hibernators and the preponderance of research in hibernation has been carried out using species from this group. Their small size, availability in the wild, and tolerance to laboratory conditions make them suitable subjects for experimental work and it might be argued that research in hibernation, particularly in the United States, is so "rodent oriented" that a biased overall picture results. In the conventional division of rodent subgroups, the Sciuromorpha or squirrel-like rodents encompass the majority of deep hibernators, including the marmots, which are the largest mammals known to enter deep hibernation. There are also hibernators in the Myomorpha or mouse-like rodent group, but no hystricomorph rodent has been reported to hibernate.

In addition to the deep hibernators, various species of rodents undergo some type of torpidity, and it is tempting to speculate that the increasing depth of this torpidity, as found in different species, represents the various stages in the physiological progression to the final condition of deep hibernation. The descriptions that follow develop such a sequence, but by no means include all the rodents known to enter torpor.

Baiomys is one of the smallest of the world's rodents, with the adults attaining a weight of about 9 g. Its origin is subtropical, and it ranges north to the southern part of the western United States. When adequate food and water are available, this species maintains a steady T_b of about 38°C, with variations that can be regarded as typical for very small mammals. If denied food they enter a state of torpor. Lack of water may also produce this condition, but it is not clear whether this is a direct effect of water deficit or whether it is because the thirsty animals also do not eat. If they are physically disturbed during the period of torpidity, they are capable of increasing their metabolic rate and warming to their active T_b using only endogenously generated heat. However, if T_b drops below 22°C, the animal cannot rewarm and will eventually die in torpor unless it is warmed artificially. It apparently can sense that its T_b is approaching the danger point, for oxygen consumption increases as T_b declines. The periods of torpidity are not seasonal in occurrence and the duration of the torpid period depends on the length of the period of starvation—the longest recorded bout of torpidity being 20 hr (Hudson, 1965).

Baiomys is remarkably coordinated at low T_b's for it can eat at a T_b of 25°C. As with the bears, the reaction to starvation or thirst results in a somnolent state rather than a period of hyperactivity. The dormancy must result in a metabolic saving in spite of the energy that is expended during arousal. As far as survival is concerned, the benefit of this brief type of torpor must be tenuous, for the availability of food cannot be expected to improve much over so short a period.

Hudson's (1967) studies of the deer or white-footed mice (*Peromyscus*) indicate that all members of this group probably undergo some sort of torpor, for he has observed this state in seven species of the genus. In all cases the torpor occurs diurnally and has not been observed to last more than 11.2 hr. The lowest T_b from which a species could recover was found to be 13.4°C in *P. maniculatus,* but *P. eremicus,* the desert mouse, could not rewarm itself from T_b's lower than 16°C (MacMillen, 1965) or 19°C (Hudson, 1967). In some cases a reduced food supply induces the onset of torpor, but this has not been reported for all species.

The deer mice tend to huddle together in their winter nest, thus decreasing the relative proportion of body surface to body mass. This behavioral reaction must reduce the chance of having the T_b drop to a level from which recovery is impossible. Actually, the posture that rodents assume during periods of torpor is not necessarily the curled posture, which is so typical of deep hibernators, and for this reason investigators overlooked the occurrence of torpor in many species until the animals were moved to the laboratory and subjected to physiological measurements.

Lynch *et al.* (1978) have shown that torpor occurs naturally in free ranging deer mice (*Peromyscus leucopus*), but was limited in Connecticut to late December, January, and February, which are the coldest months of the year. Furthermore, it was observed only when the ambient temperature was 3°C or below. Torpor was more apt to occur when a group of mice were huddled together, and in this situation the whole group tended to be torpid. In wild-caught mice confined to cages and in free ranging mice, torpor occurred even if the animals were supplied with ample food throughout the season. Gaertner *et al.* (1973) reported that *P. leucopus* remained alive below a T_b of 13°C and found that there were physiological differences between spontaneous and starvation-induced torpor, with an even heart rate in the former compared to an irregular rate in the latter. From a detailed study of *P. leucopus noveboracensis,* Hill (1977) concluded that lability of T_b may be present in this subspecies, but that this is not a routine occurrence. Clearly there are variations in the presence and depth of torpor within this species, and this may apply to the subspecific level also.

Studies on the pocket mice (*Perognathus*) of mid- and southwestern North America indicate that every species is probably capable of achieving torpor. Bartholomew and Cade (1957) reported torpor in *P. xanthonotus, P. longimembris, P. formosus, P. penicillatus,* and *P. fallax;* Tucker (1962) added *P. baileyi, P. flavus* and *P. californicus,* with the study on *P. hispidus* by Wang and Hudson (1970) and the work on *P. intermedius* by Bradley *et al.* (1975) completing the list. This comprises more than one-third of all the species in the genus, so that it is reasonable to assume that the rest of the group have the same physiological capabilities.

Torpor in *P. hispidus* most closely resembles that of *Baiomys,* for *P. hispidus* undergoes daily torpor if it is maintained on a starvation diet. However, it can tolerate lower body temperatures than can *Baiomys* and is reported to be able to warm itself from a T_b of 8°–11°C. In some of the other species of *Perognathus,* torpor occurs spontaneously, or it can be induced by a restricted food supply. Wolff and Bateman (1978) have shown that weight loss and the length of the period of torpor increase in *P. flavus* as the food supply and T_a decrease. The lowest temperature from which an animal can warm itself varies from 16°C in *P. californicus* to 4°C in *P. longimembris* (Tucker, 1962). Actually, this latter species resembles a deep hibernator, for its periods of dormancy have been observed to last about a week (Bartholomew and Cade, 1957). The daily time of arousal from torpor of this species can be entrained by the T_a (Lindberg and Hayden, 1974).

Another heteromyid, *Microdipodops pallidus,* from the deserts of western Nevada, has been shown to undergo periods of torpor lasting up to 24 hr. This species can tolerate a T_b of 6°–7°C and can arouse from T_a's in this temperature range. Starvation is not necessary to produce dormancy, but Bartholomew and MacMillen (1961) suggest that summer dormancy may be more useful than dormancy in the winter and that conservation of water rather than energy may be paramount. Daily declines in T_b to 15°C with full recovery also may occur in the kangaroo rat, *Dipodomys merriami,* when deprived of food, and this also may be an adaptation for life in dry areas (Yousef and Dill, 1971).

Recently Hudson and Scott (1979) have shown that the common albino laboratory mouse, *Mus musculus,* also may undergo periods of torpor. Although spontaneous torpor occurred in only one animal fed *ad libitum,* torpor took place frequently in animals whose food had been restricted. Below a T_a of 16°–19°C, the animals increased their oxygen consumption and maintained a gradient between T_b and T_a. The increased oxygen consumption resulted in shorter periods of torpor. As with *Baiomys,* the animals were remarkably coordinated, for they could gather millet seeds and eat them at a T_b of 24°C. This ability to function at low T_b's may be typical of animals in torpor, for *P. intermedius* can move slowly and gather and ingest food at a rectal temperature of 22°C (Bradley *et al.,* 1975).

The birch mouse, *Sicista betulina* (family Zapodidae), is a small rodent (wt. 7–16 g), which is found in northern and eastern Europe to Siberia and China. It undergoes daily torpor and wakes spontaneously at night to feed. The T_b may drop to a low of 4°C, the oxygen consumption is 1/25–1/30 of the warm-blooded resting rate, and the heart rate declines from 550–600 beats per minute to 30 beats per minute. The study of these animals was carried out from June to November when seasonal, long-term hibernation would probably not occur, but the investigators state that *Sicista subtilis* is known to be a deep hibernator and indicate that *S. betulina* is also (Johansen and Krog, 1959).

With this animal, the diurnal torpor closely resembles deep long-term hibernation as found in typical deep hibernators such as ground squirrels or hamsters. In both cases the animals can warm from hibernation using only endogenous heat, and the anterior part of the body warms faster than the posterior due to differential blood flow. The reduction in oxygen consumption during diurnal torpor in *Sicista* is comparable to that of deeply hibernating rodents, though the heart rate of thirty beats per minute is more than three times as fast as the rate of typical deep hibernators. As with many of the microchiropteran bats, daily torpor and deep hibernation are physiologically almost indistinguishable except for the duration of the dormant state.

From the starvation-induced daily torpor of *Baiomys* and the much deeper torpor of *Perognathus,* one can construct a progression that goes through the daily torpor and deep hibernation of *Sicista* and culminates in the deep hibernation of animals such as ground squirrels and hamsters. It must be recognized that this is not an evolutionary progression, for many of the species that are used as examples are only remotely related. Furthermore, from the physiological point of view, the progression may not be a real one, for deep hibernation involves physiological adjustments that are not necessary in an animal which may undergo light daily torpor. As previously mentioned, the deeply hibernating rodent enters that state from the curled position of sleep, and its T_b may decline to a low of $2°-3°C$. Many of the rodents that undergo torpor do not assume the sleeping position as their T_b declines and they will die at the low T_b's that are tolerated by hibernators. *Baiomys* and *Mus* are coordinated enough to eat at a T_b of $24°-25°C$, while the deep hibernators, if disturbed at any time during entrance into hibernation, must undergo the whole complex process of arousal before they resume their usual patterns of behavior. Differences such as these point out that our present knowledge does not give a neat, clear-cut set of transitions in living species of rodents from light torpor to deep hibernation. We simply do not know whether torpor in these animals is closely akin to hibernation or whether it is a quite separate type of temperature regulation.

Because of this, the chapters that follow will be almost exclusively concerned with the deep hibernators, not because the intermediate types are not interesting, but because they may not involve the same physiological mechanisms.

Among the rodents, the standards for hibernation research in North America are the ground squirrels. Unfortunately, the scientific generic nomenclature for these animals is confusing, and one is caught in the crossfire between the experimentalist and the classical taxonomist. For many years, this group of rodents was known as "Citellus," but Hershkovitz (1949) emphasized that this name was incorrect and that the proper name should be "Spermophilus." The result has been that some research papers refer to "Citellus" and some to "Spermophilus" and the uninitiated may not realize that it is the same species. In this volume, the scientific name as presented by the author of the paper under discussion will be quoted.

The ground squirrels are, in general, deep hibernators, but there are some, such as the desert-living antelope ground squirrel (*Citellus* (*Spermophilus*) *leucurus*), which are incapable of hibernating (Lyman, 1964). In the North American continent, *C. tridecemlineatus*, the thirteen-lined ground squirrel, has been used for research in hibernation for many years. In Europe, the European ground squirrel *C. citellus* is sometimes used, but the main object of research there has been the hedgehog, *Erinaceus,* of the order Insectivora. Other ground squirrels, particularly the golden-mantled (*C. lateralis*), Richardson's (*C. richardsonii*), the arctic (*C. parryi*), and the Californian (*C. beecheyi*) have provided important information. The largest deep hibernators are the woodchucks or marmots, and in the New World, *Marmota monax*, the woodchuck, from the eastern United States and the yellow-bellied marmot, *M. flaviventris*, from the west are used, whereas Dubois' (1896) nineteenth century classic on the comparative physiology of the marmot had the European marmot, *M. marmota*, as the subject. Both the western chipmunk (*Eutamias*) and the eastern chipmunk (*Tamias*) have been studied, but the difficulty of maintaining them in captivity and their brief periods of hibernation make them more useful for comparative purposes than for chronic physiological studies.

Hamsters are not indigenous to the Americas, but the golden or Syrian hamster (*Mesocricetus auratus*) has been studied both in Europe and North America principally because it is an easily obtainable laboratory animal. Except for its availability, it has little merit, for many individuals do not hibernate when exposed to cold, and the bouts of hibernation are short. A colony of Turkish hamsters (*Mesocricetus brandti*) was established in the United States in the mid 1960s and groups have been distributed to other laboratories. These animals are about the same size as Syrian hamsters, but they hibernate more readily (Lyman and O'Brien, 1977). The European hamster, *Cricetus cricetus*, is a much larger animal and its intractable nature perhaps explains why it has escaped more intensive research. Remarkably, the Djungarian hamster, *Phodopus sungorus*, which inhabits Siberia and is one of the smallest of the subfamily Cricetinae, is not a deep hibernator, though it may undergo periods of daily torpor with T_b as low as 19°C (Heldmaier, 1975). Figala *et al.* (1973) report that some of their animals turned white in winter, and that the animals that remained brown did not become torpid.

Hibernation in dormice was made famous by Lewis Carroll's mad tea party in "Alice in Wonderland." Since the edible dormouse, *Glis glis*, became established in England after the book was written, the sleepy dormouse was probably the hazel "mouse," *Muscardinus avellanarius*. The habits of this species are unusual among deep hibernators, for two animals, and possbly more, may hibernate in the same nest (Walhovd, 1976). Both of these dormice hibernate readily, as does the lerot or garden dormouse, *Eliomys quercinus*. According to Montoya *et al.* (1979) starvation induces torpor in this species, but torpor may also be induced in animals fed a calorically adequate diet which lacks protein. No dormouse is native to the New World, so that most of the literature with these animals as the subjects has been from Europe.

REFERENCES

Allison, T., and Van Twyver, H. (1972). Electrophysiological studies of the echidna, *Tachyglossus aculeatus*. II. Dormancy and hibernation. *Arch. Ital. Biol.* **110**, 185–194.

Andriantsiferana, R., and Rahandraha, T. (1974). Effets du séjour au froid sur le Microcèbe (Microcebus murinus, Miller 1777). *C. R. Hebd. Seances Acad. Sci., Ser. D* **278**, 3099-3102.

Augee, M. L., and Ealey, E. H. M. (1968). Torpor in the echidna, *Tachyglossus aculeatus. J. Mammal.* **49**, 446-454.

Augee, M. L., Ealey, E. H. M., and Spencer, H. (1970). Biotelemetric studies of temperature regulation and torpor in the echidna, *Tachyglossus aculeatus. J. Mammal.* **51**, 561-570.

Baldwin, S. P., and Kendeigh, S. C. (1932). Physiology of the temperature of birds. *Sci. Publ. Cleveland Mus. Nat. Hist.* **3**, 1-196.

Bartholomew, G. A., and Cade, T. J. (1957). Temperature regulation, hibernation, and aestivation in the little pocket mouse, *Perognathus longimembris. J. Mammal.* **38**, 60-72.

Bartholomew, G. A., and Hudson, J. W. (1962). Hibernation, estivation, temperature regulation, evaporative water loss, and heart rate of the pigmy possum, *Cercaertus nanus. Physiol. Zool.* **35**, 94-107.

Bartholomew, G. A., and MacMillen, R. E. (1961). Oxygen consumption, estivation, and hibernation in the kangaroo mouse, *Microdipodops pallidus. Physiol. Zool.* **34**, 177-183.

Bartholomew, G. A., and Trost, C. H. (1970). Temperature regulation in the speckled mouse bird, *Colius striatus. Condor* **72**, 141-146.

Bartholomew, G. A., Howell, T. R., and Cade, T. J. (1957). Torpidity in the white-throated swift, Anna hummingbird and poor-will. *Condor* **59**, 145-155.

Bartholomew, G. A., Dawson, W. R., and Lasiewski, R. C. (1970). Thermoregulation and heterothermy in some of the smaller flying foxes (*Megachiroptera*) of New Guinea. *Z. Vgl. Physiol.* **70**, 196-209.

Beuchat, C. A., Chaplin, S. B., and Morton, M. L. (1979). Ambient temperature and the daily energetics of two species of hummingbirds, *Calypte anna* and *Selasphorus rufus. Physiol. Zool.* **52**, 280-295.

Boswell, J. (1927). "Life of Samuel Johnson," Vol. 1, pp. 371-372. Oxford Univ. Press, London and New York.

Bourlière, F., Petter, J. J., and Petter-Rousseaux, A. (1956). Variabilité de la température centrale chez les lémuriens. *Mem. Inst. Sci. Madagascar, Ser. A* **10**, 303-304.

Bradley, W. G., Yousef, M. K., and Scott, I. M. (1975). Physiological studies on the rock pocket mouse, *Perognathus intermedius. Comp. Biochem. Physiol. A* **50A**, 331-337.

Brown, D. C., Mulhausen, R. O., Andrew, D. J., and Seal, U. S. (1971). Renal function in anesthetized dormant and active bears. *Am. J. Physiol.* **220**, 293-298.

Calder, W. A. (1971). Temperature relationships and nesting of the calliope hummingbird. *Condor* **73**, 314-321.

Calder, W. A., and Booser, J. (1973). Hypothermia of broad-tailed hummingbirds during incubation in nature with ecological correlations. *Science* **180**, 751-753.

Calder, W. A., and King, J. R. (1974). Thermal and caloric relations of birds. *In* "Avian Biology" (D. S. Farner and J. R. King, eds.). Vol. 4, pp. 259-413. Academic Press, New York.

Carpenter, F. L. (1974). Torpor in an Andean hummingbird: its ecological significance. *Science* **183**, 545-547.

Chaplin, S. B. (1974). Daily energetics of the black-capped chickadee, *Parus atricapillus,* in winter. *J. Comp. Physiol.* **89**, 321-330.

Chatfield, P. O., Lyman, C. P., and Irving, L. (1953). Physiological adaptation to cold of peripheral nerve in the leg of the herring gull (*Larus argentatus*). *Am. J. Physiol.* **172**, 639-644.

Cheke, R. A. (1971). Temperature rhythms in African montane sunbirds. *Ibis* **113**, 500-506.

Dawson, T. J., and Wolfers, J. M. (1978). Metabolism, thermoregulation and torpor in shrew sized marsupials of the genus *Planigale. Comp. Biochem. Physiol. A* **59A**, 305-309.

Dawson, W. R., and Hudson, J. W. (1970). Birds. *In* "Comparative Physiology of Thermoregulation" (G. C. Whittow, ed.), Vol. I, pp. 223-310. Academic Press, New York.

Dimple, H., and Calaby, J. (1972). Further observations on the mountain pigmy possum (*Burramys parvus*). *Victoria nat.* **89**, 101-106

Dubois, R. (1896). "Physiologie comparée de la Marmotte." Masson, Paris.

Dwyer, P. D. (1971). Temperature regulation and cave-dwelling in bats: an evolutionary perspective. *Mammalia* **35**, 424-455.

Eisenberg, J. F., and Gould, E. (1970). The tenrecs, a study in mammalian behavior and evolution. *Smithson. Contrib. Zool.* **27**, 1-137.

Eisentraut, M. (1955). A propos de la température de quelques mammifères de type primitif. *Mammalia* **19**, 437-443.

Figala, J., Hoffmann, K., and Golden, G. (1973). Zur Jahresperiodik beim Dsungarischen Zwerghamster *Phodopus sungorus*. Pallas. *Oecologia* **12**, 89-118.

Folk, G. E., Jr. (1967). Physiological observations of subarctic bears under winter den conditions. *In* "Mammalian Hibernation III" (K. C. Fisher, A. R. Dawe, C. P. Lyman, E. Schönbaum, and F. E. South, eds.), pp. 75-85. Oliver & Boyd, Edinburgh.

Gaertner, R. A., Hart, J. S., and Roy, O. Z. (1973). Seasonal spontaneous torpor in the white-footed mouse, *Peromyscus leucopus*. *Comp. Biochem. Physiol. A* **45A**, 169-181.

Gebczynski, M. (1971). Oxygen consumption in starving shrews. *Acta Theriol.* **18**, 288-292.

Godfrey, G. K. (1968). Body temperatures and torpor in *Sminthopsis crassicaudata* and *S. larapinta* (Marsupialia-Dasyuridae). *J. Zool.* **156**, 499-511.

Gould, E., and Eisenberg, J. F. (1966). Notes on the biology of the Tenrecidae. *J. Mammal.* **47**, 660-686.

Grant, T. R., and Dawson, T. J. (1978a). Temperature regulation in the platypus, *Ornithorhynchus anatinus:* maintenance of body temperature in air and water. *Physiol. Zool.* **51**, 1-6.

Grant, T. R., and Dawson, T. J. (1978b). Temperature regulation in the platypus, *Ornithorhynchus anatinus:* production and loss of metabolic heat in air and water. *Physiol. Zool.* **51**, 315-332.

Griffiths, M. (1978). "The Biology of the Monotremes." Academic Press, New York.

Hainsworth, F. R., and Wolf, L. L. (1970). Regulation of oxygen consumption and body temperature during torpor in a hummingbird, *Eulampis jugularis*. *Science* **168**, 368-369.

Hainsworth, F. R., Collins, B. G., and Wolf, L. L. (1977). The function of torpor in hummingbirds. *Physiol. Zool.* **50**, 215-222.

Heldmaier, G. (1975). Metabolic and thermoregulatory responses to heat and cold in the Djungarian hamster, *Phodopus sungorus*. *J. Comp. Physiol.* **102**, 115-122.

Henshaw, R. E. (1970). Thermoregulation in bats. *In* "About Bats" (B. H. Slaughter and D. W. Walton, eds.), pp. 188-232. South. Methodist Univ. Press, Dallas, Texas.

Herreid, C. F., II (1963). Temperature regulation and metabolism in Mexican freetail bats. *Science* **142**, 1573-1574.

Hershkovitz, P. (1949). Status of names credited to Oken, 1816. *J. Mammal.* **30**, 289-301.

Herter, K. (1962). Untersuchungen an lebenden Borstenigeln (*Tenrecinae*). 1. Über temperaturregulierung und Aktivitätsrythmik bei dem Igeltanrek *Echinops telfairi* Martin. *Zool. Beitr.* **7**, 239-292.

Hickman, V. V., and Hickman, J. L. (1960). Notes on the habits of the Tasmanian dormouse phalangers, *Cercaertus nanus* (Desmarest) and *Eudromicia lepida* (Thomas). *Proc. Zool. Soc. London* **135**, 365-374.

Hildwein, G. (1964). Evolution saisonnière de la thermorégulation chez l'Ericulus (*Setifer setosus*) (Mammifère, Insectivore). *C. R. Seances Soc. Biol. Ses Fil.* **158**, 1580-1583.

Hildwein, G. (1970). Capacités thermorégulatrices d'un mammifère insectivore primitif, le tenrec; leurs variations saisonnières. *Arch. Sci. Physiol.* **24**, 55-71.

Hill, R. W. (1977). Body temperature in Wisconsin Peromyscus leucopus: a reexamination. *Physiol. Zool.* **50**, 130-141.

Hock, R. J. (1960). Seasonal variations in physiologic functions of arctic ground squirrels and black bears. *Bull. Mus. Comp. Zool.* **124,** 155–171.

Hudson, J. W. (1965). Temperature regulation and torpidity in the pigmy mouse, *Baiomys taylori. Physiol. Zool.* **38,** 243–254.

Hudson, J. W. (1967). Variations in the patterns of torpidity of small homeotherms. *In* "Mammalian Hibernation III" (K. C. Fisher, A. R. Dawe, C. P. Lyman, E. Schönbaum, and F. E. South, eds.), pp. 30–46. Oliver & Boyd, Edinburgh.

Hudson, J. W. (1978). Shallow, daily torpor: a thermoregulatory adaptation. *In* "Strategies in Cold: Natural Torpidity and Thermogenesis" (L. C. H. Wang and J. W. Hudson, eds.), pp. 67–108. Academic Press, New York.

Hudson, J. W., and Scott, I. M. (1979). Daily torpor in the laboratory mouse, *Mus musculus* Var. Albino. *Physiol. Zool.* **52,** 205–218.

Jaeger, E. C. (1949). Further observations on the hibernation of the poorwill. *Condor* **51,** 105–109.

Johansen, K., and Krog, J. (1959). Diurnal body temperature variations and hibernation in the birch mouse, *Sicista betulina. Am. J. Physiol.* **196,** 1200–1204.

Kayser, C. (1961). "The Physiology of Natural Hibernation." Pergamon, Oxford.

Kulzer, E. (1965). Temperaturregulation bei Fledermäusen (*Chiroptera*) aus verschiedenen Klimazonen. *Z. Vgl. Physiol.* **50,** 1–34.

Lasiewski, R. C., and Dawson, W. R. (1964). Physiological responses to temperature in the common nighthawk. *Condor* **66,** 477–490.

Lignon, J. D. (1970). Still more responses of the poor-will to low temperatures. *Condor* **72,** 496–498.

Lindberg, R. G., and Hayden, P. (1974). Thermoperiodic entrainment of arousal from torpor in the little pocket mouse, *Perognathus longimembris. Chronobiologia* **1,** 356–361.

Lyman, C. P. (1964). The effect of low temperature on the isolated hearts of *Citellus leucurus* and *C. mohavensis. J. Mammal.* **45,** 122–126.

Lyman, C. P. (1970). Thermoregulation and metabolism in bats. In "Biology of Bats" (W. Wimsatt, ed.), pp. 301–330. Academic Press, New York.

Lyman, C. P., and O'Brien, R. C. (1977). A laboratory study of the Turkish hamster *Mesocricetus brandti. Breviora, Mus. Comp. Zool. Harv.* No. 442.

Lynch, G. R., Vogt, F. D., and Smith, H. R. (1978). Seasonal study of spontaneous daily torpor in the white footed mouse, *Peromyscus leucopus. Physiol. Zool.* **51,** 289–299.

McAtee, W. L. (1947). Torpidity in birds. *Am. Midl. Nat.* **38,** 191–206.

MacMillen, R. E. (1965). Aestivation in the cactus mouse, *Peromyscus eremicus. Comp. Biochem. Physiol.* **16,** 227–248.

MacMillen, R. E., and Trost, C. H. (1967). Nocturnal hypothermia in the Inca dove, *Scardafella inca. Comp. Biochem. Physiol.* **23,** 243–253.

McNab, B. K. (1969). The economics of temperature regulation in neotropical bats. *Comp. Biochem. Physiol.* **31,** 227–268.

Montoya, R., Ambid, L., and Agid, R. (1979). Torpor induced at any season by suppression of food proteins in a hibernator, the garden dormouse (*Eliomys quercinus* L.). *Comp. Biochem. Physiol. A* **62A,** 371–376.

Morrison, P. (1960). Some interrelations between weight and hibernation function. *Bull. Mus. Comp. Zool.* **124,** 75–92.

Morrison, P., and McNab, B. K. (1962). Daily torpor in a Brazilian murine opossum (*Marmosa*). *Comp. Biochem. Physiol.* **6,** 57–68.

Morton, S. R., and Lee, A. K. (1978). Thermoregulation and metabolism in *Planigale maculata* (Marsupialia: Dasyuridae). *J. Therm. Biol.* **3,** 117–120.

Müller, E. F. (1979). Energy metabolism, thermoregulation and water budget in the slow loris (*Nycticebus coucang* Boddaert 1785). *Comp. Biochem. Physiol. A* **64A,** 109–119.

Nelson, R. A., Wahner, H. W., Jones, J. D., Ellefson, R. D., and Zollman, P. E. (1973). Metabolism of bears before, during, and after winter sleep. *Am. J. Physiol.* **224,** 491-496.

Nelson, R. A., Jones, J. D., Wahner, H. W., McGill, D. B., and Code, C. F. (1975). Nitrogen metabolism in bears: urea metabolism in summer starvation and in winter sleep and role of urinary bladder in water and nitrogen conservation. *Mayo Clin. Proc.* **50,** 141-146.

Pagels, J. F. (1975). Temperature regulation, body weight and changes in total body fat of the free-tailed bat, *Tadarida brasiliensis cynocephala* (Le Conte). *Comp. Biochem. Physiol. A* **50A,** 237-246.

Pearson, O. P. (1948). Metabolism of small mammals, with remarks on the lower limit of mammalian size. *Science* **108,** 44.

Pearson, O. P. (1951). Mammals in the highlands of southern Peru. *Bull. Mus. Comp. Zool.* **106,** 117-174.

Petter, J. J. (1962a). Ecological and behavioral studies of Madagascar lemurs in the field, *Ann. N.Y. Acad. Sci.* **102,** 267-281.

Petter, J. J. (1962b). Recherches sur l'écologie et l'éthologie des Lémuriens Malgache. *Mem. Mus. Natl. Hist. Nat., Ser. A (Paris)* **27,** 1-146.

Rand, A. L. (1935). On the habits of some Madagascar mammals. *J. Mammal.* **16,** 89-104.

Rausch, R. L. (1961). Notes on the black bear, *Ursus americanus* Pallas, in Alaska, with particular reference to dentition and growth. *Z. Saeugetierkd.* **26,** 65-128.

Richter, C. P., and Rice, K. K. (1954). Comparison of the effects produced by fasting on gross bodily activity of wild and domesticated Norway rats. *Am. J. Physiol.* **179,** 305-308.

Rogers, L. L. (1974). Shedding of foot pads by black bears during denning. *J. Mammal.* **55,** 672-674.

Russell, R. J. (1975). Body temperatures and behavior of captive cheirogaleids. *In* "Lemur Biology" (I. Tattersall and R. W. Sussman, eds.), pp. 193-206. Plenum, New York.

Schmidt-Nielsen, K., Schmidt-Nielsen, B., Jarnum, S. A., and Houpt, T. R. (1957). Body temperature of the camel and its relation to water economy. *Am. J. Physiol.* **188,** 103-112.

Scholl, P. (1974). Temperaturregulation beim madegassischen Igeltanrek *Echinops telfairi* (Martin, 1838). *J. Comp. Physiol.* **89,** 175-195.

Shortridge, G. C. (1934). "The Mammals of Southwest Africa," Vol. 1. Heinemann, London.

Simpson, G. G. (1945). The principles of classification and a classification of mammals. *Bull. Am. Mus. Nat. Hist.* **85,** 1-350.

Slonim, A. D., and Shcherbakova, O. P. (1949). Metabolism and physiological peculiarities of hibernation in the badger. *In* "Experience in the Study of the Periodic Changes of Physiological Functions in the Organism" (K. M. Bykov, ed.), pp. 167-186. Acad. Med. Sci. Press, Moscow. (Engl. transl.)

Steen, J. (1958). Climatic adaptation in some small northern birds. *Ecology* **39,** 625-629.

Studier, E. H., and Wilson, D. E. (1979). Effects of captivity on thermoregulation and metabolism in *Artibeus jamaicensis* (Chiroptera: Phyllostomatidae). *Comp. Biochem. Physiol. A* **62A,** 347-350.

Svihla, A., and Bowman, H. S. (1954). Hibernation in the American black bear. *Am. Midl. Nat.* **52,** 248-252.

Taylor, C. R., and Lyman, C. P. (1967). A comparative study of the environmental physiology of an East African antelope, the eland, and the Hereford steer. *Physiol. Zool.* **40,** 280-295.

Tucker, V. A. (1962). Diurnal torpidity in the California pocket mouse. *Science* **136,** 380-381.

Twichell, A. R., and Dill, H. H. (1949). One hundred raccoons from one hundred and two acres. *J. Mammal.* **30,** 130-133.

Vogel, P. (1974). Kälteresistenz und reversible Hypothermie der Etruskerspitzmaus (*Suncus etruscus,* Soricidae, Insectivora). *Z. Saeugetierkd.* **39,** 78-88.

Wald, G., and Jackson, B. (1944). Activity and nutritional deprivation. *Proc. Natl. Acad. Sci. U.S.A.* **30**, 255-263.

Walhovd, H. (1976). Partial arousals from hibernation in a pair of common doormice *Muscardinus avellanarius* (Rodentia, Gliridae) in their natural hibernaculum. *Oecologia* **25**, 321-330.

Wallis, R. L. (1976). Torpor in the dasyurid marsupial *Antechinus stuartii*. *Comp. Biochem. Physiol. A* **53A**, 319-322.

Wang, L. C.-H., and Hudson, J. W. (1970). Some physiological aspects of temperature regulation in the normothermic and torpid hispid pocket mouse, *Perognathus hispidus*. *Comp. Biochem. Physiol.* **32**, 275-293.

Warren, J. W. (1960). Temperature fluctuation in the smooth-billed ani. *Condor* **62**, 293-294.

Wolf, L. L., and Hainsworth, F. R. (1972). Environmental influence on regulated body temperature in torpid hummingbirds. *Comp. Biochem. Physiol. A* **41A**, 167-173.

Wolff, J. O., and Bateman, G. C. (1978). Effects of food availability and ambient temperature on torpor cycles of Perognathus flavus (Heteromyidae). *J. Mammal.* **59**, 707-716.

Yousef, M. K., and Dill, D. B. (1971). Daily cycles of hibernation in the kangaroo rat, *Dipodomys merriami*. *Cryobiology* **8**, 441-446.

3

Entering Hibernation

Although a detailed analysis of physiological changes that occur during hibernation is extremely important in identifying the mechanisms involved in the whole hibernating process, it is unfortunately true that our knowledge of this subject is quite limited. One reason for this lies in the difficulty of obtaining data from an animal as it shifts from the active state to hibernation. Since the processes that control the onset of hibernation are unknown, one can not predict when hibernation will occur, and each experiment must use long-term recording with the assumption that hibernation will eventually take place. Until the invention of apparatus that could perform this function, the study of this aspect of hibernation was perforce neglected. Even now it is beset with difficulties, for animals encumbered with wires and tubes tend not to hibernate, or to destroy the paraphernalia before entering hibernation. Furthermore, studies with very small mammals are limited because of the proportionally large size of the wires and tubes that connect the animal to the recording apparatus. Miniature, transistorized units that can be chronically implanted offer a partial solution to these problems, for both T_b and heart rate can be monitored using this technique, and this is certainly the method of choice when studying hibernators under natural conditions in the field. However, there is no known method for accurately measuring metabolic rate without restraining an animal with a face mask, or confining it in a metabolic chamber. Oxygen consumption may be estimated from the relationship of heart rate and metabolic rate, but, as Wang (1978) points out, this assumes an unchanging stroke volume and arterial–venous difference in oxygen concentration and is really limited in its usefulness to steady-state conditions. Since entrance into hibernation, and arousal from it, are far from being steady states, the method has questionable value in the study of these aspects of hibernation.

Another major problem in the study of the process of entering hibernation and the hibernating state itself concerns the sensitivity of the animal to any external stimulus. Unless experiments are designed to minimize disturbance, the animal may start the process of arousal, which is a very different physiological state from either entering hibernation or the hibernating state. Thus, many of the standard acute physiological procedures that are used on anesthetized animals can not be employed in the study of hibernation. Some of the earlier investigators did not realize that the change from hibernation to arousal could be as rapid as it is, and hence, they came to erroneous conclusions. These difficulties are now recognized by most investigators, but a description of some of the major techniques used in the study of hibernation may serve to emphasize their limitations.

Immediate sampling of tissues or organs of animals entering hibernation or in the hibernating state is an acceptable procedure provided that the sampling is done rapidly. If the heart rate is monitored immediately prior to the sampling process, the investigator can be assured of the condition of the animal. A high heart rate in a "hibernating" animal that was active some hours before usually indicates that the animal is still entering hibernation. A high rate in an animal that has been hibernating for a day or more indicates that the animal has begun the arousal process.

The use of anaesthetics either to simulate the entrance into hibernation or to stop arousal in the hibernating animal distorts the balanced physiological condition, and the observed results can not be equated with natural hibernation. Arousal from hibernation can be permanently stopped by transection of the spinal cord at C_1 immediately followed by artificial respiration. This technique has been useful in studying spinal reflexes in hibernation and in screening drugs that were to be used later in the natural hibernating state. It does not completely simulate natural hibernation, though the technique could be greatly improved by precise control of the artificial respiration and maintenance of normal pH and P_{CO_2}. This is difficult to accomplish in a small mammal with a limited blood supply (Lyman and O'Brien, 1969).

In general, hibernators will tolerate indwelling catheters and wires provided that the exteriorized portion is not within easy reach. The technique developed by

Still and Whitcomb (1956) for cannulation of the aorta has been used in the study of hibernation since the late 1950s (Lyman and O'Brien, 1960, 1963). Popovic and Popovic (1960) have developed techniques for permanent cannulation of the vena cava and the aorta and have successfully used this preparation with animals in hibernation. Permanent cannulation of artery or vein permits the introduction of pharmacological agents without physically disturbing the animal which is entering hibernation or is in the hibernating state. The pharmacological approach is useful, but has definite limitations. The matter of dosage is important, and the effect of a drug on a rat with a T_b of 37°C can not be extrapolated to the effect of the same drug on a different species in hibernation with a T_b of 5°C. It is not possible to predict whether the animal in hibernation will be more or less sensitive to a drug than the same animal in the active state. Furthermore, many drugs have side effects that may distort the specific response. Cardioacceleration, often followed by arousal from hibernation, may occur after infusion of drugs which have little cardioacceleratory effect in the active animal. Thus, the use of infused pharmacologic agents must be carefully controlled.

In spite of their sensitivity to external stimuli, some hibernators will tolerate quite bulky apparatus when it is firmly attached to the skull, and will hibernate with the device in place. Heller and Hammel (1972) were first to use thermodes to change the temperature of specific areas of the brain during the hibernating cycle. This technique was originally developed to study temperature regulation in larger, active animals, but its modification for studying hibernation in small animals such as ground squirrels has been an important one. The thermode assembly consists of three stainless steel tubes, sealed at one end, which are mounted perpendicularly on a Plexiglas block and positioned so that the sealed ends may be lowered into the brain of the experimental animal. Two of the tubes are 1 mm in diameter and are the thermodes. Each contains a smaller tube through which water is pumped at a controlled temperature and returned to a reservoir via the surrounding, larger tube. The third, thinner tube receives a thermocouple for measuring the temperature of the brain. The assembly is mounted firmly on the skull, using a stereotaxic instrument, so that the sealed ends of the thermodes straddle the preoptic–anterior hypothalamic area (POAH), with the smaller tube slightly rostral and off the midline (Fig. 3-1). The thermode technique is a difficult approach, but the results are rewarding. Its main disadvantage is that the bulk and position of the head cap may prevent the experimental animal from reaching the very deepest stages of hibernation. It must also be realized that the thermode does not change the temperature of the POAH area alone, for its temperature must spread to other areas of the brain.

Simultaneous recording of various physiological measurements is of great importance when studying hibernation because it offers clues to the sequence of the changes that take place. The usefulness of this approach is typified in the study of the process of entering hibernation.

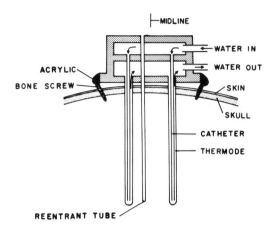

Fig. 3-1.   Schematic diagram of the thermode assembly used to heat and cool the POAH. Note that the assembly rests on the scalp rather than on the bone of the skull. (After Heller and Hammel, 1972. Reprinted with permission from *Comparative Biochemistry and Physiology*, copyright 1972, Pergamon Press, Ltd.)

The manner in which a hibernator enters the hibernating state is still the subject of argument and speculation. Part of this disagreement stems from differing experimental observations that have led to different conclusions, and the crux of the matter lies in the relationship of metabolic rate to T_b. It is universally agreed that an animal entering hibernation is not simply a chilled homeotherm. In the latter condition, metabolic rate is higher than it is when the animal is at thermal neutrality and T_b is maintained until the animal is overcome by cold and body temperature starts to drop. As T_b declines, metabolic rate and heart rate also drop and the animal enters a state of hypothermia from which it can not recover unless warmed by heat from exogenous sources. In contrast, entrance into hibernation is such a natural process that many (South *et al.*, 1969; Heller and Glotzbach, 1977; Walker *et al.*, 1977) maintain that the hibernator enters the hibernating state from the condition of normal sleep. In woodchucks (Lyman, 1958) body temperature, heart rate, and oxygen consumption may decline for a short period, followed by a burst of oxygen consumption and a sudden increase in heart rate. Body temperature rises to the normal level of the active animal and remains there for a few minutes when oxygen consumption, heart rate, and T_b again decline. The decrease in T_b may then form a smooth curve as T_b approaches T_a. Often, however, the entrance into hibernation is interrupted by sudden increases in oxygen consumption and heart rate. Bursts of muscle action potentials appear on the electromyograph and T_b may slow its decline or actually start to rise. After a brief period, heart and metabolic rate again are reduced, the bursts of muscle action potentials no longer occur, and T_b starts to drop again. The occurrence of these periodic "resistances" to hibernation appears to vary with the condition of the individual animal and with the species. In woodchucks entering hibernation for the first few times during the start of the winter season, the descent into hibernation was always interrupted by brief periods of warming, but animals that had entered hibernation several times did not interrupt the decline in T_b (Lyman, 1958). In contrast, we have found no evidence that the descent into hibernation in golden hamsters is ever interrupted except when the animal is physically disturbed.

Changes in blood pressure during entrance into hibernation have been measured only in the thirteen-lined and the golden mantled ground squirrel (Lyman and O'Brien, 1960). Using indwelling aortic cannulae, it was possible to show that, though a decline in systolic and diastolic pressure occurred as the animals started to enter the hibernating state, still the mean blood pressure remained within the range of the active animal. As hibernation deepened, the blood pressure decreased further, the pulse pressure increased, and the diastolic runoff time lengthened. If observed from the same systolic pressure, an estimate of the peripheral resistance may be made by measuring the slope of the angle which the diastolic runoff trace makes with the perpendicular. The increase in this angle is linear when plotted against T_b, indicating that peripheral resistance continually increases as T_b declines. Part of this change must be due to the effect of cold itself, such as increase in the viscosity of blood or shifts in the elasticity of the vascular walls. However, an important part must also be caused by increased vascular tone, for, if the ground squirrel is disturbed or if a vasodilatory drug is introduced via the cannula, a dramatic steepening of the runoff curve occurs without any change in T_b. We concluded that a generalized mild vasoconstriction takes place in these animals as they enter the hibernating state, and that this is maintained during hibernation. This differs from animals in hypothermia, and may be one of the reasons for the short survival time of hypothermed animals. Popovic (1960) showed that rats and ground squirrels which maintain their blood pressure during the first hours of hypothermia had an increased prognosis for survival. The heart rate in hypothermia is much faster than the rate in hibernation at the same T_b, and the blood pressure in the former must be maintained at the expense of a rapidly beating, cold heart.

In the chilling hibernator, some clues concerning the distribution of the blood can be obtained by recording sequential temperatures from various parts of the body. Strumwasser (1959) has reported in detail these changes as they take place in the California ground squirrel, *Citellus beecheyi*. He demonstrated that there is an interplay between the temperature of the brain and that of the skin of the back, with the temperature of one area rising while the temperature of the other area declines. The changes in the temperature of the back were interpreted as evidence of vasodilation and vasoconstriction, with concomitant heat loss or heat retention in that exposed area of the body. Bouts of shivering slowed the drop in

T_b and the waxing and waning of muscle tonus also influenced this decline. Entrance into hibernation was visualized as a series of controlled plateaus and "step-downs" of T_b. Body temperature declined with increasing rapidity in four consecutive entrances into hibernation. This was due to shorter plateaus of T_b and a greater amplitude or faster rate of decline in T_b during the step-downs, or a combination of the last two factors.

Using thermocouples located near the heart, on the back, and in the abdomen, we have recorded no such small steplike declines in T_b in thirteen-lined ground squirrels (Lyman and O'Brien, 1960) or woodchucks (Lyman, 1958). However, the bursts of muscle action potentials and increase in oxygen consumption and heart rate that have been reported in the woodchuck may transiently stop the decline, or actually result in an increase, of T_b. If T_b rises, the skin of the dorsum warms faster than the abdomen, presumably due to differential blood flow. When shivering ceases and cooling resumes, the back temperature slowly approaches the abdominal temperature. It seems probable that this sequence of events occurs in other rodent hibernators when the decline of T_b is interrupted by shivering or overt muscular movement. With thirteen-lined ground squirrels, it has been shown that the increase in heart rate and the appearance of muscle action potentials are accompanied by an increase in blood pressure which may result in changes in the vascular bed. Brief periods of tachycardia also may occur from time to time accompanied by an increase in blood pressure, but without muscle action potentials (Lyman and O'Brien, 1960).

These changes differ from the plateaus and "step-downs" reported by Strumwasser in that the declines in temperature in *C. beecheyi* apparently occur in a stepwise fashion throughout the whole process of entering hibernation and the "step-downs" were observed every time this species entered hibernation. In contrast, the changes observed in thirteen-lined ground squirrels and woodchucks had no stepwise pattern and often the entrances into hibernation were not interrupted at all by increased muscular activity and a plateau in T_b. To our knowledge, *C. beecheyi* is the only hibernator that has been shown to exhibit these precise stepwise changes in T_b during entrance into hibernation, and this may be a refinement which has not been developed in other ground squirrels. Using the brain thermode assembly, Heller *et al.* (1977) have convincingly demonstrated that the "set point" (T_{set}) threshold for temperature regulation slowly drops as the golden mantled ground squirrel enters hibernation. By manipulating the temperature (T_{hy}) of the brain in the preoptic anterior hypothalamic area, they were able to establish the T_{set} within a relatively narrow range of temperature at various times as the animals entered hibernation. T_{set} lay between the lowest T_{hy} which would not cause an increase in metabolic rate and the highest T_{hy} which would cause an increase (Fig. 3-2). If the T_b formed a smooth curve as the animal entered hibernation, then T_b was always higher than T_{set} as both T_b and T_{set} slowly declined. If T_b dropped faster than T_{set} and actually passed it, the

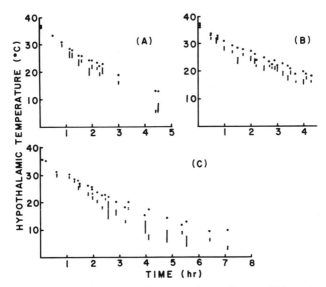

Fig. 3-2. Continuous decline in T_{set} as three ground squirrels entered hibernation. (A) is taken from two entrances into hibernation, (B) is from three entrances, and (C) is from four entrances. Dots are unmanipulated temperatures of the POAH and vertical lines indicate the range of T_{set} as determined by manipulating the POAH temperature. Note that dots are always above T_{set}. If temperature of POAH were to drop to T_{set} or below, metabolic heat production would increase, thus slowing entrance into hibernation. (After Heller et al., 1977.)

animal reacted by increasing its metabolic rate and raising its T_b above the set point. Presumably at least some of the plateaus or increases in T_b observed by us in thirteen-lined ground squirrels and woodchucks were the result of T_b passing T_{set} in their downward paths. The California ground squirrels used by Strumwasser were surgically deafened prior to the experiments. With this deficit in afferent input, it is possible that T_b kept dropping faster than the low threshold for set point and the entrance into hibernation consisted of a series of corrections which slowed down the cooling process.

The difference of the step-down method of entering hibernation is reflected in the length of time required to enter the hibernating state. Strumwasser (1960) illustrates this with *C. beecheyi* in a series of sequential entrances into hibernation of a single animal, with T_b dropping more rapidly during each entrance due to shorter plateaus and steeper declines of T_b.

A further refinement in the onset of hibernation has been reported for *C. beecheyi*. After the animals had been moved to cold, declines in T_b took place in a precise manner. Strumwasser (1960) illustrated a series of drops in brain temperature over a period of 6 days (Fig. 3-3). On a 12-hr light–dark schedule, the T_b of *C. beecheyi* dropped during every dark period and rose again before the

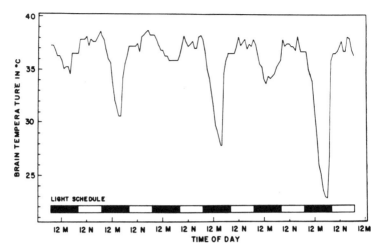

Fig. 3-3. Test drops as shown by the brain temperature (T_{br}) of a California ground squirrel over a 6-day period. Note that the test drops occur every other day, and that the decline in T_{br} begins at about the same time each day. M, midnight; N, noon. (After Strumwasser, 1960.)

period of light. On the first, third, and fifth "nights" T_b dropped only 3°–4°C, but on the second, fourth, and sixth nights T_b dropped successively to 30.8°, 27.9°, and 22.9°C. Strumwasser called these changes "test drops" and the lowest temperature reached on a drop was referred to as the "critical point." He presented evidence that, at each critical point, the animal resisted a further decline in T_b by shivering, and that this critical point was controlled very precisely—perhaps to less than 0.1°C. Each test drop was regarded as indicating further metabolic and perhaps neuronal preparation for deep hibernation, with the animal lowering its T_b to a specific safe level and no further. As illustrated, the lowest test drop is to 22.9°C; it is not stated whether test drops continue in this manner until the hibernator reaches the low temperatures that are typical of deep hibernation.

Test drops have been reported for Richardson's ground squirrel, *Spermophilus richardsonii,* but these test drops do not fulfill Strumwasser's criterion, for each sequential drop does not reach a lower T_b (Scott *et al.,* 1974).

Pivorun (1976) used radiotelemetered temperature recording in studying entrance into hibernation in two species of chipmunk, *Tamias striatus* and *Eutamias minimus.* He found that test drops took place in *T. striatus,* but not in *E. minimus,* which indicates that test drops may be confined to certain species. Test drops have not been reported for many of the intensively studied species of hibernators, and certainly would not appear at all with animals in the laboratory that enter deep hibernation at once when moved to the cold. It has been suggested that animals may actually experience test drops while in the warm and are thus prepared for hibernation, so that these drops are unnecessary in the cold. However, in a series of unpublished experiments, we monitored the T_b of thirteen-lined ground squirrels in the autumn. Some of the animals experienced test drops in the warm, though none showed a progressive decline of T_b in sequential test drops. Other animals maintained a steady T_b varying around 37°C. When moved to the cold room, some animals of each group entered hibernation at once, with no evidence of test drops. With other individuals, test drops occurred, but, as with the situation in the warm room, there was no evidence of progressive decline in T_b in sequential drops. Thus, test drops may be necessary for one species, but they do not appear to be a universal characteristic of rodents entering hibernation.

The slowing of the heart prior to the decline in body temperature is typical of entrance into hibernation in ground squirrels (Lyman and O'Brien, 1960; Strumwasser, 1960), woodchucks (Lyman, 1958), and in the eastern chipmunk (Wang and Hudson, 1971). Actually, T_b may be rising while the heart rate is slowing (Fig. 3-4). The appearance of skipped beats on the electrocardiogram seems to signal the onset of hibernation in most rodent hibernators, but slowing of the heart is also accomplished by a lengthening of the time between the even beats.

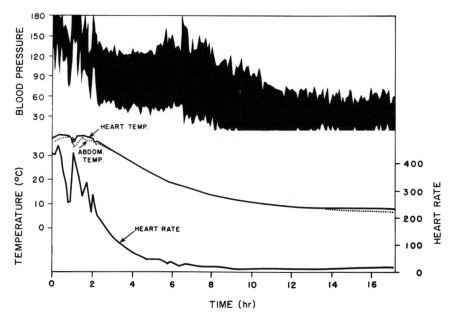

Fig. 3-4. Thirteen-lined ground squirrel entering hibernation. Drop in blood pressure and heart rate always precede decline in body temperature. Blood pressure in dark area is highest systole and lowest diastole recorded every 4 min for a 1-min period. Note drop in heart rate while body temperatures are rising. (After Lyman, 1965.)

As T_b drops, the skipped beats change to periods of asystole occurring at regular intervals. These are followed by a series of systoles that raise the blood pressure to a higher level (Fig. 3-5). In the thirteen-lined ground squirrel, evenly spaced electrical depolarizations of the heart often appear on the electrocardiogram with little or no change in pulse pressure (Fig. 3-5). In addition, sporadic electrical depolarizations also occur with no recorded change in pulse pressure, and these continue in deeply hibernating animals. Electrical depolarization without visible contraction has also been observed in cold, isolated hearts from animals that hibernate (Lyman and Blinks, 1959), and it seems probable that the effective heart rate of animals in hibernation is always slower than the electrically recorded rate would indicate.

During a single entrance into hibernation, the relationship of temperature to heart rate is quite precise, and this is accomplished in part by the occurrence of the skipped beats. Thus, if the heart rate is calculated by counting only the periods of even beats and omitting the effect of the asystoles, the temperature–heart rate curve is very uneven when compared to the relationship of the actual heart rate and temperature (Fig. 3-6). Ths presence of skipped beats suggests parasympathetic activity, and this can be tested by the use of atropine, which blocks vagal action. We have rarely succeeded in having a fully atropinized animal enter hibernation, but atropinization can be induced by infusing the drug via a chronically implanted aortic cannula after the ground squirrel has started

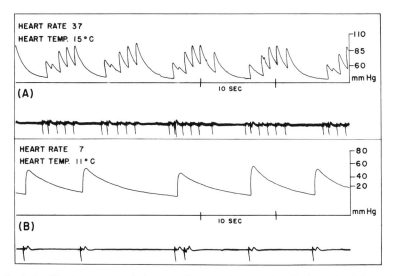

Fig. 3-5. Blood pressure and electrocardiogram of a ground squirrel entering hibernation. (A) At the start of each group of beats, there is an interpolated premature beat accompanied by little or no pulse pressure. (B) Extra systole with no pulse pressure. (After Lyman, 1965.)

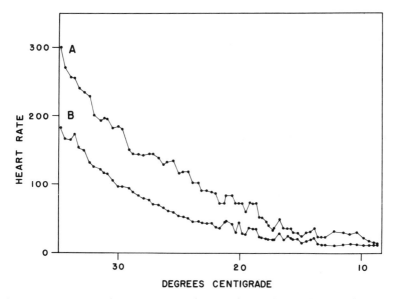

Fig. 3-6. Temperature-heart rate curves of a normal ground squirrel entering hibernation. Heart rate registered for a 1-min period at 5-min intervals. (A) Rate calculated by counting only the evenly occurring beats. (B) Actual heart rate; note fewer irregularities in the curve in (b) and increase in irregularities in this curve as animal reaches lower temperatures, possibly indicating loss of vagal influence. (After Lyman and O'Brien, 1963.)

entrance into hibernation. If this is done, the animal continues into hibernation, but the skipped beats are abolished and the heart rate is faster at any given temperature than in the untreated animal (Fig. 3-7a,b). The parasympatholytic action of atropine in the awake animal is quite specific and the doses used in these experiments were higher than those which blocked vagal action in active ground squirrels with high T_b's. These observations indicate that the heart rate is being modulated and slowed by parasympathetic action as the animal enters hibernation by inducing skipped beats and asystoles as well as contributing to the slowing of the even beats. This modulation serves to maintain an even rate-temperature curve as hibernation deepens. The correlation of heart rate with T_b is less exact as the woodchuck (Lyman, 1958) and the California ground squirrel (Strumwasser, 1960) enter hibernation, but the effect of atropinization has not been tested in these species.

The heart rate in the atropinized animal may offer further information of the forces at work during entrance into hibernation. At unpredictable intervals the heart may stop for periods of over 80 sec (Fig. 3-7C). As the period of asystole continues, respiratory rate increases and a flinch of the body often occurs before the heart resumes beating. These asystoles can not be caused by parasympathetic

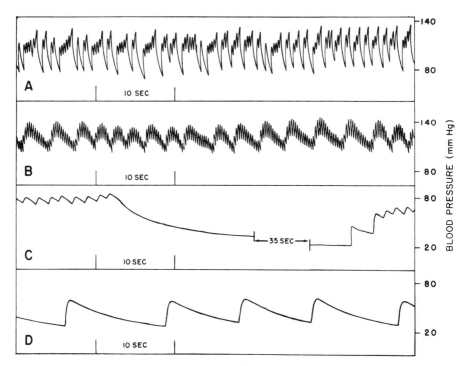

Fig. 3-7.   (A) Typical pulse pressure pattern of thirteen-lined ground squirrel entering hibernation, showing skipped beats. Thoracic temperature 29.5°C. (B) Same animal 10 min after treatment with atropine, which abolishes skipped beats. Thoracic temperature 29.0°C. (C) Long asystole in atropinized animal entering hibernation. Asystole lasted 58 sec in this case. Thoracic temperature 14.5°C. (D) Atropinized animal (18.5 mg/kg) with pulse pressure pattern that is typical of untreated ground squirrel in deep hibernation. Thoracic temperature 10.6°C. (After Lyman and O'Brien, 1963.)

activity, for they occur when the animals are treated with doses of atropine which are known to block any vagal influence. We have suggested that the heart rate in the atropinized animal indicates the force of sympathetic influence devoid of parasympathetic activity, and that the prolonged period of asystole is a result of withdrawal of sympathetic activity. Alternatively, the high heart rate at low temperature may use up some substrate or cause the accumulation of a metabolite that may interfere with the inherent rhythmicity of the heart, and the heart does not resume beating until the situation is corrected.

As the atropinized ground squirrel approaches the deeply hibernating state, the periods of asystole become more frequent and the periods of tachycardia become shorter until the animal attains the typical heart rate and pulse pressure of deep hibernation. This suggests that parasympathetic influence is at a minimum level in the last stages of entering hibernation and in the hibernating state.

This pharmacological evidence indicates that the heart is being slowed by parasympathetic influence as the animal starts to hibernate, and this influence continues to regulate the heart rate as the animal cools. Sympathetic influence is manifest when parasympathetic influence is blocked and the heart rate then accelerates. However, parasympathetic activity does not force the animal into hibernation, for the fully atropinized animal will enter this state. Furthermore, the balance between sympathetic and parasympathetic influence is not always the same, for heart rate–temperature curves differ with the same animal entering hibernation at different times. The concept of a controlled heart rate during the whole entrance period is buttressed by the comparison of heart rate in the anesthetized chilled woodchuck with that of the hibernator. In the former, the decline in T_b is much faster, and the heart rate is much more rapid at any given temperature during the first part of the entrance process (Lyman, 1958). The effect of vagotomy on entrance into hibernation has not been successfully tested. Cutting the vagi in the neck region also divides the recurrent laryngeal nerve and the animal becomes choked with mucus.

Obviously, the metabolic rate must drop as the animal enters hibernation, but the timing of this drop has been the subject of some controversy. Using relatively crude apparatus, we reported that the oxygen consumption of hamsters reached a minimum level before T_b had dropped to the temperature of deep hibernation (Lyman, 1948). With more sophisticated apparatus, Robertson et al. (1968) demonstrated that oxygen consumption declined before T_b in the Syrian hamster. In woodchucks, oxygen consumption dropped before any measurable decline in body temperature and, as with the hamster, reached a minimal level before T_b had completed its decline (Lyman, 1958).

Landau (1956) reported that respiratory rate was first to decline as thirteen-lined ground squirrels entered hibernation, and that this was followed by a decrease in heart rate and, finally, a drop in T_b. These results were interpreted to indicate that an active depression of metabolism was taking place as the animal entered hibernation (Lyman and O'Brien, 1963). Later, Wang and Hudson (1971) found that the oxygen consumption and heart rate of the eastern chipmunk, *Tamias striatus*, declined before T_b as the animal entered torpor. Skipped beats occurred during this time, and the Q_{10} at the beginning of the entrance was higher than later, suggesting that there was active inhibition of the heart rate. As with woodchucks and Syrian hamsters (Lyman, 1948, 1958), the oxygen consumption reached a minimum value before T_b achieved its lowest level. The eastern chipmunk is apparently not a typical deep hibernator, for it usually maintains its T_b between 10° and 24°C at a T_a of 1.5°–9.5°C. Its minimum heart rate varies from 23 to 46 beats per minute, and periods of torpor do not extend beyond 5 days. Nevertheless, its manner of entering hibernation appears to be very similar to that of woodchucks and ground squirrels. Hammel et al. (1973) found that heat loss increased and metabolic heat production decreased prior to the drop in hypothalamic and cervical temperature as the golden mantled ground squirrel entered hibernation. Furthermore, heat production reached basal levels well before body temperature. Recently, Wang (1978) has reported that the decline in oxygen consumption, CO_2 output, and heat loss precede the decline in T_b in *Spermophilus* (*Citellus*) *richardsonii*.

Using data from the California pocket mouse, *Perognathus californicus,* Tucker (1965) constructed theoretical curves depicting the decline in T_b and heat production over time as the animal entered torpor. One set of curves assumed that minimum heat production and maximum heat loss were taking place over time, whereas another set assumed that heat production and heat loss were both minimal. With actual measurements at given times on animals entering torpor, he found that actual heat production and T_b might exceed the theoretical curves for minimum heat production and heat loss but not for minimum heat production and maximum heat loss. He concluded that conductance must increase (i.e., insulation decrease) with the onset of torpor, and that conductance decreased as the animal reached the minimal T_b level of 15°C. It was his view that no active suppression of metabolism need occur when the animal entered torpor. If the pocket mouse was exposed to a T_a which was below the temperature of thermal neutrality, and if it did not respond to this lower temperature by increasing its heat production, then heat production would be less than heat loss and the animal would start to cool. Body temperature and heat production must change simultaneously and the animal enters hibernation by "sliding down a temperature-metabolism function."

This view is only partially at variance with our conclusions. There seems to be no doubt that animals entering either torpor or deep hibernation must change their "temperature setting" so that they permit heat loss to exceed heat production. It is agreed that in many mammals, including *P. californicus,* the abandonment is pulsatile, and heat production can exceed heat loss from time to time, thus slowing the decline, or actually causing a transitory rise in T_b, followed by a return to the original rate of cooling. The evidence for an active depression of metabolism depends on the sequence of changes in heart, respiratory, and metabolic rate in relation to body temperature. Tucker believed that declines in T_b were too small to be measured conveniently and, therefore, the reported relationship between metabolic events and T_b was not dependable. In our opinion, this view is untenable. It is well known that representative body temperatures in very small mammals are difficult to obtain and are apt to change quickly with shifts of position, activity, and other factors, but this is not true with properly placed temperature sensors in animals the size of ground squirrels or woodchucks, and the sequence of events in these animals as they enter hibernation is very clear. The rapidly beating heart and the absence of skipped beats in the atropinized ground squirrel as it enters hibernation is further evidence that the heart rate is being slowed by parasympathetic influence in the untreated animal.

Furthermore, the observation that atropinized animals rarely hibernate suggests that vagal control may be important, though not absolutely necessary, for the entry process.

It must be realized that *P. californicus* is not a typical deep hibernator, for it dies if T_b is below 15°C and only remains in natural torpor for a few hours. Thus, a comparison between this species and deep hibernators such as woodchucks and thirteen-lined ground squirrels may not be valid. In a more recent paper, Wang and Hudson (1970) demonstrated with actual measurements that the hispid pocket mouse, *Perognathus hispidus,* also underwent a decrease in oxygen consumption and heart rate prior to the decline in T_b. *P. hispidus* remains alive in torpor at T_b's as low as 8.1°C and hence has a greater tolerance to low body temperatures than has *P. californicus,* but otherwise the pattern of torpor in the two species is very similar. It seems reasonable to expect that the sequence in the decline of metabolic functions and T_b would also be the same. Furthermore, the white-footed mouse, *Peromyscus leucopus,* undergoes a torpor similar to *Perognathus californicus* and Gaertner *et al.* (1973) report that the heart rate drops before T_b also in this species as it enters torpor.

In our present state of knowledge it is difficult to explain how metabolic function can be actively depressed and thus initiate and control the entrance into hibernation. The "father of all hibernators," Raphael Dubois (1896), espoused the concept of "autonarcosis" and believed that the European marmot was forced into the hibernating state by the buildup of expired carbon dioxide in its burrow. This concept has not stood the test of time, and it is generally agreed that the high concentrations of carbon dioxide used in Dubois' experiments produced narcosis followed by hypothermia. Scholander (1963) has suggested that the physiological changes that take place during a dive in aquatic mammals are comparable to the changes during entrance into hibernation. In the former case, bradycardia and peripheral vasoconstriction are profound, apnea is complete, and metabolism is anaerobic during the dive. There is no evidence that these striking changes take place during entrance into hibernation. Metabolism can be depressed in mammals, under certain conditions, both on a long-term (Hudson, 1969) or on a short-term basis. In the latter case, it was found that the depression of metabolic rate and heart rate in rats occurred even in atropinized animals, so parasympathetic activity appeared to be unnecessary for depressed metabolism (Spielman and Lyman, 1971). As previously mentioned, Heller and his group have presented evidence that hibernation is an extension of slow wave sleep and believe that mammals enter hibernation from that state (Heller and Glotzbach, 1977). Sleep itself may be caused by the accumulation of a "sleep substance" in the brain of the awake animal, but its mechanism of action is not known nor has it been determined that it actually reduces metabolic rate (Pappenheimer, 1979).

The description given here of entrance into hibernation demonstrates that in rodents and in the insectivore *Erinaceus* the process of entering hibernation is a controlled one. The amount of control may vary from species to species, but control appears to be always present. It seems reasonable to assume that similar controls take place when species from other orders of mammals enter hibernation, but there is virtually no evidence that this is so, for the measurements have not been made. We believe that the controlled entrance into hibernation permits the animal to remain in hibernation for a given period and arouse from that state using self-generated heat after a definite period or when physically disturbed. The evidence to date backs this view, but the evidence is not yet complete.

REFERENCES

Dubois, R. (1896). "Physiologie comparée de la Marmotte." Masson, Paris.

Gaertner, R. A., Hart, J. S., and Roy, O. Z. (1973). Seasonal spontaneous torpor in the white footed mouse, *Peromyscus leucopus*. *Comp. Biochem. Physiol. A* **45A,** 169-181.

Hammel, H. T., Heller, H. C., and Sharp, F. R. (1973). Probing the rostral brainstem of anesthetized, unanesthetized, and exercising dogs and of hibernating and euthermic ground squirrels. *Fed. Proc., Fed. Am. Soc. Exp. Biol.* **32,** 1588-1597.

Heller, H. C., and Glotzbach, S. F. (1977). Thermoregulation during sleep and hibernation. *Int. Rev. Physiol.* **15,** 147-188.

Heller, H. C., and Hammel, H. T. (1972). CNS control of body temperature during hibernation. *Comp. Biochem. Physiol. A* **41A,** 349-359.

Heller, H. C., Colliver, G. W., and Beard, J. (1977). Thermoregulation during entrance into hibernation. *Pfluegers Arch.* **369,** 55-59.

Hudson, J. W. (1969). Comparative aspects of depressed metabolism in homeotherms. *In* "Depressed Metabolism" (X. J. Musacchia and J. F. Saunders, eds.), pp. 231-263. Am. Elsevier, New York.

Landau, B. R. (1956). Physiology of mammalian hibernation. *Diss. Abstr.* **16,** 2195-2196.

Lyman, C. P. (1948). The oxygen consumption and temperature regulation of hibernating hamsters. *J. Exp. Zool.* **109,** 55-78.

Lyman, C. P. (1958). Oxygen consumption, body temperature and heart rate of woodchucks entering hibernation. *Am. J. Physiol.* **194,** 83-91.

Lyman, C. P. (1965). Circulation in mammalian hibernation. *Handb. Physiol., Sect. 2: Circ.* **3,** 1967-1989.

Lyman, C. P., and Blinks, D. C. (1959). The effect of temperature on the isolated hearts of closely related hibernators and non-hibernators. *J. Cell. Comp. Physiol.* **54,** 53-63.

Lyman, C. P., and O'Brien, R. C. (1960). Circulatory changes in the thirteen-lined ground squirrel during the hibernating cycle. *Bull. Mus. Comp. Zool.* **124,** 353-372.

Lyman, C. P., and O'Brien, R. C. (1963). Autonomic control of circulation during the hibernating cycle in ground squirrels. *J. Physiol. (London)* **168,** 477-499.

Lyman, C. P., and O'Brien, R. C. (1969). Hyperresponsiveness in hibernation. *Symp. Soc. Exp. Biol.* **23,** 489-509.

Pappenheimer, J. R. (1979). "Nature's soft nurse": a sleep-promoting factor isolated from brain. *Johns Hopkins Med. J.* **145,** 49-56.

Pivorun, E. B. (1976). A biotelemetry study of the thermoregulatory patterns of *Tamias striatus* and *Eutamias minimus* during hibernation. *Comp. Biochem. Physiol. A* **53A,** 265-271.

Popovic, V. (1960). Physiological characteristics of rats and ground squirrels during prolonged lethargic hypothermia. *Am. J. Physiol.* **199,** 467-471.

Popovic, V., and Popovic, P. (1960). Permanent cannulation of aorta and vena cava in rats and ground squirrels. *J. Appl. Physiol.* **15,** 727-728.

Robertson, W. D., Yousef, M. K., and Johnson, H. D. (1968). Simultaneous recording of core temperature and energy expenditure during the hibernating cycle of *Mesocricetus auratus*. *Nature (London)* **219,** 742-743.

Scholander, P. F. (1963). The master switch of life. *Sci. Am.* **209(6),** 92-106.

Scott, G. W., Fisher, K. C., and Love, J. A. (1974). A telemetric study of the abdominal temperature of a hibernator, *Spermophilus richardsonii* maintained under constant conditions of temperature and light during the active season. *Can. J. Zool.* **52,** 653-658.

South, F. E., Breazile, J. E., Dellman, H. D., and Epperly, A. D. (1969). Sleep, hibernation and hypothermia in the yellow-bellied marmot (*M. flaviventris*). *In* ''Depressed Metabolism'' (X. J. Musacchia and J. F. Saunders, eds.), pp. 277-312. Am. Elsevier, New York.

Spielman, R. S., and Lyman, C. P. (1971). Thermal bradycardia in the mildly stressed rat. *Am. J. Physiol.* **221,** 948-951.

Still, J. W., and Whitcomb, E. R. (1956). Technique for permanent long-term intubation of rat aorta. *J. Lab. Clin. Med.* **48,** 152-154.

Strumwasser, F. (1959). Regulatory mechanisms, brain activity and behavior during deep hibernation in the squirrel *Citellus beecheyi. Am. J. Physiol.* **196,** 23-30.

Strumwasser, F. (1960). Some physiological principles governing hibernation in *Citellus beecheyi. Bull. Mus. Comp. Zool.* **124,** 285-320.

Tucker, V. A. (1965). The relation between the torpor cycle and heat exchange in the California pocket mouse *Perognathus californicus. J. Cell. Comp. Physiol.* **65,** 405-414.

Walker, J. M., Glotzbach, S. F., Berger, R. J., and Heller, H. C. (1977). Sleep and hibernation in ground squirrels (*Citellus* spp.): electrophysiological observations. *Am. J. Physiol.* **233,** R213-R221.

Wang, L. C. H. (1978). Energetic and field aspects of mammalian torpor: the Richardson's ground squirrel. *In* ''Strategies in Cold: Natural Torpidity and Thermogenesis'' (L. C. H. Wang and J. W. Hudson, eds.), pp. 109-145. Academic Press, New York.

Wang, L. C.-H., and Hudson, J. W. (1970). Some physiological aspects of temperature regulation in the normothermic and torpid hispid pocket mouse, *Perognathus hispidus. Comp. Biochem. Physiol.* **32,** 275-293.

Wang, L. C.-H., and Hudson, J. W. (1971). Temperature regulation in normothermic and hibernating eastern chipmunk, *Tamias striatus. Comp. Biochem. Physiol. A* **38A,** 59-90.

4

The Hibernating State

The study of animals in the deeply hibernating state is not as difficult as the study of entrance into hibernation. Unlike the transient entrance, the deeply hibernating animal stays in this condition for days at a time and, therefore, is available for experimentation during that period. Clips may be attached to permanently implanted electrodes and splices to thermodes or exteriorized catheters from arteries or veins, and, with skill and luck, the animal will not start the arousal process.

When one attempts to separate the process of entering hibernation from the hibernating state itself, one runs head-on into the need for a precise definition of deep hibernation. This condition should not be visualized as a static state in which standard measurements such as T_b, heart rate, or oxygen consumption simply reach a minimum level and remain there. Instead, it must be realized that the animal in hibernation is constantly reacting to internal and external stimuli, and, though changes may take place slowly, they are nonetheless real.

A few examples may illustrate this and help define the deeply hibernating state. The relationship between T_a and T_b under steady-state conditions is usually taken as a criterion of deep hibernation, and it is accepted that, between a T_a of approximately 5° and 15°C, the T_b remains less than 2°-3°C above T_a. However, a steady T_a for long periods of time is seldom attained except in a laboratory and some animals hibernate in places where T_a fluctuates widely. It would generally be agreed that a dormouse was in deep hibernation if T_a was 9° and T_b was 10°C. However, if T_a shifted to 5°C, the T_b would follow only slowly and, during this time, the criteria for deep hibernation would not be fulfilled even though oxygen consumption and heart rate might be slower than before. Similarly, even with an unchanging T_b, oxygen consumption and heart rate may vary, either together or separately. For example, the heart rate of a thirteen-lined ground squirrel in hibernation may vary between three and nine beats a minute during a day's recording, with no concurrent variation in T_a or T_b. Tripling of the heart rate in awake, active animals is a notable physiological change, yet it may occur during hibernation with so little effect that its physiological significance is not yet understood. Oxygen consumption may also vary from time to time in the absence of any detectable external stimulus. This is usually, but not invariably, accompanied by an increase in heart rate, though T_b often remains unchanged.

The sensitivity to external stimuli during hibernation varies from species to species and, in an individual animal, from day to day. Syrian hamsters are more "nervous" than Turkish hamsters and thirteen-lined or golden mantled ground squirrels and seem to be poised for arousal at all times. It has been reported that golden mantled ground squirrels become habituated and can be tossed and caught like a ball without waking from hibernation (Pengelley and Fisher, 1968), yet at other times the slight vibration caused by the starting of a refrigeration unit may result in an increase in heart rate with this same species (Lyman and O'Brien, 1969). Twente *et al.* (1970) categorize animals in this nervous condition as "stressed" hibernators and consider that this is a different physiological state from deep hibernation. In our experience, there appear to be gradations between the least and the most sensitive state. Furthermore, the sensitivity can shift during a single bout of hibernation, though the reason for this change is obscure.

Until we know more about these changes, it is perhaps best to recognize that they exist and categorically state that an animal is in deep hibernation if its T_b is below 10°C, with a greatly reduced heart and metabolic rate, while maintaining the ability to rewarm from this state using only endogenously generated heat.

With the exception of bats, all hibernators are curled in a ball when in hibernation and usually have the same position in natural sleep. The head is tucked under the abdomen, and the back is usually exposed above the insulating material of the nest. If the tail is long and bushy, it is curved over the back, but species with bare tails usually curl the tail beneath them. Very occasionally an animal in the laboratory hibernates outside its nest in a prone position as if suddenly overcome by hibernation before it was properly prepared. Most, if not all, nonvolant hibernators prepare nests when exposed to cold, and the insulative properties of these nests plus their curled position may protect the animal from sudden changes in T_a. Bats hang by their hind feet when sleeping or hibernating and their wings and tail membrane may be curled close to the body for added insulation. Many species of bats huddle together when hibernating, thus reducing the amount of body surface per unit mass. Unlike the other hibernators that are confined to a single nest, bats hibernate in caves and other secluded, dark places. The temperature is not the same throughout the cave, and bats wake from time to time and seek more suitable resting places (Twente, 1955; McManus, 1974).

In animals that hibernate curled in a ball, the region of the heart is slightly warmer than the rest of the body. The position must curtail blood flow to the abdomen (Lyman, 1958), but deep rectal temperature is only slightly lower than the heart. In bats, no differences in temperature have been detected in various parts of the body.

It has been reported that the skin of the exposed back of the California ground squirrel (*Citellus beecheyi*) remained two or more degrees Celsius colder than the temperature of the brain. At a steady T_a, there was no change in the temperature of the brain but the temperature of the dorsum underwent a slow rise of about 0.5°C with a faster fall, in a wave lasting about 15 min. Superimposed on the wave of declining temperature were smaller oscillations of about 0.2°C which lasted about 2.5 min. Strumwasser (1959) concluded that these rhythmic changes indicated alterations in blood flow and were manifestations of a delicate balancing of T_b with flushes of slightly warmer blood being chilled by the cooler temperature of the exposed back. If the hibernating ground squirrel was given a lethal dose of pentobarbital, T_b slowly declined toward T_a. During this decline there were regular oscillations of the temperature of the skin lasting 5 min and never greater than 0.2°C. Since these oscillations continued after the heart had stopped, they presumably were artifacts, but it is not clear why they were not present during normal hibernation. Even oscillations of the temperature of the back have not been seen in *Citellus tridecemlineatus* or other species in which suitable measurements have been made. In *C. tridecemlineatus* variations in blood pressure do occur, which may reflect shifts in blood flow, but the changes take place in hours instead of minutes and they are not evenly spaced in time (Lyman and O'Brien, 1960). *C. beecheyi* does not appear to be typical of the genus *Citellus* in its reactions as a hibernator, but the measurements made on this species were all above a T_b of 7°C, whereas the measurements on other species were usually obtained at T_b's of 5°C or lower. Perhaps there are more homeostatic mechanisms at work at higher T_b's in all hibernators.

At times during a bout of hibernation the heart rate may be very even, but at other times no heart beat can be detected for as long as a minute. These periods of asystole are followed by a series of relatively rapid beats, often preceded by one or more deep respirations. Strumwasser (1959) reported that the heart rate of the

California ground squirrel was more regular and slower as the animal approached deep hibernation than at a steady T_b of the deeply hibernating state. This has not been observed in the hedgehog, the arctic ground squirrel (*Citellus parryi*), or in Franklin's ground squirrel (*Citellus franklini*) (Dawe and Morrison, 1955), and does not occur in the thirteen-lined ground squirrel (Lyman and O'Brien, 1960).

In ground squirrels, woodchucks, and marmots, heart rates over ten beats or under three beats a minute are rare. Heart rates in the hibernating Syrian hamster are somewhat faster, with the lowest reported rate being four beats a minute (Chatfield and Lyman, 1950). In general, a rate of over 15 beats a minute at a T_b of 5°C indicates that the rodent is undergoing some physiological change. It may still be in hibernation but it is not in a steady state. With deeply hibernating bats, the heart rates are somewhat higher. For example, Davis (1970) reports rates between 24 and 32 beats a minute in *Myotis lucifugus*, the little brown bat, at a T_b of 5°C.

The electrocardiogram of the hibernating animal differs from the active animal in that the electrical events take place much more slowly. The principal cause for the slowing of the heart rate is the lengthening of the T–P interval, i.e., the time between individual beats. In the hedgehog (Sarajas, 1954; Dawe and Morrison, 1955) and in Franklin's and the arctic ground squirrel (Dawe and Morrison, 1955) it has been reported that the lengthening of the various intervals in the depolarization and recovery process contributes an important fraction of the overall slowing of the heart rate. The lengthening of the intervals in this P–T complex increases disproportionately at the lower temperatures of hibernation and curves of interval length plotted against temperature break quite sharply at about the temperature that hearts of nonhibernators cease to beat completely.

A notched R wave was reported by Nardone (1955) for *Citellus barrowensis* in hibernation and was interpreted as left bundle branch block. The heart rates of these animals were extremely high for animals in deep hibernation and it seems probable that they actually were arousing from the hibernating state. Notching of the R wave has been seen in hamsters as they started to arouse from hibernation (Chatfield and Lyman, 1950), but this notching does not normally occur during deep hibernation. Auricular–ventricular dissociation was reported by some of the early investigators (Buchanan, 1911a,b), though this may have been due to the difficulty of recording the low voltage P wave. Alternatively, the observations may have been made on animals that had started the process of arousal, for the heart rates that are illustrated are high for animals in deep hibernation. With more modern amplifiers and chronically implanted electrodes, the recording of A–V dissociation is a rarity in the healthy hibernating animal.

Cardiac output in deep hibernation has been measured only in the thirteen-lined ground squirrel. Using chronically implanted aortic and right ventricular cannulae, Popovic (1964) measured the arteriovenous difference of oxygen content in the blood and calculated the cardiac output using the direct Fick principle.

In resting, awake ground squirrels the cardiac output was 69 ± 5.5 ml/min or 313 ± 28 ml/min kg (means ± standard deviation). In contrast, the output during deep hibernation was 1.04 ± 0.1 ml/min or 4.6 ± 0.6 ml/min kg at T_b's of 6.6° ± 0.4°C. The value in hibernation was about 1/65 of that found in the euthermic animal, whereas the oxygen consumption was reduced to only about 1/61. The slight discrepancy is due to an increase in the A–V difference in the hibernating animal, in spite of a moderate decrease in the oxygen content of the arterial blood. Popovic suggested that there was no firm evidence leading to the conclusion that the blood in hibernating animals was highly enriched with oxygen. The values for cardiac output were the same throughout a bout of hibernation. Eliassen (1960) reported much higher cardiac outputs per unit body weight in two hedgehogs (*Erinaceus*) at T_b's below 9°C. However, the acute techniques used to obtain the A–V differences also started the arousal process, so the results would not be comparable even if the animals were closely related phylogenetically.

Blood pressure during hibernation has been measured only in thirteen-lined and golden mantled ground squirrels and most of the observations have been made on the former species (Lyman and O'Brien, 1960, 1963). The slow diastolic run-off and the increase in pulse pressure that develop during entrance into hibernation continue in the deeply hibernating state so that diastolic pressures are relatively high (7–40 mmHg) with systolic pressures varying between 40 and 90 mmHg. Usually the blood pressure is higher when the heart rate is rapid, but this is not invariably the case. As with the process of entering hibernation, a complete electrical depolarization of the heart with little or no change in blood pressure may occur from time to time.

The slow diastolic run-off time has been taken to indicate an increase in peripheral resistance brought about by an increased viscosity of the blood and a mild, but generalized vasoconstriction. The presence of vasoconstriction is confirmed by comparing the hibernating animal with chilled, nembutalized ground squirrels. In the latter case, peripheral resistance rises only slightly as the animals cool and the pulse pressure does not notably increase. The slight rise in peripheral resistance may be attributed to the increased viscosity of the cold blood, for there is no question that the blood becomes more viscous as both the hibernator and the hypothermic animal become cooler, though Maclean (1981) has found that change in viscosity is less in the chipmunk and ground squirrel than in the rat. Furthermore, infusion of a vasodilatory or adrenergic blocking drug via an intra-arterial cannula will quickly reduce peripheral resistance in the hibernating animal before there is a change in T_b (Lyman and O'Brien, 1963).

Respiration during hibernation is, of course, greatly reduced in comparison to the active animal. It may take place at quite evenly spaced intervals or long periods of apnea may occur followed by several deep respirations. The deep respirations are often referred to as Cheyne–Stokes, but true Cheyne–Stokes respiration is a pathological condition consisting of a gradual increase in the rapidity of breathing followed by a gradual decrease and cessation. In the hibernating animal, respiration usually starts and stops suddenly and there is little change in the rapidity of respiration when the animal is breathing (Fig. 4-1). Very long periods of apnea have been reported. In the Chinese hedgehog, *Erinaceus dealbatus*, Chao and Yeh (1950) timed an apneic period lasting 65 min, but this was outdone by Kristoffersson and Soivio (1964) with a period of 150 min in the European hedgehog, *Erinaceus europaeus*. In the garden dormouse, *Eliomys quercinus*, the longest apneic period was 112 min at a T_a of 4.2°C (Pajunen, 1970), and Twente *et al.* (1970) have observed apnea lasting 55 min in the golden mantled ground squirrel. At a T_a of 3.7° and a T_b of 4.2°C, the respiratory rate of this animal averaged 0.4/min. The latter authors point out that the overall respiratory rate is about the same whether respiration is of the even or the apneic type. In the type of respiration involving very long periods of apnea, the periods of rapid breathing are correspondingly long, and 50 or more respirations may occur during that time (Fig. 4-1). In our laboratory we have not observed these very prolonged periods of apnea, the record being 20 min in a European dormouse (*Glis glis*) hibernating with a T_b of 5°C. Usually four to eight respirations bring to an end a breathless period of this length.

It has been shown that hibernating Syrian hamsters and thirteen-lined ground squirrels increase their respiratory rate when exposed to more than 2.5% CO_2. Respiratory volume could not be measured without causing arousal, but it is clear that these animals are sensitive to relatively low concentrations of CO_2. The heart rate of the ground squirrels usually increased with the respiratory rate, but the

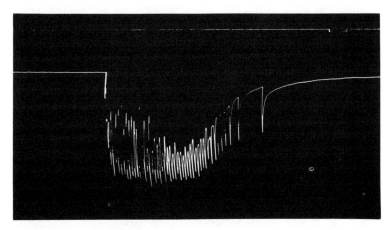

Fig. 4-1. Fifty respiratory movements after a period of apnea lasting about 56 min in a European hedgehog hibernating at T_b 4.7°C. Time marker interval is 7.5 min. (After Kristoffersson and Soivio, 1964.)

heart rate of hamsters did not increase until the concentration of CO_2 was over 5%. This concentration was a stimulus for arousal in the hamsters, for if they were exposed for several minutes they awoke from hibernation (Lyman, 1951). Tähti (1975) found that the apneic periods in the hibernating hedgehog were shorter when the ambient CO_2 was 1% and disappeared completely when CO_2 reached 5%. Also, if the concentration of ambient O_2 dropped below 16%, the apneic periods became shorter. Thus, sensitivity of the respiratory mechanisms appears to be similar during hibernation in a rodent and an insectivore. It has been reported that the CO_2 concentrations in the occupied burrows of the California ground squirrel reach about 2.5% when the animals are active (Baudinette, 1974). It seems probable that lower concentrations than this occur during the hibernating period and hence the animals avoid the possibility of stimulation by CO_2, but see Chapter 13.

Using an extracorporeal circulation technique, Goodrich (1973) was able to pump blood of hibernating marmots (*Marmota monax* and *M. flaviventris*) from the aortic arch through a blood–gas measuring cuvette and return it through the vena cava. She found that P_{CO_2} increased during the periods of apnea and fell almost immediately when respiration occurred. As shown in Fig. 4-2, the level of P_{CO_2} was not precisely correlated with the onset of respiration. In these experiments T_b averaged 11°C and T_a was about 5°C lower. Marmots have been reported to remain in normal hibernation at T_a's of 2.2°C and T_b's as low as 3.6°C (Benedict and Lee, 1938). It seems probable that periods of apnea under the latter conditions would be more prolonged and the correlation between the build up of P_{CO_2} and the onset of breathing might be more striking. With the hedgehog, Tähti and Soivio (1975) found a similar result, for the P_{CO_2} was lowest at the start of the apneic period, and climbed until another series of breaths

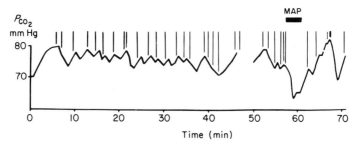

Fig. 4-2.   Variations in P_{CO_2} with respirations. Vertical bars indicate respiration recorded on EKG with P_{CO_2} read simultaneously. Spontaneous muscle action potentials (MAP) at horizontal bar. Decreases in P_{CO_2} follow a breath by 1–2 min. Tubing dead space and lag in the electrode and meter cause a delay in response of slightly less than 2 min, so that the drop in P_{CO_2} actually occurs almost immediately after a respiration. (After Goodrich, 1973.)

occurred. The P_{O_2} was higher at the beginning of the apneic period than it was in euthermic animals, but fell to very low levels as apnea continued.

In Goodrich's experiments the number of respirations in each period of breathing did not closely relate to the depth of decline of P_{CO_2} that followed. This observation suggests that tidal volume varies from breath to breath during hibernation. Other indirect evidence points to short-term and daily variations in tidal volume. Kayser (1940) published the results of a series of long-term experiments on European marmots in which he found that the respiratory frequency varied from 12 to 94 per hour, whereas the oxygen consumption varied only from 19.3 to 21.1 ml/kg/hr. In unpublished observations we have seen the same lack of correlation between oxygen consumption and respiratory frequency in Turkish hamsters and in thirteen-lined and golden mantled ground squirrels.

Direct measurements of tidal volume in deeply hibernating animals have not been successfully accomplished because the necessary apparatus causes the animal to arouse from hibernation. Endres (1930) developed a technique that involved a face mask and an inlet and outlet valve external to an airtight box which held the animal. Endres and Taylor (1930) measured the change of volume within the box that occurred with each respiration and referred to this as "depth of respiration." This was not a true measurement of tidal volume because expiration produced a contraction of the chest to a volume smaller than either the beginning of inspiration or than during the pause between respirations. They reported that at least one marmot remained in torpor overnight when held in this apparatus, but it is doubtful that they measured changes which are typical of deep hibernation, for they recorded no prolonged periods of apnea though they knew that these had been reported previously by Pembrey (1901–1902). In one experiment the "depth of respiration" varied from 11.3 to 14.8 ml at a T_b of 5.6°C, and this relatively steady depth of respiration may be typical of the first stages of arousal rather than deep hibernation.

Kayser made an exhaustive study of the metabolic rates of various species of hibernators, including hedgehogs, dormice, ground squirrels, hamsters, and bats, and summed up his results in 1961 (Kayser, 1961). In making these comparisons he did not use the same T_a or T_b for each species, but rather an optimum T_a which he derived by observation and which involved the duration of uninterrupted hibernation, the loss of weight during hibernation, and the length of the period between cold exposure and hibernation. This T_a varied from 7°C in the European hamster and 5°C for *C. citellus,* to below 5°C for bats. The minimum metabolic rate of the various species, hibernating at their optimum T_a's, was almost directly proportional to their weight. In contrast, the metabolism of these species when they were not in hibernation was more nearly proportional to their body surface, and Tähti (1978) came to the same conclusion when comparing hibernating and euthermic hedgehogs. It is reasonable to assume that this would be the case, for the animal hibernating at its "preferred" T_a is maintaining a very small gradient between T_a and T_b, and the amount of heat lost from its surface area is minimal. Kayser also firmly established in his long series of painstaking experiments that the respiratory quotient (RQ) in hibernation was slightly more than 0.7, of which more in Chapter 8.

Although changes in respiratory rate, heart rate, blood pressure, and oxygen consumption have now been recorded in the deeply hibernating animal, very little is known about the physiological mechanisms that orchestrate these changes. The investigator is strictly limited in any experimental approach because of the sensitivity of the hibernating animal to any arousal stimulus. However, changes in the heart rate are relatively easily monitored and indwelling vascular cannulae may be used to introduce drugs or measure blood pressure without causing the animal to arouse from hibernation.

As detailed in the previous chapter, intra-arterial infusion of parasympatholytic drugs will cause a striking increase in heart rate in the ground squirrel entering hibernation, and will abolish the skipped beats that are so typical of the entry process (Lyman and O'Brien, 1963). We interpret this to indicate that parasympathetic influence is slowing the heart, and, without this influence, the heart accelerates. Hudson (1973) has suggested that the infused parasympatholytic drugs may cause vasodilation with a resulting compensatory cardioacceleration. However, we have observed the effects of various vasodilatory drugs infused via an aortic cannula while monitoring the blood pressure. In this situation, there is a transient drop in blood pressure, followed by an increase in heart rate, with the whole episode lasting a matter of minutes. In the atropinized animal entering hibernation the heart rate is much faster at any given T_b than it is when compensatory acceleration occurs, and this acceleration lasts for hours, punctuated by the peculiar prolonged periods of asystole.

Once deep hibernation has been attained, we reported that the effect of infused atropine is strikingly different. Doses which would result in a marked cardioacceleration in the animal during the induction of hibernation, or completely block the bradycardia which accompanies the startle reflex in the active animal, often did not increase the rate at all. If there were uneven periods of asystole, these no longer occurred, causing a more even rate that was sometimes, but not always, faster than before the infusion. Extremely slow heart rates of two to four beats a minute showed no more tendency to increase than faster rates. The parasympatholytic drug methantheline bromide usually resulted in some slight cardioacceleration, but in several experiments the rate remained unchanged. Since infusion of similar volumes of isotonic saline often was followed by cardioacceleration in the deeply hibernating ground squirrel, the evidence suggested that the parasympathetic system had little influence on the heart rate during deep hibernation.

To examine this concept further, shielded stainless steel electrodes were implanted on the right vagus of ground squirrels and the terminals were exteriorized. When these animals hibernated, the vagus was stimulated with biphasic shocks of 25 msec duration at a frequency of 25/sec and up to 140 V for as long as 60 sec. Some stimuli were so strong that they caused contraction of the neck muscles followed by cardioacceleration and arousal, but never bradycardia. After the cardioacceleration had proceeded for some minutes, electrical stimulation of the vagus at much lower voltages resulted in bradycardia, even though heart temperature had changed only slightly.

Manually choking euthermic hibernators results in reflex bradycardia (Chatfield and Lyman, 1950) and this can be blocked by atropinization, indicating that it is parasympathetically induced. When hibernating ground squirrels were manually choked, using a chilled glove in order not to heat the neck region, the heart rate was not slowed. Choking started arousal and, 15 min or more after the initial stimulus, choking resulted in profound bradycardia. Temperature was not precisely related to vagal function induced by choking, for the heart of one animal slowed at a T_b of 8°C, whereas another failed to slow at a T_b of 10.6°C.

The influence of the parasympathetic system on heart rate was complicated by the anomalous effect of acetylcholine and methacholine on ground squirrels in hibernation. When acetylcholine, in a wide variety of doses, was infused via an aortic cannula, an almost immediate cardioacceleration occurred. Infusion of methacholine produced a similar cardioacceleration, but there was a delay of at least 24 sec before acceleration occurred. Detailed examination of this effect will be undertaken in the next chapter, but it is important to emphasize here that neither drug produced the expected bradycardia upon infusion. After the cardioacceleration had proceeded for several minutes, infusion of either drug resulted in bradycardia. Immediate slowing of the heart sometimes occurred when acetylcholine was forced upstream in the aorta in animals entering into or arousing from hibernation, but this never occurred in the many experiments with undisturbed hibernating animals, even though the slower blood flow would presumably make the upstream flow more likely. In animals entering or arousing from hibernation, the heart slowed some seconds after infusion of the drug. This delay was taken to indicate the circulation time from the end of the cannula to the heart. Again, this was not observed in deeply hibernating animals (Lyman and O'Brien, 1963).

Biewald and Raths (1959) reported similar results with European hamsters that were anesthetized in hibernation and then artificially warmed. Bilateral electrical stimulation of the vagi failed to produce a chronotropic effect on the heart until T_b reached 10° or 12°C. This lack of chronotropic effect was not observed in artificially cooled animals, and for this reason it was suggested that some unexplained change took place in the receptivity of the heart during hibernation. The conditions of the experiment were not the same as with the unanesthetized hibernating ground squirrels and did not permit stimulation of the vagi at different heart rates while at the same T_b.

The increase in threshold to parasympathetic influence has also been observed in the hypothermic hedgehog anesthetized with chlorolose (Johansen *et al.,* 1964). It was found that electrical stimulation of the vagus or arterial infusion of acetylcholine failed to slow the heart at T_b's of 5°–7°C, though the expected bradycardia took place at higher temperatures. There are several reports on the limitations of vagal activity by cold in animals that do not hibernate and, though the experiments are complicated by the inability of the hearts of nonhibernators to beat at low temperatures, the evidence is clear that vagal influence is blocked at low temperatures (Adolph and Nail, 1960).

The observed lack of vagal parasympathetic influence in the deeply hibernating ground squirrel differs from the experiments with hypothermic hedgehogs and nonhibernators in that the effect is not simply one of low temperature, for it also involves a period of cardioacceleration, even without a change in T_b, before the parasympathetic influence is restored. Presumably, some shift in the internal milieu or changes in some component in the heart itself causes this increase in threshold. It was observed that the return to a lower threshold occurred after a shorter period of acceleration in ground squirrels which had just entered hibernation than in animals which had hibernated for longer periods, so that the increase in threshold may build up through the first few days of a bout of hibernation (Lyman and O'Brien, 1963).

If these observations can be extended to all other hibernators, the observed changes in heart rate during deep hibernation probably can not be ascribed to parasympathetic influence. The mechanisms involved in this block are not known, though the experiments with parasympathomimetics indicate that more is involved than a simple failure of nerve conduction.

Changes in the intrinsic rhythmicity of the heart would be reflected in changes of heart rate, and this has been tested by infusion of the alkaloid veratramine, which slows the heart and antagonizes the cardioaccelerator action of epinephrine by direct action on the pacemaker in nonhibernating mammals (Krayer, 1949; Kosterlitz *et al.,* 1955). Although respiratory rate increased, the heart rate of ground squirrels infused with this drug did not change even with doses that produce convulsions if the animals were then stimulated to arouse from hibernation (Lyman and O'Brien, 1963). Since the amount of drug which affects the heart in nonhibernators is much lower than the dose which produces convulsions (Krayer, 1949), it appears that inherent changes in the pacemaker have little effect on heart rate in hibernation. Furthermore, if the pacemaker were the sole determinant of the heart rate, one would expect that rate and T_b would be closely correlated during deep hibernation, and this is not the case.

The influence of the sympathetic system on the hibernating ground squirrel is particularly difficult to test by infusion of pharmacological agents because of the side effects of available drugs. The effect of the catecholamines epinephrine and norepinephrine will be discussed in more detail in Chapter 5. However, in unpublished observations we find that arterial infusion of either drug in very small amounts (0.004–0.005 mg/kg) results in an increase in heart rate, oxygen consumption, and T_b. In high doses, arterial infusion of the α-adrenergic blocking agent piperoxan (benodaine) was followed by a decrease in peripheral resistance and an increase in heart rate which may have been compensatory (Lyman and O'Brien, 1960). Infusion of the ganglionic blocking agent hexamethonium chloride resulted in a drop in blood pressure and peripheral resistance, and an increase in heart rate which was also interpreted as being compensatory (Lyman and O'Brien, 1963).

Intravenous infusion of adrenergic neuron blocking agents in nonhibernating mammals results first in sympathomimetic effects which are followed by adrenergic depression (Boura and Green, 1965). A similar reaction occurred when hibernating ground squirrels were infused intra-arterially with the adrenergic neuron blocker, β-TM10, for the heart rate increased and many of the animals aroused from hibernation. In a few cases, however, the heart rate slowed after 1–3 hr, and peripheral resistance and blood pressure declined to levels that were lower than before the infusion.

Further considerations concerning the control of heart rate and circulation will be detailed in Chapter 7, but some conclusions can be drawn from the evidence presented here. It appears that the influence of the parasympathetic system is at a low ebb during the deeply hibernating state, and it is probable that this holds true for both rodent and insectivore hibernators. The variation in heart rate that occurs during deep hibernation is not due to changes in the inherent rhythmicity of the heart and is not strictly dependent on temperature. At least some of the physiological control mechanisms that relate heart function to the rest of the circulatory system in active mammals are retained during hibernation. Thus, a decline in peripheral resistance and a drop in blood pressure will usually be followed by a compensatory increase in heart rate and, presumably, an increase in cardiac output. The heart in the hibernating animal is sensitive to sympathomimetic stimulation and responds by an increase in rate. On the other hand, some sympathetic influence must be present during hibernation, because blocking of that influence is followed by a decrease in peripheral resistance and a decline in blood pressure and heart rate. Recently, Twente and Twente (1978) have advanced a theory concerning the control of hibernation, which involves a functioning and effective parasympathetic system. This is described and discussed in Chapter 14.

The T_b of the animal in deep hibernation is usually slightly above T_a over a range of temperature which may be as great as 4°–20°C. Very little investigation has been done on the upper limits at which an animal will hibernate, but it seems to vary both with the season and the species and possibly even with the individual. Wang and Hudson (1971) have shown that the eastern chipmunk (*Tamias striatus*) will hibernate at ambient temperatures between about 2° and 25°C. At lower T_a's both the mean and the range of the difference between T_b and T_a were greater than they were at higher T_a's. Though there is little information on hibernation at high temperatures, a great deal of research has been carried out at low T_a's, particularly in the last few years, as investigators attempt to discover the extent of temperature regulation that can take place in the deeply hibernating mammal. Wyss (1933) appears to be the first to report that the hibernating dormouse (*Glis glis*) could respond to subzero ambient temperatures by increasing its metabolic rate, and continue in hibernation. Since then, the same response has been described in many hibernators such as hamsters (Lyman, 1948), chipmunks (Wang and Hudson, 1971), thirteen-lined and golden mantled ground squirrels (Lyman and O'Brien, 1972), marmots (Mills and South, 1972), hedgehogs (Kristoffersson and Soivio, 1964), and including bats (Davis and Reite, 1967) and a hummingbird (Hainsworth and Wolf, 1970).

It appears reasonable to assume that this is a characteristic of hibernators as a whole. At the outset it must be realized that the reaction does not always occur. As was shown with the golden hamster many years ago (Lyman, 1948), some animals may undergo a complete arousal from hibernation when challenged by a low T_a, and others may die. In the latter case, we have never seen a hibernating animal that did not respond to a low T_a by increasing heart and respiratory rate, but sometimes the response is not brisk enough to forestall death. In the majority of cases, the animal remains in hibernation with an augmented rate of heat production and this state provides an important aspect in the study of the mechanisms of hibernation. Whether the animal entering hibernation simply shuts off its temperature regulating mechanisms and slides down into hibernation, or whether it actually suppresses these mechanisms and controls the descent in hibernation, still it is clear that some drastic change must take place that permits a decrease in body temperature over a range of 30°C. Why, then, is the change limited, so that decline in temperature does not continue until the animal reaches a T_b from which it cannot recover? In its response to low temperatures, is the animal using the same mechanisms for regulating its body temperature as it used in the euthermic state, or is it using a different set of mechanisms that are not involved in maintenance of T_b during euthermia? If the former, how can the "thermostat" be shifted from 37°-39°C to 3°-5°C? Once changed, does the "thermostat" regulate the T_b of the hibernating animal within a narrow range of both high and low T_b's, or does the "thermostat" only protect the animal from possibly lethal low temperatures? If the "thermostat" is not functioning during hibernation, can the observed response be explained as the result of a nonspecific stimulus brought on by cold, just as an animal in hibernation may respond to a mechanical stimulus with a burst of muscle action potentials, an increase in heart rate, and a rise in metabolic rate?

In attempting to answer these questions, a major approach has been to compare the methods of temperature regulation in hibernation with those of a euthermic animal. In the latter case, it is widely accepted that temperature sensors from various parts of the body, including the skin, the extremities, and the spinal cord all contribute to supplying information to the temperature regulating center in the preoptic–anterior hypothalamic area of the brain (POAH). The response of this area, in turn, is influenced by its own temperature. By altering the temperature of the peripheral sensors while holding the temperature of the POAH area steady, or vice versa, the effect of these sensors can be determined.

For experiments such as these, thermodes which can change the temperature of the brain have been used extensively (Fig. 3-1) and a simpler device consisting of a spoon-shaped copper thermode which could be positioned against the head of the hibernating animal was used on Turkish hamsters (*Mesocricetus brandti*), thirteen-lined (*Citellus tridecemlineatus*) and golden mantled (*C. lateralis*) ground squirrels, and the common dormice (*Glis glis*) (Lyman and O'Brien, 1972). Most hibernating rodents curl in a ball, so it is difficult to manipulate the temperature of the extremities without causing arousal. An exception is the dormouse, which hibernates on its back with its hind legs exposed (Fig. 4-3). Double-layered, jacketed half-boots or Wellingtons (T_w) can be slipped over these extended feet, and a temperature-controlled liquid pumped between the two layers of the boot. With this device the temperature of the hind feet can be altered without affecting the deep body or the brain temperature (Lyman and O'Brien, 1974).

The experiments of Heller and Colliver (1974) with golden mantled ground

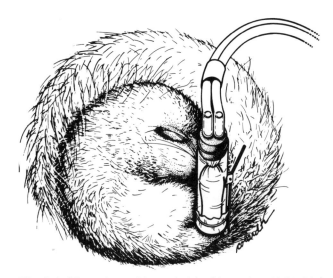

Fig. 4-3. *Glis glis* in hibernating position typical for this species with hind feet at level of dorsal surface and one of the two jacketed half-boots in place. (After Lyman and O'Brien, 1974.)

squirrels, and those of Lyman and O'Brien (1972, 1974) with golden mantled and thirteen-lined ground squirrels and Turkish hamsters indicated that input from peripheral temperature sensors had no detectable effect on the central temperature regulator. When the temperature of the POAH area in the case of the golden mantled ground squirrel, or the whole head in all three species, was held steady and T_b was reduced by lowering T_a, there was no change which would indicate that metabolic heat production had been altered and increased. Such was not the case with the common dormouse or the yellow-bellied marmot, *Marmota flaviventris* (South *et al.*, 1975). With the dormouse, any change in the temperature of the spoon-shaped thermode resulted in an immediate increase in heart and respiratory rates and in muscular activity, indicating that the temperature of the thermode was stimulating receptors in the skin of the head. Furthermore decrease of the temperature of the hind feet by means of the Wellingtons produced the same result. In contrasting the response of hamsters and ground squirrels with that of the dormouse it was emphasized that the former hibernated in burrows which would not be exposed to sudden changes of T_a. For this reason, quick-acting receptors of changes in T_a would not be necessary for survival. On the other hand, dormice hibernate in hollow logs and other exposed places, where there is little protection against sudden changes in T_a. Here a rapid response to a drop in T_a might save the life of the animal.

The response of the marmot was similar to that of the dormouse. Rapid increases or decreases in T_a resulted in alterations of electromyographic activity and changes in heart rate without any change in the temperature of the brain. This indicated that peripheral sensors must be effective, yet the marmot hibernates in well-protected burrows as do hamsters and ground squirrels. The yellow-bellied marmot is one of the largest of the deep hibernators, and its head is tucked beneath its body during hibernation. In this position the brain as a whole, and specifically the POAH area, might be the last part of the body to sense the penetration of lethal cold. South *et al.* (1975) postulate that functioning peripheral temperature sensors may act as an early warning system to protect the marmot from a drop in T_b that might be fatal.

Experiments with euthermic mammals, both hibernators and nonhibernators, have shown that metabolic heat production will remain unchanged over a small range of POAH temperatures. If the POAH temperature drops below this range, metabolic heat production will increase. Heller and group have designated the temperature at which this increase takes place as "T_{set}", and below T_{set} the change in metabolic rate is inversely proportional to the actual POAH temperature. The slope of the straight line generated by these data is the proportionality constant for the response (α). Figure 4-4 (Heller and Colliver, 1974) demonstrates this concept.

Mills and South (1972) showed that direct cooling of the POAH area using a single thermode resulted in an increase in heart rate and heat production in the

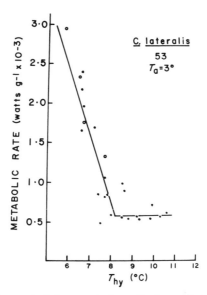

Fig. 4-4.   Changes in metabolic heat production with changes in temperature of the POAH area in a hibernating golden mantled ground squirrel. Animal was maintaining T_b above T_a by bursts of thermogenesis. Open circles, heat production following naturally occurring declines in POAH temperature. Closed circles, response to manipulations of POAH area. (After Heller and Colliver, 1974.)

hibernating marmot. Similarly, Lyman and O'Brien (1972, 1974) found that heart and respiratory rates increased when the whole brain of hibernating thirteen-lined and golden mantled ground squirrels, and Turkish hamsters was chilled. In the thirteen-lined ground squirrel there was an increase in the heart and respiratory rates soon after the brain temperature (T_{br}) began to drop. This reaction could be repeated by holding the brain at one temperature until the heart and respiratory rates returned to basal levels and then cooling the brain again (Fig. 4-5). Because of the rapidity of the response when the brain was chilled, and because the response occurred at brain temperatures which varied by almost 5°C, it was concluded that it was the decline and change in T_{br} rather than the temperature reaching a precise threshold which resulted in an increase of heart and respiratory rates. Alternatively, changes in threshold could produce the same result, but, as illustrated by Fig. 4-5, this would involve both decreases and increases in threshold over periods that were shorter than 1 hr. Ground squirrels appeared to be unable to tolerate large differences between T_b and T_{br}, for apnea followed by death was apt to occur under these circumstances.

Although Heller and group reported in 1972 that manipulations of the temperature of the POAH area did not result in temperature regulating responses (Heller

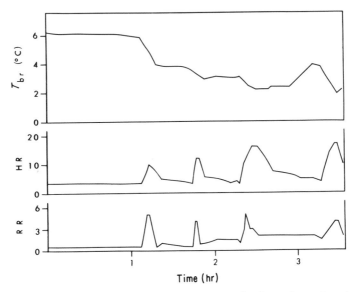

Fig. 4-5.   Repeated decreases of brain temperature (T_{br}) by thermode result each time in transient increases in heart and respiratory rates in hibernating thirteen-lined ground squirrel. Note that the fourth response occurs at a higher brain temperature than does the third. T_a varied between 2.2° and 3.9°C. (After Lyman and O'Brien, 1972.)

and Hammel, 1972), they were later able to show that metabolic heat production increases proportionately as the POAH dropped below T_{set} in hibernating golden mantled ground squirrels (Heller and Colliver, 1974). This relationship occurred whether the POAH temperature was manipulated by the bilateral thermodes in the POAH or whether it occurred naturally (Fig. 4-4). This is very convincing evidence that the POAH temperature regulating system is functioning during hibernation though the response is greatly reduced when compared to the euthermic animal.

   The experiments of Lyman and O'Brien (1972) in which the temperature of the whole brain was manipulated are not comparable to experiments in which the temperature of a discrete area is altered, for in the former case the response appeared to be due solely to a decrease in temperature. It is also worth noting that chilling the whole brain has more widespread effects than chilling with thermodes, for differences between brain and body temperature in the former situation were apt to cause death.

   Using the hibernating yellow-bellied marmot and a single thermode, Mills and South (1972) shifted the temperature of the POAH area both above and below its natural temperature of hibernation. In one case an animal responded to hypothalamic heating by moving from its bed and lying prone outside its nest.

Unfortunately, this response could not be elicited in later experiments, but it strongly suggests that the reaction is an attempt to lose heat from the whole body when the POAH area alone is heated. In a later paper, South *et al.* (1975) observed that nonspecific body movements occurred when the POAH was heated, and these were interpreted as attempts to leave the nest without arousing from hibernation. In this report the POAH area was insensitive to thermal manipulations until the animal had been in hibernation for 4 or more days. This was taken as evidence that T_b was plastic at the initial phase of a bout of hibernation, and that regulation of T_b developed as the bout progressed.

A great variation was evident when the open loop gain was calculated from the metabolically generated rise in T_b resulting from the stimulus of chilling the POAH area. This variability appeared to be related to the temperature of the brain before the stimulus was applied, for there was an obvious relationship when the logarithm of the open loop gain was plotted against the prestimulus T_{br}, with the larger gains occurring at the lower temperatures. This is the opposite of an Arrhenius effect, and implies active regulation rather than a simple physicochemical reaction. South *et al.* (1975) point out that, with such a system in operation, the open loop gain increases markedly as the T_b approaches 0°C and hence the animal is able to respond and rewarm itself in spite of the low T_a. If the open loop gain did not increase with decreasing temperature, the response might be inadequate and the animal might die in hibernation.

Florant and Heller (1977) also studied the reactions of the yellow-bellied marmot, using bilateral thermodes with euthermic and hibernating animals as well as animals entering hibernation. As with the ground squirrels in the earlier study, the change in metabolic heat production was inversely proportional to the temperature of the POAH area once this temperature had dropped below T_{set}. Unlike the results of South *et al.* (1975) they found that the POAH area was responsive to changes in temperature at all times during a bout of hibernation though it appeared to be least responsive when the POAH area was between 15° and 11°C as the animal entered the hibernating state. They concluded from their results that responsiveness was low at the end of entrance into hibernation, and that it then rose during the first day of deep hibernation only to decline until 1–2 days before the end of the bout. At this point sensitivity again rose, followed by arousal from the hibernating state. The T_b of the hibernating marmot increased gradually throughout the day preceding arousal which suggested that the threshold of the POAH area rose above the actual temperature of this area long before arousal took place. In contrast, it was predicted that ground squirrels would have the lowest sensitivity on the first day of hibernation and that it would increase as the bout progressed. Thus, differences have been found in the sensitivity of the POAH area during hibernation, not only between species, but also with the same species using somewhat different methods.

As will be discussed at length in Chapter 5, animals in hibernation appear to become increasingly sensitive to various types of stimuli as a bout of hibernation progresses. The response to these stimuli is usually a burst of muscular activity, and an increase in heart and respiratory rates and in oxygen consumption. Various types of stimuli have been employed including mechanical stimulation of the periphery, intraperitoneal injection, and intraventricular infusion of drugs with known pharmacological effects. Heller and Hammel (1972) had suggested in an earlier paper that the responses to change of temperature in the hibernating mammal were the result of stimulating the lower centers of the brain, possibly the reticular activating system. In spite of the elegant demonstration that the POAH is functional during hibernation, the possibility remains that the reticular activating system may be involved in temperature regulation in certain situations. For example, we suggested that the response to low temperature in the dormouse might involve two systems (Lyman and O'Brien, 1974). When the hind feet or the whole animal was rapidly cooled a burst of muscle action potentials, an increase in heart and respiratory rates, and an increase in oxygen consumption (unpublished observations) occurred at a very specific temperature (Fig. 4-6). This response is indistinguishable from the reaction to a mechanical or other type of stimulus and may well be the same as a response to a stimulus mediated by the

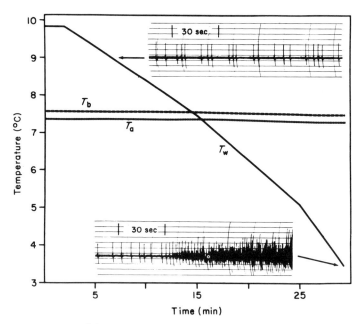

Fig. 4-6. Response of dormouse to rapid chilling of hind feet (T_w). Upper insert is typical undisturbed ECG-EMG. Lower insert shows burst of muscle action potentials, with increase in cardiac and respiratory rates. (After Lyman and O'Brien, 1974.)

reticular activating system. Its threshold may change during a bout of hibernation, and its final effect is clearly thermoregulatory even if the initial response is not followed by arousal from hibernation.

On the other hand, if the dormouse is slowly cooled a response may occur, again at a very specific temperature, consisting of an increase in heart and respiratory rates and in oxygen consumption, but with a delay in the appearance of muscle action potentials. Changes in threshold occur from experiment to experiment, with the range of T_b falling between 4.5°C and 1.3°C (Fig. 4-7). This reaction is clearly different from the previous one, and may well be mediated by the POAH, though proof of this is lacking. Therefore, it can be argued that this animal in hibernation has two defenses against potentially lethal cold. The defense that is perhaps the most basic consists of the usual, though sluggish, responses of the temperature regulating mechanisms which apparently function in all euthermic mammals. A fundamental and unanswered question in the field of hibernation research concerns the mechanisms by which the set point of this regulator can be changed to permit the animal to hibernate.

The second defense is typical of animals, such as dormice and marmots, that respond to the peripheral stimulus of cold as other hibernators respond to various nonspecific stimuli. Even though it may involve only the briefest increase in

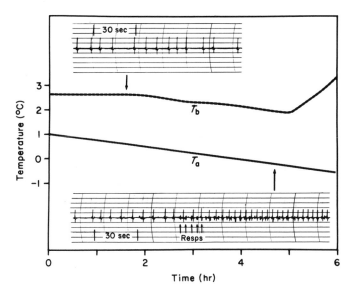

Fig. 4-7. Response of dormouse to slowly declining T_a and T_b. Upper insert is typical undisturbed ECG-EMG. At 4.7 hr cardiac and respiratory rates suddenly increase as shown in lower insert with no accompanying burst of muscle action potentials. Resps, respirations. (After Lyman and O'Brien, 1974.)

muscular activity and increase in respiratory, heart and metabolic rates, still it is thermoregulatory in the sense that it causes an increase in T_b and if the response continues it results in arousal from hibernation. It is conceivable that a hibernator could regulate its temperature above the lethal point by periodic bursts of activity followed by quiet periods. Thus, regulation of T_b might take place without involving the usual temperature regulating system at all. The response of animals in hibernation to various other stimuli will be discussed in Chapter 5.

REFERENCES

Adolph, E. F., and Nail, R. L. (1960). Vagal inhibition of the heart in deep hypothermia. *J. Appl. Physiol.* **15,** 911–913.

Baudinette, R. V. (1974). Physiological correlates of burrow gas conditions in the California ground squirrel. *Comp. Biochem. Physiol. A* **48A,** 733–743.

Benedict, F. G., and Lee, R. C. (1938). Hibernation and marmot physiology. *Carnegie Inst. Washington Publ.* No. 497.

Biewald, G.-A., and Raths, P. (1959). Die chronotrope Vaguswirkung auf das Hamsterherz unter dem Einfluss von Winterschlaf und Hypothermie. *Pfluegers Arch. Gesamte Physiol. Menschen Tiere* **268,** 530–544.

Boura, A. L. A., and Green, A. F. (1965). Adrenergic neurone blocking agents. *Annu. Rev. Pharmacol.* **5,** 183–212.

Buchanan, F. (1911a). Dissociation of auricles and ventricles in hibernating dormice. *J. Physiol. (London)* **42,** xix–xx.

Buchanan, F. (1911b). The frequency of the heart-beat in bats and hedgehogs and the occurrence of heart-block in bats. *J. Physiol. (London)* **42,** xxi–xxii.

Chao, I., and Yeh, C. J. (1950). Hibernation of the hedgehog. II. Respiratory patterns. *Chin. J. Physiol.* **17,** 379–390.

Chatfield, P. O., and Lyman, C. P. (1950). Circulatory changes during process of arousal in the hibernating hamster. *Am. J. Physiol.* **163,** 566–574.

Davis, W. H. (1970). Hibernation: ecology and physiological ecology. *In* "Biology of Bats" (W. A. Wimsatt, ed.), Vol. I, pp. 265–300. Academic Press, New York.

Davis, W. H., and Reite, O. B. (1967). Responses of bats from temperate regions to changes in ambient temperature. *Biol. Bull. (Woods Hole, Mass.)* **132,** 320–328.

Dawe, A. R., and Morrison, P. R. (1955). Characteristics of the hibernating heart. *Am. Heart J.* **49,** 367–384.

Eliassen, E. (1960). Cardiovascular pressures in the hibernating hedgehog, with special regard to the pressure changes during arousal. *Arbok Univ. Bergen, Mat.-Naturvitensk. Ser.* No. 6, 1–27.

Endres, G. (1930). Recording of respiration of small animals. *J. Physiol. (London)* **70,** 218–220.

Endres, G., and Taylor, H. (1930). Observations on certain physiological processes of the marmot. II. The respiration. *Proc. R. Soc. London, Ser. B* **107,** 231–240.

Florant, G. L., and Heller, H. C. (1977). CNS regulation of body temperature in euthermic and hibernating marmots (*Marmota flaviventris*). *Am. J. Physiol.* **232,** R203–R208.

Goodrich, C. A. (1973). Acid-base balance in euthermic and hibernating marmots. *Am. J. Physiol.* **224,** 1185–1189.

Hainsworth, F. R., and Wolf, L. L. (1970). Regulation of oxygen consumption and body temperature during torpor in a hummingbird, *Eulampis jugularis*. *Science* **168,** 368–369.

Heller, H. C., and Colliver, G. W. (1974). CNS regulation of body temperature during hibernation. *Am. J. Physiol.* **227**, 583-589.

Heller, H. C., and Hammel, H. T. (1972). CNS control of body temperature during hibernation. *Comp. Biochem. Physiol. A* **41A**, 349-359.

Hudson, J. W. (1973). Torpidity in mammals. *In* "Comparative Physiology of Thermoregulation" (G. C. Whittow, ed.), Vol. 3, pp. 98-165. Academic Press, New York.

Johansen, K., Krog, J., and Reite, O. (1964). Autonomic nervous influence on the heart of the hypothermic hibernator. *In Ann. Acad. Sci. Fenn., Ser. A4* **71**, 243-255.

Kayser, C. (1940). Les échanges respiratoires des hibernants à l'état de sommeil hibernal. *Ann. Physiol. Physicochim. Biol.* **16**, 127-221.

Kayser, C. (1961). "The Physiology of Natural Hibernation." Pergamon, Oxford.

Kosterlitz, H. W., Krayer, O., and Matallana, A. (1955). Studies on veratrum alkaloids. XXII. Periodic activity of the sino-auricular node of the denervated cat heart caused by veratramine. *J. Pharmacol. Exp. Ther.* **113**, 460-469.

Krayer, O. (1949). Studies on veratrum alkaloids. VIII. Veratramine, an antagonist to the cardioaccelerator action of epinephrine. *J. Pharmacol. Exp. Ther.* **96**, 422-437.

Kristoffersson, R., and Soivio, A. (1964). Hibernation in the hedgehog (*Erinaceus europaeus* L.). Changes of respiratory pattern, heart rate and body temperature in response to gradually decreasing or increasing ambient temperature. *Ann. Acad. Sci. Fenn., Ser. A4* **82**, 1-17.

Lyman, C. P. (1948). The oxygen consumption and temperature regulation of hibernating hamsters. *J. Exp. Zool.* **109**, 55-78.

Lyman, C. P. (1951). Effect of increased CO_2 on respiration and heart rate of hibernating hamsters and ground squirrels. *Am. J. Physiol.* **167**, 638-643.

Lyman, C. P. (1958). Oxygen consumption, body temperature and heart rate of woodchucks entering hibernation. *Am. J. Physiol.* **194**, 83-91.

Lyman, C. P., and O'Brien, R. C. (1960). Circulatory changes in the thirteen-lined ground squirrel during the hibernating cycle. *Bull. Mus. Comp. Zool.* **124**, 353-372.

Lyman, C. P., and O'Brien, R. C. (1963). Autonomic control of circulation during the hibernating cycle in ground squirrels. *J. Physiol. (London)* **168**, 477-499.

Lyman, C. P., and O'Brien, R. C. (1969). Hyperresponsiveness in hibernation. *Symp. Soc. Exp. Biol.* **23**, 489-509.

Lyman, C. P., and O'Brien, R. C. (1972). Sensitivity to low temperature in hibernating rodents. *Am. J. Physiol.* **222**, 864-869.

Lyman, C. P., and O'Brien, R. C. (1974). A comparison of temperature regulation in hibernating rodents. *Am. J. Physiol.* **227**, 218-223.

Maclean, G. S. (1981). Blood viscosity of two mammalian hibernators: *Spermophilus tridecemlineatus* and *Tamias striatus*. *Physiol. Zool.* **54**, 122-131.

McManus, J. J. (1974). Activity and thermal preference of the little brown bat, *Myotis lucifugus*, during hibernation. *J. Mammal.* **55**, 844-846.

Mills, S. H., and South, F. E. (1972). Central regulation of temperature in hibernation and normothermia. *Cryobiology* **9**, 393-403.

Nardone, R. M. (1955). Electrocardiogram of the arctic ground squirrel during hibernation and hypothermia. *Am. J. Physiol.* **182**, 364-368.

Pajunen, I. (1970). Body temperature, heart rate, breathing pattern, weight loss and periodicity of hibernation in the Finnish garden dormouse, *Eliomys quercinus* L. *Ann. Zool. Fenn.* **7**, 251-266.

Pembrey, M. S. (1901-1902). Observations upon the respiration and temperature of the marmot. *J. Physiol. (London)* **27**, 66-84.

Pengelley, E. T., and Fisher, K. C. (1968). Ability of the ground squirrel, *Citellus lateralis*, to be habituated to stimuli while in hibernation. *J. Mammal.* **49**, 561-562.

Popovic, V. (1964). Cardiac output in hibernating ground squirrels. *Am. J. Physiol.* **207,** 1345–1348.

Sarajas, H. S. S. (1954). Observations on the electrocardiographic alterations in the hibernating hedgehog. *Acta Physiol. Scand.* **32,** 28–38.

South, F. E., Hartner, W. C., and Luecke, R. H. (1975). Responses to preoptic temperature manipulation in the awake and hibernating marmot. *Am. J. Physiol.* **229,** 150–160.

Strumwasser, F. (1959). Regulatory mechanisms, brain activity and behavior during deep hibernation in the squirrel, *Citellus beecheyi. Am. J. Physiol.* **196,** 23–30.

Tähti, H. (1975). Effects of changes in CO_2 and O_2 concentrations in inspired gas on respiration in the hibernating hedgehog (Erinaceus europaeus L.). *Ann. Zool. Fenn.* **12,** 183–187.

Tähti, H. (1978). Seasonal differences in O_2 consumption and respiratory quotient in a hibernator (Erinaceus europaeus L.) *Ann. Zool. Fenn.* **15,** 69–75.

Tähti, H., and Soivio, A. (1975). Blood gas concentrations, acid-base balance and blood pressure in hedgehogs in the active state and in hibernation with periodic respiration. *Ann. Zool. Fenn.* **12,** 188–192.

Twente, J. W., Jr. (1955). Aspects of a population study of cavern-dwelling bats. *J. Mammal.* **36,** 379–390.

Twente, J. W., and Twente, J. (1978). Autonomic regulation of hibernation by *Citellus* and *Eptesicus. In* "Strategies in Cold: Natural Torpidity and Thermogenesis" (L. C. H. Wang and J. W. Hudson, eds.), pp. 327–373. Academic Press, New York.

Twente, J. W., Twente, J., and Giorgio, N. A. (1970). Arousing effects of adenosine and adenine nucleotides in hibernating *Citellus lateralis. Comp. Gen. Pharmacol.* **1,** 485–491.

Wang, L. C.-H., and Hudson, J. W. (1971). Temperature regulation in normothermic and hibernating eastern chipmunk, *Tamias striatus. Comp. Biochem. Physiol. A* **38A,** 59–90.

Wyss, O. A. M. (1932). Winterschlaf und Wärmehaushalt, untersucht am Siebenschläfer (Myoxus glis). *Pfluegers Arch. Gesamte Physiol. Menschen Tiere* **229,** 599–635.

5

Sensitivity to Arousal

A mammal in natural hibernation is capable of arousing from that state at any time using only self-generated heat. As will be detailed in Chapter 6, animals in hibernation spontaneously arouse from time to time even if they are maintained in unchanging external conditions. In addition, arousal may be evoked at any time by an adequate external stimulus. The strength of the stimulus necessary to start the arousal may vary from species to species and from day to day with individual animals, and the mechanisms involved in these changes have not yet been completely clarified.

The physiological changes that occur at the start of a spontaneous arousal have not been reported in detail. Although Twente and Twente (1965) showed that spontaneous arousals occurred in predictable pattern in the golden mantled ground squirrel and in other ground squirrels (Twente et al., 1977), still recordings of spontaneous arousals are usually made by chance during long-term monitoring of animals in deep hibernation. Once spontaneous arousal is well under way it is indistinguishable from an induced arousal, so that not much attention has been focused on the first few minutes of spontaneous arousal.

The changes that occur during an induced arousal are relatively easy to record. Fitting the hibernator to the recording devices can serve as the arousal stimulus, and the experiment can proceed smoothly until the animal is almost fully aroused and starts to struggle purposefully. If the first few minutes of arousal are of interest, the techniques can be the same as those used in the studies of deep hibernation. Attaching the various recording leads may cause some temporary disturbance but will usually not result in arousal. Once the animal has settled back into hibernation, a stimulus may be applied and arousal initiated.

If the sensitivity of an animal to stimuli during hibernation is the object of study, graded stimuli may be applied which may initiate arousal responses but which are not sufficiently potent to start a full arousal. After a variable period the excitatory phase terminates, the animal returns to deep hibernation and its sensitivity can be tested again. Obviously, some stimuli are so disturbing that arousal is inevitable. In these cases the "spinal hibernator" (Lyman and O'Brien, 1969) has some usefulness. The hibernating animal is prepared by quickly dividing the spinal cord at C_1 and applying artificial respiration. The operation stops the arousal process, but sensitivity to peripheral stimuli remains and the preparation may be used for many hours. It is particularly useful when screening drugs for the proper dosage that can then be used on the intact hibernator.

Our interest was focused for some time on a reaction which is associated with evoked arousal and appears to be ubiquitous in deeply hibernating animals (Lyman and O'Brien, 1969). In hibernators fitted with dermal electrodes, an externally applied stimulus results in a long-lasting burst of muscle action potentials (MAP) which is immediately followed by an increase in respiratory and cardiac rate and a rise in oxygen consumption. These changes may be followed by a complete arousal from hibernation, or the heart and respiratory rate and oxygen consumption may decline and the animal return to the condition of deep hibernation. We have observed this reaction in *Mesocricetus auratus* and *brandti*, in *Citellus tridecemlineatus* and *lateralis*, in *Marmota monax* and *flaviventris*, in *Glis, glis*, and Kulzer (1967) reports the burst of muscle action potentials in the bat *Myotis myotis*. The response in the dormouse (*Glis*) is particularly striking. This arboreal rodent hibernates with its bushy tail curled over its back (Fig. 4-3). If the piloerected hairs are gently displaced, the typical burst of muscle action potentials occurs (Fig. 5-1) with a concurrent increase in respiratory rate and heart rate, though the stimulus does not usually result in arousal.

The burst of MAP and its sequelae appeared to offer a means of testing changes in sensitivity of the hibernator over the period of one bout of hibernation and, therefore, it has been examined in some detail. It was found that many types of stimuli resulted in the burst of MAP. These included vibration, pressure, locally applied heat or cold and electrical shocks. Infusion of a variety of substances via chronically implanted aortic cannuli also caused the MAPs. Acetyl-

| 20 sec |

Touch

Fig. 5-1. Burst of muscle action potentials and increase in heart rate after touching the tail hair of a hibernating dormouse. (From Lyman, unpublished observations.)

choline and various mild acids and bases that cause no observable effect in the active animal produce the MAPs in animals during hibernation.

Using a heated disk applied to the shaved back of hibernating thirteen-lined ground squirrels, we observed that the duration and the voltage of the MAPs decreased as the T_b of the animal was increased by artificial warming (Fig. 5-2). The increase in specificity of the reflex with increasing T_b was demonstrated in an experiment with golden mantled ground squirrels. At a T_b of 10°C, heating of the disk caused a long-lasting burst of MAPs and a slight rocking of the body. At a T_b of 17°C the local thermal stimulation resulted in a well-defined scratching movement of one hind leg, which knocked the disk from the back of the animal.

In order to trace this reflex to the spinal cord, spinal hibernating ground squirrels were shown to respond with MAPs to a variety of stimuli. The spinal cord was then destroyed by passing a roughened wire down the spinal canal. After this, MAPs could not be elicited with stimuli of any magnitude. We concluded that the burst of MAPs was the manifestation of a spinal reflex which became less specific as the temperature of the hibernating animal decreased.

Koizumi *et al.* (1959) have examined the effect of cold on the spinal reflex in cats. They found that excitability decreased but responsiveness increased as the temperature of the spinal cord was lowered from 37° to 20°C. Below 18°-20°C the response ceased—a result that might be expected in a species that does not hibernate. They attributed the hyperresponsiveness to "(1) repetitive discharge, probably contributed to by failure of accommodation and prolongation of excitatory potentials; (2) the greater participation of interneurones in reflex action; and (3) a greater power of recruitment at synaptic junctions because of the changes just mentioned." They considered that this probably caused a greater spread of activation, overflow of excitation into systems that were not ordinarily invaded, and general failure of confinement of activity to normal pathways.

It has been shown that nervous tissue from animals that hibernate can function at much lower temperatures than can homologous tissue from nonhibernators (Chatfield *et al.*, 1948; South, 1961). The hyperresponsiveness which has been

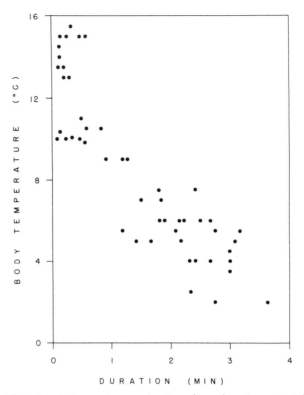

Fig. 5-2. Effect of body temperature on duration of muscle action potentials in thirteen-lined and golden mantled ground squirrels. Stimulus of heated disk which induced muscle action potentials was removed within 5 sec after the beginning of the response. (After Lyman and O'Brien, 1969.)

reported in the cat as temperatures were lowered to 20°C appears to continue and increase in the hibernator with decreasing temperatures down to nearly 0°C. This means that the hibernator must become more responsive as its body temperature decreases. This increase in responsiveness may compensate for a loss of coordination of various physiological systems at the low temperatures of deep hibernation, so that the animal has an ever-present alarm which is strong enough to activate the mechanisms that are involved in the arousal process.

The hyperresponsiveness that occurs in hibernating ground squirrels should not be equated with hypersensitivity, and the sensitivity of hibernators to stimuli which may cause arousal still presents a somewhat confusing picture. This picture is intertwined with the puzzle of the periodic arousals that occur in all hibernating mammals and which is examined in detail in Chapter 6.

The response of the animal in hibernation to acetylcholine (ACh) infused via an aortic cannula gave rise to speculations concerning changes in sensitivity to stimuli during a bout of hibernation. As little as 0.001 mg/kg acetylcholine may result in a burst of MAPs which is indistinguishable from the response to a mechanical or electrical stimulation. It seemed possible that the ACh was acting directly on the muscle and that its nicotinic effect was causing the burst of MAPs. Further, the sensitivity of the animal in hibernation might be akin to the sensitivity that develops in the skeletal muscle of an active animal when that muscle is denervated. If this were the case, it would be expected that the muscle becomes increasingly sensitive as a bout of hibernation progresses, just as the denervated muscle increases its sensitivity with the passage of time. Furthermore, it has been shown that the sensitivity of the denervated muscle fiber to acetylcholine moves from the area of the motor end plate and toward the ends of the muscle fiber as denervation sensitivity proceeds (Axelsson and Thesleff, 1959), and this change would also be expected in the muscles of hibernating animals if the parallelism were to hold. It was attractive to postulate that the periodic spontaneous arousals so characteristic of all hibernators might be caused by the bursts of activity from muscles which had become increasingly sensitive over a period of continuous hibernation.

In order to test this concept, a long series of *in vitro* experiments was performed at 10°C to determine the sensitivity to ACh of the *extensor digiti quarti* muscles taken from active, hypothermed, and hibernating thirteen-lined ground squirrels, as well as from active animals in which this muscle had been previously denervated. Dose–response curves of the amount of tension produced by addition of ACh to the perfusion bath were inconsistent, but some information was obtained on the dose necessary to produce a threshold response, although there was large variation from preparation to preparation. It was found that the threshold response of muscles from artificially chilled and from active animals was the same and, therefore, these two categories were combined. In comparison, the threshold response of muscles from animals in hibernation was significantly lower ($p < 0.01$), in spite of the large standard deviation (585 ± 591 versus 127 ± 155 μg ACh/100 ml). However, there was no evidence that the muscles became more sensitive as the days in hibernation increased, for muscles removed from animals in the first day of hibernation were as sensitive as muscles from animals which had hibernated continuously for 6 days. Furthermore, the muscles from active animals whose foreleg had been denervated 14 days earlier were significantly more sensitive to ACh than were the muscles from hibernating animals.

In an attempt to obtain more consistent results, the technique of close arterial infusion of the gastrocnemius muscle was used. Again, the results were variable, with the threshold of the muscle in the hibernating, spinal cold animals somewhat lower than the threshold from nonhibernating control animals which were first made hypothermic, then made spinal and tested at the same temperature (5°C).

The lowest threshold in the former group was 0.0088 μg ACh compared to a low of 0.22 in the latter group and $p < 0.05$, which tended to reinforce the concept that the muscles from animals in hibernation were sensitized to ACh.

A method, first developed by Fatt (1950), measures thresholds and detects the migration of sensitivity to acetylcholine from the region of the motor end plate. For this method it is necessary to use a long, thin muscle with the end plates located in discrete bands. In his experiments, Fatt used the *sartorius* or the *rectus abdominis* from the frog. Unfortunately, rodents do not possess a discrete *sartorius,* and the *rectus abdominis* is too bulky, so the *omohyoideus* was used instead. This necessitated a long dissection, using "hibernating," chilled, anesthetized animals, so that it is questionable whether the excised muscle was in its normal state. Reliable thresholds and dose–response curves could not be established, and we obtained no evidence that sensitivity shifted from the motor end plate (Lyman and O'Brien, 1969).

More recently, Moravec *et al.* (1973) reexamined this problem using the phrenic nerve–diaphragm preparation in the golden hamster, and iontophoretic infusion of ACh. With this elegant technique, minute amounts of ACh may be deposited in discrete areas along a muscle fiber and the sensitivity of each part of the muscle membrane can be measured. These investigators also used microelectrodes to measure frequency of miniature end plate potentials. The preparations were tested at 20°C and it was found that the zones of sensitivity of the muscle fibers from hibernating animals were approximately twice as long as in the active control animals, and the maximal sensitivity at the motor end plate zone also was about double that of the controls. The frequency of miniature end plate potentials was reduced more than seven times in muscles from the hibernating animals compared to the controls. In addition, South (1961) found that the diaphragm muscle of hibernating hamsters was more excitable to direct stimulation than were the muscles from euthermic controls. Moravec *et al.* (1973) point out that the muscle from the hibernating animal reacts in a manner similar to a muscle in early stages of denervation, with decrease in spontaneous release of ACh at the motor end plates and an increase in sensitivity occurring in both cases. They also mention that the decrease in release of ACh occurs in the muscles of old animals and that in all three situations the changes are associated with disuse.

In spite of intensive research on the phenomenon of denervation sensitivity, it has not been firmly established that disuse is the cause of the changes in the end plate and the muscle fiber. It is generally agreed, however, that the increase in sensitivity and its movement away from the motor end plate are progressive over a period of days. The evidence to date indicates that these changes are not progressive during a bout of hibernation. Furthermore, denervation results in atrophy of the muscle, yet Moravec *et al.* (1973) observed no atrophy of the muscle in hibernating hamsters and South (1961) found that the maximum tension produced at low temperatures in diaphragm muscle from active and hibernating hamsters was essentially the same. Thus, the analogy between denervation sensitivity and increased sensitivity in hibernation is not complete.

If skeletal muscle is more sensitive to ACh in the hibernating animal, an alternate explanation may be offered. A reduction in the concentration or the effectiveness of acetylcholinesterase could result in the apparent increase in sensitivity of skeletal muscle to ACh. With this thought in mind, we compared the activity of acetylcholinesterase in blood and muscle of hibernating and active ground squirrels tested at 37°C and found that there was no significant difference. The Q_{10} of acetylcholinesterase over the range of 7°–37°C was found to be very low, confirming the results of others. Eserinization of the *extensor digiti quarti* muscles from active and hibernating ground squirrels resulted in a similar proportional increase in sensitivity to ACh. Moravec *et al.* (1973) also found that the difference in sensitivity between the muscle fibers from hibernating and active animals still obtained when the preparations were treated with an anticholinesterase.

Thus, the bulk of the evidence to date indicates that skeletal muscle of animals in hibernation is more sensitive to the neuromuscular transmitter ACh than is the muscle of active animals, but the reason for this increase in sensitivity is not known. Whatever the reasons for the increase in sensitivity, it is apparently dependent on the hibernating state rather than the effect of cold on the whole animal, for the muscles from artificially chilled animals are comparable to controls.

We had originally assumed that the burst of MAPs that occurred after intra-arterial infusion of ACh was due to its direct nicotinic action on the muscle. When we found later that many chemicals, including weak acids and bases, produced the same result, we examined the problem further. It has been known for some time that ACh will stimulate receptors and this drug has been implicated as the neurochemical transmitter between the adequate stimulus to the receptor and the initiation of the nervous discharge. In examining this concept, Douglas and Ritchie (1960) were able to stop the effect of ACh on receptors by using the ganglionic blocking agent hexamethonium. In spite of this block, the receptors in question still responded to their respective adequate stimuli. The same kind of experiment was carried out on spinal and normal cannulated hibernators, with the same results in both types of preparation (Lyman and O'Brien, 1969). It was first shown that the animals responded to intra-arterial infusion of ACh with a burst of MAPs and that the same result was produced by an adequate physical stimulus. After infusion of hexamethonium, a much larger dose of ACh produced virtually no effect, but the reaction to the same adequate physical stimulus was unchanged (Fig. 5-3). This, of course, does not prove that infused ACh, acid, and base are

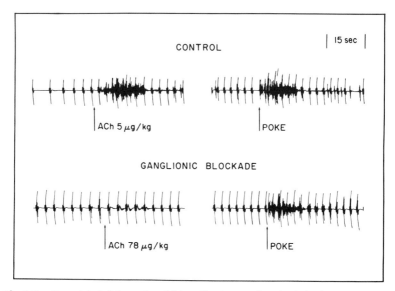

Fig. 5-3. Cannulated, hibernating thirteen-lined ground squirrel, body temperature 7°C, dermal electrode recording. Upper panels: muscle action potentials in the normal animal caused by infusion of acetylcholine (left), or mechanical stimulation by poking with a stiff wire (right). Lower panels: 18 min after infusion of 10.8 mg/kg hexamethonium chloride to produce ganglionic blockade. Effect of acetylcholine is blocked (left), whereas effect of mechanical stimulation is not (right). (After Lyman and O'Brien, 1969.)

acting on peripheral receptors such as those for touch, pain, and temperature, which, in turn, produce the burst of MAPs, but the parallelism between these results and those of Douglas and Ritchie strongly suggests that this is the case.

The bursts of MAPs appeared to offer a unique opportunity to chart the change in sensitivity in the intact hibernating animal. The MAPs which followed a stimulus were easily recorded and an entire series of experiments could be carried out in a single day. For this reason many tests were conducted using a variety of stimuli, including point pressure, electric shock, local heat and cold, and intra-arterial infusion of ACh and weak acids and bases. It was hoped that these studies would reveal a shift in sensitivity either with a change in the T_b of the hibernator or with the passage of time during a single bout of hibernation, but consistent results could not be obtained. One technical difficulty was never solved to complete satisfaction. In spite of the many types of stimuli that would elicit the response, there was no assurance that stimuli were repeatably identical over a period of time. Local application of heat, cold, or pressure depended on the exact replacement of stimulating apparatus, and this could not be assured. Electrical stimulus will vary with moisture content of the skin and other factors that can not be controlled. The stimulation of specific receptors by intra-arterial infusion of drugs or chemicals assumes that the circulatory bed remains unchanged throughout the test period, and it is generally recognized that changes in the circulation take place during hibernation.

Toth (1977) has confirmed the lack of repeatability in the response of the hibernating animal to graded stimuli and expanded the study to include spinal ground squirrels in hibernation and hypothermia, using as a stimulus the tip of a pressure transducer on a marked, shaved area of the mid-dorsum. She found little correlation between the T_b in the hibernating animal and the pressure necessary to produce a burst of MAPs, nor was there evidence that the sensitivity increased over a bout of hibernation. However, in spinal, hibernating golden mantled and thirteen-lined ground squirrels, the threshold to tactile stimuli was shown to be consistently correlated with T_b, with higher T_b's resulting in lower thresholds. In spite of this correlation, the absolute threshold varied from animal to animal, which might be expected, since it would be unlikely that the identical receptors would be stimulated in the same way in each animal. In euthermic animals which were then rendered hypothermic, peripheral stimuli did not result in a burst of MAPs, but if the spinal cord was divided, peripheral stimulation resulted in MAPs. In anesthetized intact, euthermic animals there was also no response, but after the animal was rendered spinal a short burst of MAPs occurred after each peripheral stimulus.

Toth postulated that the reflex is inhibited by higher centers in the brain, and that this inhibition is released in the euthermic anesthetized or in the hypothermic ground squirrel when the spinal cord is cut at C_1. The variability of the threshold to peripheral stimuli in the intact hibernating animal may thus be due to an interplay between the spinal reflex and inhibition from higher centers. The fact that Toth obtained such reproducible results with spinal animals indicates that the technique used in applying the stimulus was repeatable, and, therefore, the variability of the response found in the intact hibernator was not due to methodological artifacts.

In experiments with this spinal reflex, the difference between the sensitivity to a peripheral stimulus and the sensitivity to arousal is emphasized again and again. An animal may be stimulated repeatedly during a day's experiments and may respond many times with MAPs and accelerated heart rate and oxygen consumption without arousing from hibernation. At some unpredictable point, which may occur after several days of sequential experiments, an applied stimulus will be followed by the usual burst of MAPs and cardioacceleration, but the MAPs and rapid heart rate will continue instead of returning to their original states, and arousal from hibernation will begin. Obviously, a massive stimulus will result in an arousal, but in the realm of smaller peripheral stimuli which produce recordable muscle action potentials, there appears to be no correlation between the strength of the stimulus, or the voltage and duration of the response, and the sudden appearance of the start of arousal.

An evoked arousal from hibernation can, thus, be regarded as a two-step system. The applied stimulus of sufficient magnitude results in a response which is manifest by a burst of muscle action potentials, an increase in heart and respiratory rate, and a rise in oxygen consumption. In the intact hibernator, the threshold for this response changes unpredictably over time, possibly due to inhibition from higher centers. If the stimulus is sufficient to cause a response, arousal from hibernation may follow, but this in turn depends largely on the sensitivity of the animal to the arousal stimulus, and this also changes over time. This second threshold has been examined in considerable detail and some of the aspects of these studies will be discussed in Chapter 6, but the experimental procedures are outlined here and emphasize the difference between the initial and the final response to a stimulus.

Twente and Twente (1968a) showed that hibernating *Citellus lateralis* would not arouse during the first half of a bout of hibernation when given an intraperitoneal injection of isotonic saline. During the second half of the bout, the animals became increasingly responsive to the stimulus, so that progressively more animals aroused from hibernation as they approached the expected termination of their bout. Kristoffersson and Soivio (1964) also noted that pricking hedgehogs with a needle would cause arousal only at the end of a bout of hibernation. These studies were extended to the effect of epinephrine and norepinephrine on the arousal response (Twente and Twente, 1968b). It was found that intraperitoneal injection of epinephrine resulted in a significant number of arousals in the first half of a bout of hibernation, and that larger doses resulted in arousals earlier in the bout. The injections were often followed by a "partial arousal" or a measurable change in T_b followed by a return to deep hibernation, and these partial arousals occurred more often with lower doses of epinephrine. Injections of norepinephrine were not followed by an increase in T_b, though heart rate increased transiently, as it did with injections of epinephrine. Later, this anomalous effect with norepinephrine was found to be incorrect (Twente and Twente, 1978). See also Chapter 14. These authors concluded that partial arousals were due to the calorigenic effect of epinephrine, and that the full arousals were due to the effect of epinephrine on the hypothalamic area. They suggested that the progressive increase in irritability during a bout of hibernation might be due to a gradual accumulation of epinephrine in that area. Subsequent measurements of epinephrine from various parts of the brain gave no indication of differential accumulation (Twente *et al.*, 1970a) and measurements of [^{14}C]epinephrine indicated that it did not pass the blood–brain barrier (Twente and Twente, 1971).

There seems no doubt that this painstaking work has established that arousal from hibernation after application of a stimulus is increasingly more likely as a bout of hibernation continues in *C. lateralis*. However, the mechanisms of the action of injected epinephrine remain obscure. The injection of a liquid into the animals must have involved a variety of stimuli, including tactile, pressure, and possibly pain, yet it is clear that different injected substances produced different effects. Though no muscle action potentials were recorded and oxygen consumption was not measured, there was an immediate increase in heart rate following the injection even if T_b remained unchanged. It seems certain that this was the response which we have studied and which gives no indication of increasing in sensitivity or frequency as a bout of hibernation continues. Therefore, the substances which the Twentes and co-workers injected must be doing more than evoking a burst of muscle action potentials and an increase in cardiac and metabolic rate.

In their search for the critical substance that was responsible for the arousal reaction, a great array of substances was tested (Twente et al., 1970b). The results suggested to them that the substances which evoke arousal act through the stimulation of adenylate cyclase and that cyclic AMP may be involved in naturally occurring progressive irritability during a bout of hibernation (Chapter 6). However, they recognized that the method of testing had inherent weaknesses (Twente and Twente, 1970). The distribution of a substance after intraperitoneal injection could not be uniform from animal to animal, and injections into the intestine or fat could not be avoided. They estimated that 10–15% of the injections may have been faulty, and the site of injection may have been one reason for the variability of the results. The period which transpired between the time of injection and arousal varied greatly both from animal to animal and from substance to substance. For example, the range of latency after injection of 0.075 μmoles of luteinizing–interstitial cell stimulating hormone was 1–31 hr, whereas the range for one-third of that dose was 25–47 hr. Thus, lack of uniformity of response is a major difficulty, and the fact that even injection of 0.5 ml of isotonic saline will cause arousal during the latter part of a bout of hibernation emphasizes that the site and action of the stimulus are still unknown.

Beckman and collaborators made a direct attack on the problem by infusing drugs into discrete areas of the brain via chronically implanted bilateral cannulae. Infusion of norepinephrine or 5-hydroxytryptamine into the preoptic–hypothalamic area resulted in arousal from hibernation, which was indistinguishable from arousal produced by physically stimulating the ground squirrel. Infusion of the same amount of isotonic saline never resulted in a rise in temperature that was more than that which occurred normally in the undisturbed hibernator and infusion of drugs into the region of the putamen had no effect (Beckman and Satinoff, 1972).

In an effort to delineate the neural pathways and mechanisms involved in the arousal process, further experiments were carried out using the same technique but with the site of the infusion changed to the midbrain reticular formation. In these experiments particular emphasis was placed on the effect of ACh and of carbachol, which is a synthetic analog of ACh that is less susceptible to hydrolysis by acetylcholinesterase, and hence remains active for a longer period of time (Beckman *et al.*, 1976a). In these experiments the results were not as clear-cut as in the previous work. Infusion of carbachol at the highest dose (1.50 μg) usually resulted in complete arousal from hibernation, but failed to cause any change in one case out of six. At the lowest dose (.75 μg) complete arousal occurred in one case, but no change took place in two other experiments. ACh usually resulted in partial warming followed by a return to the deeply hibernating state, whereas the effect of epinephrine was usually negative, though it resulted in a complete arousal in one case and partial warming in another. 5-Hydroxytryptamine was without effect. In general, infusion of control substances had no effect, but infusion of sodium bitartrate at a pH of 3.4, as a control for the norepinephrine infusion, resulted in complete arousal in one case and partial warming in another. In euthermic animals, carbachol was not tested, but ACh resulted in an increase in T_b. Norepinephrine sometimes caused an increase in T_b, and control substances usually did not, but there were some exceptions.

In further work using the same technique, they studied the effect of ACh using various concentrations of the drug and infusing at various times during a bout of hibernation (Beckman *et al.*, 1976b; Beckman and Stanton, 1976). Of particular interest to us here is the finding that the animals became increasingly sensitive to the infusion as a bout of hibernation progressed. With the larger doses, complete arousal usually occurred during the last quarter of a bout of hibernation, whereas infusion of smaller amounts was apt to be followed by partial warming during the same period, but not before. They concluded that the response to ACh when infused into the midbrain reticular formation is diminished during the entrance and early phases of hibernation. They point out that this finding is very similar to the "progressive irritability" of Twente and Twente (1968a), and so it seems to be, but it must be realized that the Beckman technique is a more direct approach to the problem. They have shown that infusion of a drug into an area of the central nervous system produces a specific effect after a short period of time, and that the threshold for this effect becomes lower as a bout of hibernation progresses. They did not test the preoptic anterior hypothalamic area with epinephrine and norepinephrine for "progressive irritability." The arousal response is similar to the response to peripheral injection of epinephrine or norepinephrine reported by Twente and Twente (1978) in their later paper. Twente and Twente concluded in their earlier papers that their intraperitoneal injections of epinephrine were not reaching the brain, and this conclusion seems justified. This indicates that the hibernating animal experiences "progressive irritability" in two areas—the brain and the periphery—as hibernation progresses.

Some criticism may be warranted toward the method of Beckman *et al.* (1976a,b) of quantitating the effect of a drug using only the change in brain and interscapular body temperature. We have often observed in thirteen-lined and golden mantled ground squirrels and in the Turkish hamster that heart and respiratory rate and oxygen consumption may increase markedly without increasing body temperature. They also ascribe the rise in interscapular temperature to the thermogenic effect of brown fat. The golden mantled ground squirrel has a bilateral subscapular deposit of brown fat, but no midline interscapular deposit, so an interscapular temperature would not necessarily reflect the temperature of the brown fat. Nevertheless, the evidence is persuasive that, in this species at least, a progressive lowering of threshold to ACh takes place in the midbrain reticular formation during a bout of hibernation.

It is tempting to assume that the midbrain reticular formation is crucial to the arousal process, but not enough evidence has accumulated to conclude that this is the case, and the results from electrical recording from this area are, in the main, negative. Chatfield and Lyman (1954) concluded that the limbic system was first to show electrical activity as the golden hamster woke from hibernation, and South et al. (1969) found that an increase in asynchronous cortical activity accompanied by increased activity in the midbrain reticular formation and hypothalamus occurred before or with the rise in brain temperature in the yellow bellied marmot (Marmota flaviventris). Shtark (1970) reported that the first sign of electrical activity in tentative arousal in the European hedgehog was in the limbic system. In contrast, Raths (1958) found early electrical activity in the area of the pons in the European hamster (Cricetus cricetus), but not in all regions above the midbrain, especially the hypothalamus, rhinencephalon, thalamus, and neocortex.

Thus, the information on electrical activity is contradictory and it is difficult to know how to evaluate it. In our hands, the cerebral cortex of the golden hamster is electrically silent during hibernation, whereas that of the woodchuck is not (Chatfield and Lyman, 1954; Lyman and Chatfield, 1953). One would hope that there would be no differences between species in such a basic entity as the functioning of the brain, but such does not appear to be the case.

REFERENCES

Axelsson, J., and Thesleff, S. (1959). A study of supersensitivity in denervated mammalian skeletal muscle. J. Physiol. (London) **147**, 178-193.

Beckman, A. L., and Satinoff, E. (1972). Arousal from hibernation by intrahypothalamic injections of biogenic amines in ground squirrels. Am. J. Physiol. **222**, 875-879.

Beckman, A. L., and Stanton, T. L. (1976). Changes in CNS responsiveness during hibernation. Am. J. Physiol. **231**, 810-816.

Beckman, A. L., Satinoff, E., and Stanton, T. L. (1976a). Characterization of midbrain component of the trigger for arousal from hibernation. Am. J. Physiol. **230**, 368-375.

Beckman, A. L., Stanton, T. L., and Satinoff, E. (1976b). Inhibition of the CNS trigger process for arousal from hibernation. Am. J. Physiol. **230**, 1018-1025.

Chatfield, P. O., and Lyman, C. P. (1954). Subcortical electrical activity in the golden hamster during arousal from hibernation. Electroencephalogr. Clin. Neurophysiol. **6**, 403-408.

Chatfield, P. O., Battista, A. F., Lyman, C. P., and Garcia, J. P. (1948). Effects of cooling on nerve conduction in a hibernator (golden hamster) and a non-hibernator (albino rat). *Am. J. Physiol.* **155**, 179–185.

Douglas, W. W., and Ritchie, J. M. (1960). The excitatory action of acetylcholine on cutaneous non-myelinated fibres. *J. Physiol. (London)* **150**, 501–514.

Fatt, P. (1950). The electromotive action of acetylcholine at the motor end-plate. *J. Physiol. (London)* **111**, 408–422.

Koizumi, K., Brooks, C. McC., and Ushiyama, J. (1959). Hypothermia and reaction patterns of the nervous system. *Ann. N.Y. Acad. Sci.* **80**, 449–456.

Kristoffersson, R., and Soivio, A. (1964). Hibernation of the hedgehog (*Erinaceus europaeus* L.). The periodicity of hibernation of undisturbed animals during the winter in a constant ambient temperature. *Ann. Acad. Sci. Fenn., Ser. A4* **80**, 1–22.

Kulzer, E. (1967). Die Herztätigkeit bei lethargischen und winterschlafenden Fledermäusen. *Z. Vgl. Physiol.* **56**, 63–94.

Lyman, C. P., and Chatfield, P. O. (1953). Hibernation and cortical electrical activity in the woodchuck (*Marmota monax*). *Science* **117**, 533–534.

Lyman, C. P., and O'Brien, R. C. (1969). Hyperresponsiveness in hibernation. *Symp. Soc. Exp. Biol.* **23**, 489–509.

Moravec, J., Melichar, I., Janský, L., and Vyskočil, F. (1973). Effect of hibernation and noradrenalin on the resting state of neuromuscular junction of golden hamster (*Mesocricetus auratus*). *Pfluegers Arch.,* **345**, 93–106.

Raths, P. (1958). Die bioelektrische Hirntätigkeit des Hamsters im Verlaufe des Erwachens aus Winterschlaf und Kältenarkose. *Z. Biol.* **110**, 62–80.

Shtark, M. B. (1970). "The Brain of Hibernating Animals." Natl. Aeronaut. Space Adm., Washington, D.C. [Engl. transl. of "Mozg Zimnespyashchikh." Nauka, Sib. Branch, Novosibirsk.]

South, F. E. (1961). Phrenic nerve-diaphragm preparations in relation to temperature and hibernation. *Am. J. Physiol.* **200**, 565–571.

South, F. E., Breazile, J. E., Dellman, H. D., and Epperly, A. D. (1969). Sleep, hibernation and hypothermia in the yellow-bellied marmot (*M. flaviventris*). *In* "Depressed Metabolism" (X. J. Masacchia and J. F. Saunders, eds.), pp. 277–312. Am. Elsevier, New York.

Toth, D. M. (1977). EMG responses of intact and spinal ground squirrels to tactile stimulation during hibernation, hypothermia and normothermia. *Comp. Biochem. Physiol. A* **57A**, 167–177.

Twente, J., and Twente, J. W. (1971). Distribution of [^{14}C] epinephrine and [^{131}I] albumin in regions of the brain and other tissues of hibernating *Citellus lateralis* after intraperitoneal injection. *Comp. Gen. Pharmacol.* **2**, 138–144.

Twente, J. W., and Twente, J. A. (1965). Regulation of hibernating periods by temperature. *Proc. Natl. Acad. Sci. U.S.A.* **54**, 1058–1061.

Twente, J. W., and Twente, J. A. (1968a). Progressive irritability of hibernating *Citellus lateralis*. *Comp. Biochem. Physiol.* **25**, 467–474.

Twente, J. W., and Twente, J. A. (1968b). Effects of epinephrine upon progressive irritability of hibernating *Citellus lateralis*. *Comp. Biochem. Physiol.* **25**, 475–483.

Twente, J. W., and Twente, J. (1970). Arousing effects of trophic hormones in hibernating *Citellus lateralis*. *Comp. Gen. Pharmacol.* **1**, 431–436.

Twente, J. W., and Twente, J. (1978). Autonomic regulation of hibernation by *Citellus* and *Eptesicus*. *In* "Strategies in Cold: Natural Torpidity and Thermogenesis" (L. C. H. Wang and J. W. Hudson, eds.), pp. 327–373. Academic Press, New York.

Twente, J. W., Cline, W. H., Jr., and Twente, J. A. (1970a). Distribution of epinephrine and norepinephrine in the brain of *Citellus lateralis* during the hibernating cycle. *Comp. Gen. Pharmacol.* **1**, 47–53.

Twente, J. W., Twente, J., and Giorgio, N. (1970b). Arousing effects of adenosine and adenine nucleotides in hibernating *Citellus lateralis. Comp. Gen. Pharmacol.* **1,** 485–491.

Twente, J. W., Twente, J., and Moy, R. M. (1977). Regulation of arousal from hibernation by temperature in three species of *Citellus. J. Appl. Physiol.* **42,** 191–195.

6

The Mystery of the Periodic Arousal

As described in Chapter 1, the popular view of hibernation of an animal's sinking into deep torpor at the approach of winter, not to waken again until spring, is not found in nature. All species of hibernators studied exhibit a behavior of arousing periodically, that is, spontaneously regaining a high body temperature in the face of a large thermal gradient, before resuming deep torpor. The ability so to rewarm is one of the most characteristic and definitive features of hibernators, and the physiological mechanisms which underlie it will be considered in Chapter 7.

The frequency of the arousal and the length of the euthermic periods vary widely with species, and, within species, among individuals and with the time of year. As detailed in Chapter 2, some species show daily torpor, which may be obligatory, whereas others may undergo voluntary torpor, and this pattern may be lengthened and body temperature lowered still further in longer, seasonal hibernation. Some hibernators, such as hamsters, store food for the winter in their nests and burrows, whereas other species become fat before entering hibernation. The food storers eat during their euthermic periods, and, in the laboratory, they also drink. It is not known whether they require water under natural conditions or, in fact, how the water could be obtained. Syrian hamsters, which do not store large reserves of fat, arouse at frequent intervals (2–10 days among all studied; commonly 3–5 days in the laboratory). Long-term, seasonal hibernators, such as ground squirrels, dormice, and hedgehogs, exhibit a pattern of short bouts of hibernation at the beginning of the season which gradually become longer, reaching lengths of perhaps 10–15 days, and then shortening again at the end of the hibernating season (Pengelley and Fisher, 1961; Kristoffersson and Soivio, 1964; Strumwasser et al., 1964; Twente and Twente, 1967; Wang, 1973). Small bats exhibit perhaps the longest uninterrupted bouts, often in excess of 30 days (Menaker, 1964). Species which store food do not necessarily undergo bouts of hibernation which are shorter than those that occur in animals which depend on fat for winter nourishment, for the bouts that occur in the food-storing Turkish hamster are as long as those occurring in ground squirrels or dormice (Lyman and O'Brien, 1977). At least in ground squirrels, hamsters, and hedgehogs there is a marked tendency for the animal to become progressively more sensitive to arousal by external stimulation during a bout of hibernation (Chapter 4).

Whatever the length or depth of the bouts of hibernation among various species and individuals, the existence of spontaneous arousal raises the compelling question, why do they bother to rewarm? Kayser (1953) computed for a hibernating dormouse that the energy expended in a single arousal lasting hours was equivalent to that used in 10 days of hibernation. If one views hibernation as a device for escaping high energy expenditure in the cold, it would seem that the process of arousal dissipates a significant part of that advantage. Many who have considered this question have supposed that there must be some inescapable constraint placed upon the animal at low body temperature and that rewarming is necessary for removing that constraint. On the other hand, it is possible that periodic rewarming has little to do with hypothermia per se, but rather represents the extension and the persistent expression of rhythmic processes present in the euthermic individual and indeed in euthermic individuals in all mammalian species. We shall consider the results of investigations based upon both of these points of view.

THE EMERGENCY EMERGENCE

The idea that emergence reflects an internal emergency, present or potential, has given rise to two subsidiary hypotheses, namely, that at low temperature nutrients are being depleted that must be restored or that toxic substances are accumulating in the internal environment which must be eliminated. In its grossest form this amounts to saying that the animal must rewarm in order to eat or to urinate.

Judged even in these gross terms, the hypothesis may have merit with respect to the "nutrient" aspect. As described, Tucker (1962) showed that the duration of daily torpor of pocket mice was inversely proportional to the amount of food available, and, even in normally nonhibernating species (*Mus, Baiomys*), it is possible to induce daily torpor by food deprivation. Nevertheless, that the timing of arousal in the pocket mice was still a preprogrammed event (that is, arousal was not initiated by actual depletion of metabolic reserve) is suggested by his observation that the mice sometimes did not eat all of the food available. (Unfortunately, these observations on mice have never been extended. Would they take into account the effect of ambient temperature? Could the period have been extended by intravenous feeding?) It is also possible that in the hamster, which lays on no reserve of body fat and which does cache food in its nest and eats during periods of arousal, the length of the bout may be set by metabolic reserves.

Nevertheless, most long-term hibernators do possess fat reserves and may survive an entire winter season without eating or drinking. For these, at least, actual food is not the problem. Even so, it has been suggested (Hock, 1958; Strumwasser *et al.*, 1964) that there may be blocks in metabolic pathways at low body temperatures with depletion of some critical substrate. Formally in this category, though not actually concerned with a nutrient, is the proposal of Strumwasser *et al.* (1964) that nerve cells must replenish chemical templates at high temperature. Although it is likely that major shifts may occur in preferred metabolic pathways at low temperature (Chapter 8), little evidence has been adduced to support these hypotheses of metabolic limitation.

Perhaps in part because of this idea, however, there has been a long sustained interest in blood glucose levels and carbohydrate reserves. Because of differences in species and techniques, the status of blood sugar in hibernation is complex (Chapter 8). In any case, what is required, but seldom achieved, are systematic measurements during the course of single bouts of hibernation. This stringent condition was met in the studies of Galster and Morrison (1975) on chronically catheterized arctic ground squirrels whose periods of hibernation were well established and predictable. They found that as a ground squirrel entered a bout of hibernation its blood glucose concentration was about twice that of a resting, nonhibernating ground squirrel and that it fell more or less linearly during the bout to about half that level. During arousal blood glucose was replenished from liver glycogen.

Turning to the "toxicity" side of the "emergency" hypothesis, Dubois (1896) considered the possibility of need to void the bladder as a signal for arousal, but dismissed it when he found that marmots would still arouse periodically even when their afferent pathways from the bladder had been abolished by spinal transection. This venerable hypothesis was resurrected briefly (Pengelley and Fisher, 1961; Fisher, 1964; Pengelley *et al.*, 1971) before being laid once again to rest. As with Dubois and others, they noticed that urination was the one consistent act performed by golden mantled ground squirrels during an arousal period, but they found no correlation between distended bladder volume and length of a bout of hibernation. Fisher (1964) found that arousal could be stimulated more readily by injecting a solution of urea in saline into the peritonea of the ground squirrels than was achieved with just saline alone. His attempt, however, to correlate concentration of naturally produced urea in the blood with length of bout did not lead to a convincing conclusion. Some years later Pengelley *et al.* (1971), sampling blood from individuals at different times within a single bout by means of an indwelling aortic catheter, found quite conclusively that, far from rising, the urea concentration fell during a bout.

Aside from urea, other factors in the internal environment have been nominated as involved, such as thyroxine and water, largely on the basis of arousal caused by injection of the given substance (see Lyman and Chatfield, 1955). Such observations are often of little value since injecting any substance may inflict pain and stimulate arousal, and in earlier work this possibility was not adequately controlled.

Not in this category, however, are the careful experiments of Twente and Twente on golden mantled ground squirrels (Twente and Twente, 1968a,b, 1970; Twente *et al.*, 1970). By keeping continuous records of arousal frequency for many individual squirrels, they were able to "calibrate" the sensitivity of the organism to substances injected into its peritoneum by whether and to what extent it aroused prior to its expected time of arousal. As detailed in Chapter 14, they reported that epinephrine (but not norepinephrine) in very small doses would initiate early arousal. In a later paper, norepinephrine was shown to be nearly as effective as epinephrine (Twente and Twente, 1978). They found further that cyclic adenosine monophosphate (cAMP) and certain of its precursors and congeners which would stimulate adenylate cyclase would also cause arousal. Moreover, trophic hormones whose effects are mediated by stimulation of adenylate cyclase could also induce early arousal but not other trophic hormones, proteins, or polypeptides. They proposed that activation of the adenylate cyclase could play a role in the progressive irritability that characterizes arousal. There are several gaps in the argument, however. No evidence was presented of increasing cyclase activity during a bout; there were several (nontrophic) hormones whose effects are mediated by cAMP which had no effect on arousal, and it was unexplained how adenosine triphosphate (ATP), which was one of the precursors of cAMP that did have an effect, could have got into cells to be converted to cAMP.

In short, no "injection study," even one as careful and well-controlled as those of Twente and Twente (1970) or of Galster and Morrison (1975), can by itself establish an internal cause of periodic arousal because the demonstrations that some factor has the effect of evoking arousal and that also varies in a predictable way during a bout of hibernation does not prove that it does do so in the unperturbed, naturally hibernating animal. As with the classical paradigm for establishing a "candidate" as a neurotransmitter, it must not only be shown that a factor mimics the biological effect, and that under normal circumstances it is present, but also that when its presence is prevented the effect does not occur.

Another candidate for which the first two of these criteria have been at least partially met is K^{ion}. Loss of K^{ion} from cells at low temperature, especially from excitable cells, leads to depolarization, which ultimately would result in inactivation, but which at an intermediate stage would also produce hyperexcitability. Thus, the necessity of periodic arousal could be the restoration of high cytoplasmic K, and the increase of arousability of the organism during a bout might be explained as an effect of progressive loss of K from some tissues. Fisher and Mrosovsky (1970) confirmed the expectation that a solution containing an isotonic mixture of KCl and NaCl injected in enough volume to raise the K concentration in the extracellular space by about 20%, caused arousal more frequently than the same volume (0.5 ml) of isotonic NaCl alone.

Retention of K by cells is a critical requirement for survival at low body temperature, and correspondingly, cells of hibernators are well adapted to this end. Many tissues, such as brain, heart muscle, and diaphragm, appear to retain K perfectly during the hypothermia of hibernation (Willis *et al.*, 1971). Nevertheless, some tissues, such as leg skeletal muscle in hamsters and thirteen-lined ground squirrels, do lose K when the organism becomes cold, and in erythrocytes of thirteen-lined ground squirrels there is a slow, but progressive, loss of K as well (Kimzey and Willis, 1971; Willis *et al.*, 1971).

Most careful and well-controlled studies of ions in serum of hibernating individuals have failed to show any tendency for increase of K that could be attributed to the low body temperature of a single bout. Willis *et al.* (1971) suggested that hyperkalemia resulting from the loss of K from muscle and other tissues might be forestalled by uptake in other cells, notably kidney cortex whose K content in hamsters and ground squirrels may rise by as much as 30% during a bout of hibernation. Conceivably, only when such "sinks" had been saturated would the concentration of K in the serum rise, and thus only at the very end of the bout serve as the fine signal for arousal. Several studies of serum K within single bouts of chronically catheterized individuals have failed to reveal even this "late-bout" hyperkalemia in groundhogs or ground squirrels.

Of course, membrane potential of cells is a function of the ratio of concentration of K between the inside and outside of the cell, so that rise in serum or extracellular K is not a necessary correlate of depolarization. Nevertheless, Willis *et al.* (1971) did not find any loss of K from the brain of hamsters and ground squirrels, and Bito and Roberts (1974), who did find an initial loss of K from gray matter in groundhogs which had just entered a bout, found that it was higher again in individuals which had hibernated a few days. Bito and Roberts (1974) ascribe this fall to cortical shutdown, not cold, and suggest that reaccumulation of K was preparatory to arousal—the opposite of the present proposal. If such a mechanism does play a role, therefore, it would apparently have to be confirmed to the spinal cord or more peripheral tissues. Clausen and Storesund (1971) found not a fall but a rise in skeletal muscle K and in several other tissues of hedgehogs, during a bout of hibernation, no real decline in any tissue K concentration, but, paradoxically, and in contrast to Biörck *et al.* (1956), they did find a slight but significant hyperkalemia.

The third requirement for validating a candidate stated above (i.e., removing the candidate and observing an absence of the effect) when translated into terms of this hypothesis would involve preventing loss of cytoplasmic K (or depolarization resulting therefrom) and demonstration that this lengthens bouts of hibernation or precludes arousal. Such a condition seems at present to defy execution.

THE DEEP—BUT NOT UNBROKEN—SLEEP

The first suggestion that periodic arousal from hibernation may be an extension of rhythmicity in the normothermic state seems to have been made by Folk (1957) who claimed that the arousal in hamsters occurred predominantly in what would have been the active period. The numbers, however, were small and no chi square or other statistical verification was given. Subsequent attempts of this type indicated some instances of arousal synchronizing with a previously determined diurnal rhythm in early hibernation periods of bats, dormice, California ground squirrels (Strumwasser, 1959), and in thirteen-lined ground squirrels provided with light–dark cycles (Strumwasser *et al.*, 1967). On the other hand, with longer seasonal periods of hibernation (i.e., more bouts) no synchronization was found in golden mantled ground squirrels or hedgehogs (Kristoffersson and Soivio, 1964; Twente and Twente, 1965a).

Subsequent studies have been of two types, the further statistical treatment of arousal frequency and the attempt to identify within the still hibernating individual the persistence of measurable physiological rhythms, which may or may not synchronize with arousal. Discussion and interpretation of these observations is complicated by the variety of method and expectation regarding two factors, the effect of temperature and the effect of a light cycle.

Thus, with respect to temperature, one might envision two possibilities, either that with cooling the daily activity cycle becomes lengthened (i.e., slowed down in the rest phase because of low temperature), or, alternatively, that the cycle is independent of temperature (as many endogenous rhythms are) and continues to function with about the same period in hibernation. According to the first idea, *each* full cycle of the endogenous rhythm results in an arousal, but according to the second formulation the ongoing rhythm, whether endogenous (circadian) or not (i.e., entrained), only expressed itself by initiating an arousal occasionally. As will be seen, there is evidence—and there are arguments—for both of these. Regarding light, the question is whether one is looking for the continuation of an endogenous, free-running rhythm or one that is entrained by an impinging environmental cycle. Investigators vary as to whether they use constant dark, constant light or a light–dark cycle, and the results vary as to the effect of the last.

Among the studies of the first type described above (analysis of arousal frequency), perhaps the most provocative was that of Twente and Twente (1965a,b) who recorded the duration of midwinter bouts of hibernation in a population of golden mantled ground squirrels. As mentioned above, no preferred clock time of arousal was observed, but when the animals were placed in ambient temperatures ranging from 25° to 2°C, the duration of their bouts of hibernation varied monotonically and inversely with temperature. This function appeared virtually linear as an Arrhenius plot (i.e., log of duration versus reciprocal of absolute temperature) and when this was extrapolated to 36°C, the duration of the normal sleep cycle fell close to it. The sum of these two observations, therefore, appeared to favor the first hypothesis previously described, namely, that a bout of hibernation represents a temperature-dependent prolongation of the sleep cycle. Pohl (1961), who found that while hamsters generally entered hibernation during the rest phase of their activity cycle and distributed their arousals evenly around the clock, suggested that the prolongation of the rest phase at low temperature governed the termination of a bout in this species as well.

Such a conclusion was somewhat unexpected in view of Rawson's (1960) earlier studies showing that cooling deer mice, hamsters, and bats during the rest phase of their activity cycles had only a small effect of delaying onset of activity. Strumwasser *et al.* (1967), moreover, disputed both the results of the Twentes regarding the effect of temperature and their interpretation regarding the clock time of arousals. In an attempt to test the effect of temperature in golden mantled ground squirrels, he found a log-linearity with inverse of absolute temperature only during early season bouts. Lengths of later bouts showed no systematic dependence on temperature. However, as was pointed out in the discussion of this work, the specific conditions used by Strumwasser and co-workers were far from duplicating those of the Twentes. Pengelley and Fisher (1961), whose conditions were more like those used by Twente and Twente, also saw a decrease in cycle length at higher temperatures in the same species, but they did not attempt to examine this quantitatively.

The difficulty with concluding that nonpreference of arousal time precludes a rhythm is very apparent. If the rhythm were endogenous, typically circadian (i.e., not precisely 24 hr in length) and free-running (either because of constant conditions or because of nonreceptivity to available zeitgebers), the cycle would process around the clock within an individual, and for a population of animals after many bouts would result in an apparently random assortment of arousal times. To detect a rhythm, therefore, would require resorting to more subtle and analytical statistical procedures. Strumwasser *et al.* (1967) cogently reviewed three of the pet techniques and incantations of certified circadians (periodograms, autocorrelograms, and power spectra), but neglected to apply any of these to actual arousal data.

To deal with the same problem, Pohl (1967) devised a test for periodicity based upon size of standard deviation (S.D.) of grouped data. Thus, the actual time between a number of successive arousals in individual dormice was divided by integers (1, 2, 3, . . .) to allow for the possibility of multiples of periods lying within a single bout. The data were then rank ordered and those values lying between 16 and 36 hr were parceled into groups of equal size and the standard deviation computed. The idea is that with random numbers so treated the S.D. will rise (more or less) linearly with the size of the value, but that if there is a natural period (i.e., a nonrandom grouping) a drop below this line will be seen.

Pohl (1967) applied this procedure first to activity cycles of nonhibernating dormice in which the periodicity could be clearly seen both with and without the aid of the analysis. He then applied it to the hibernation cycles both of dormice which he had studied and of hedgehogs reported by Kristoffersson and Soivio (1964). He found apparently clear latent 24-hr rhythms in at least two dormice (out of six examined) and in one of two hedgehogs. Pohl (1967) also found that light cycles at intensities sufficient to entrain activity did not produce a clear rhythm of arousal of the dormice. It should be emphasized that only those dormice that showed strong activity rhythms during their periods of arousal were treated in this way. Other dormice, which exhibited a pattern of changing cycle lengths like those of ground squirrels and which did not emerge from their next boxes during periods of arousal, were not examined.

Furthermore, this treatment, though clever, appears to have shortcomings. First, it is quite subjective, involving as it does the judgement of departure of a curve from a supposedly straight line. Even the curve connecting S.D. generated from random numbers is not entirely straight. Second, it seems likely that in such a large number of trials, chance departures from linearity large enough to be scored as positive might occur. Presumably, both of these objections could be resolved by further development and verification of the procedure.

There are several instances of internal diurnal rhythms having been detected during hibernation. Menaker (1961) reported a daily rise in temperature of re-strained "hibernating" little brown bats. Pohl (1961) found that a rise in respiration during the prior active phase persisted for several days in hibernating bats and dormice. The cycle in dormice was dependent upon a provided light cycle and these rhythms in both bats and dormice all eventually died out. Strumwasser *et al.* (1967) found a circadian periodicity of EEG spectrum in the septum of awake California ground squirrels and a 9-hr cycle in the amygdala of a hibernating indidvidual. In the latter case the peak activity slightly preceded a small rise in brain temperature, which in one case led to arousal. Lew and Quay (1973) found distinct cycles in the tissue concentrations of catecholamines of both awake and hibernating Syrian hamsters which were provided with a light cycle. Finally, in accordance with the observations on arousal in thirteen-lined ground squirrels exposed to a light–dark cycle, Pohl (1961) also observed a daily rise in thoracic temperature even on days when there was no arousal.

We have seen that there is considerable evidence favoring the view that periodic arousal represents *some kind* of continuation of normal periodicity, but that there appears to be no certainty about what the relationship is. Taking together the evidence of both internal rhythms and of frequency of arousal, it seems that so far the best evidence for a continued, temperature-independent circadian period occurs in species and individuals with short-term cycles (California ground squirrels, transitional bats, short-term dormice). The one notable exception to this is that of Pohl and Strumwasser's independent observations with the light-entrained arousal in thirteen-lined ground squirrels (Pohl, 1961; Strumwasser *et al.* 1967). On the other hand, in view of the tendency for increasing irritability with length of bout, these results could also be explained as a special case of induced arousal. Strumwasser *et al.* (1967) pointed out that for a hypothesis based on a continued circadian cycle capable of acting as a stimulus for arousal, it is necessary also to postulate a decreasing threshold to that stimulus. In the case of the light-aroused squirrel, the cycle is provided artificially. It still remains to be determined whether in long-term hibernators an internal cycle also exists that might substitute for the light.

It is obvious that the two general hypotheses discussed here (emergency emergence and extended sleep) are not mutually exclusive; a hibernating animal may arouse spontaneously for any of a variety of reasons just as a person may so arouse from sleep. Even to explain the arousal pattern as the extension of the sleep pattern is really to beg the question, since no one knows what the function of the sleep pattern is, nor what causes it. So it is possible that ultimately the two hypotheses will not even be separable. It may be that changing factors in the internal environment (glucose, K, amines), which would fall under the ''emergency'' category, are, in terms of the Strumwasser ''two factor'' concept, either part of what is setting the threshold or part of what is cycling, or both.

In reviewing this subject one is impressed by how little has been attempted to advance it in recent years. Thus, while Strumwasser and Pohl spelled out methods for analyzing free-running cycles in long-term hibernators over 10 years ago, no one has attempted to apply them in detail. Another approach should be to produce models of hibernators that do not arouse. One example was the fortuitously placed lesions of Satinoff (1967) in the forebrains of thirteen-lined ground squirrels which caused the animals to die in hibernation because they did not awake, even though their thermoregulation while awake was allegedly intact and their hibernation in other respects normal. While this particular result may be difficult to repeat, because the coordinates for making the lesions are unknown and the skulls of adult ground squirrels are not all the same shape and size, other possibilities, such as using anesthetics, other lesions, or blocking agents for effectors would seem to abound. With an animal stably in hibernation, many experiments should be possible to survey factors of the internal environment, rhythms, and the effects in artificially aroused (rewarmed) animals of enforced prolonged bouts of hibernation.

Obviously, the pattern of hibernation is a product of evolutionary processes and, while these certainly involve selective pressures leading to adaptation, they are also in part simply stochastic. This conclusion is engrained in both of the hypotheses discussed here: hibernators are seen as having "failed" to free themselves from one of two lineal traits, either rhythms or adaptation to high body temperature. While either might be true, it should be noted that the latter makes an assumption about the specific course of evolution of hibernation which may or may not be warranted.

This chapter began with the proposition that periodic arousal is a mystery because the energetic cost thereof is so high. We will now close it with the suggestion that the energetic argument is overdrawn and that periodic arousal is a mystery only because we (or so many other hiberniks) choose to make it so. As pointed out in Chapter 1, the saving of energy of even the shortest possible bout, in which the animal immediately rewarms, is of considerable significance to small mammals. The fact that this device has only been extended to days or up to a month, and not to an entire season may only mean that this failure represents an unoccupied thermal niche, just as does the fact that there are no aquatic or arctic mammals with body temperatures of 25°C.

REFERENCES

Biörck, G., Johansson, B., and Veige, S. (1956). Some laboratory data on hedgehogs, hibernating and non-hibernating. *Acta Physiol. Scand.* **37,** 281-294.

Bito, L. Z., and Roberts, J. C. (1974). The effects of hibernation on the chemical composition of cerebrospinal and intraocular fluids, blood plasma and brain tissue of the woodchuck (*Marmota monax*). *Comp. Biochem. Physiol. A* **47A,** 173-193.

Clausen, G., and Storesund, A. (1971). Electrolyte distribution and renal function in the hibernating hedgehog. *Acta Physiol. Scand.* **83,** 4-12.

Dubois, R. (1896). Physiologie comparée de la Marmotte. Masson, Paris.

Fisher, K. C. (1964). On the mechanism of periodic arousal in the hibernating ground squirrel. *Ann. Acad. Sci. Fenn., Ser. A4* **71,** 141-156.

Fisher, K. C., and Mrosovsky, N. (1970). Effectiveness of KCl and NaCl injections in arousing 13-lined ground squirrels from hibernation. *Can. J. Zool.* **48**, 595-596.

Folk, G. E., Jr. (1957). Twenty-four hour rhythms of mammals in a cold environment. *Am. Nat.* **91**, 153-166.

Galster, W., and Morrison, P. R. (1975). Gluconeogenesis in arctic ground squirrels between periods of hibernation. *Am. J. Physiol.* **228**, 325-330.

Hock, R. J. (1958). In discussion of Symposium on Metabolic Aspects of Adaptation of Warm Blooded Animals to Cold Environment. *Fed. Proc., Fed. Am. Soc. Exp. Biol.* **17**, 1066-1069.

Kayser, C. (1953). L'hibernation des mammifères. *Année Biol.* **29**, 109-150.

Kimzey, S. L., and Willis, J. S. (1971). Resistance of erythrocytes of hibernating mammals to loss of potassium during hibernation and during cold storage. *J. Gen. Physiol.* **58**, 620-633.

Kristoffersson, R., and Soivio, A. (1964). Hibernation of the hedgehog (*Erinaceus europaeus* L.). The periodicity of hibernation of undisturbed animals during the winter in a constant ambient temperature. *Ann. Acad. Sci. Fenn., Ser. A4* **80**, 1-22.

Lew, G. M., and Quay, W. B. (1973). Circadian rhythms in catecholamines in organs of the golden hamster. *Am. J. Physiol.* **224**, 503-508.

Lyman, C. P., and Chatfield, P. O. (1955). Physiology of hibernation in mammals. *Physiol. Rev.* **35**, 403-425.

Lyman, C. P., and O'Brien, R. C. (1977). A laboratory study of the Turkish hamster *Mesocricetus brandti. Breviora, Mus. Comp. Zool. Harv. No.* 442.

Menaker, M. (1961). The free running period of the bat clock; seasonal variations at low body temperature. *J. Cell. Comp. Physiol.* **57**, 81-86.

Menaker, M. (1964). Frequency of spontaneous arousal from hibernation in bats. *Nature (London)* **203**, 540-541.

Pengelley, E. T., and Fisher, K. C. (1961). Rhythmical arousal from hibernation in the golden mantled ground squirrel, *Citellus lateralis tescorum. Can. J. Zool.* **39**, 105-120.

Pengelley, E. T., Asmundson, S. J., and Uhlman, C. (1971). Homeostasis during hibernation in the golden mantled ground squirrel, *Citellus lateralis. Comp. Biochem. Physiol. A.* **38A**, 645-653.

Pohl, H. (1961). Temperaturregulation und Tagesperiodik des Stoffwechsels bei Winterschläfern. *Z. Vgl. Physiol.* **45**, 109-153.

Pohl, H. (1967). Circadian rhythms in hibernation and the influence of light. *In* "Mammalian Hibernation III" (K. C. Fisher, A. R. Dawe, C. P. Lyman, E. Schönbaum, and F. E. South, eds.), pp. 140-151. Oliver & Boyd, Edinburgh.

Rawson, K. S. (1960). Effects of tissue temperature on mammalian activity rhythms. *Cold Spring Harbor Symp. Quant. Biol.* **25**, 105-113.

Satinoff, E. (1967). Disruption of hibernation caused by hypothalamic lesions. *Science* **155**, 1031-1033.

Strumwasser, F. (1959). Factors in the pattern, timing and predictability of hibernation in the squirrel, *Citellus beecheyi. Am. J. Physiol.* **196**, 8-14.

Strumwasser, F., Gilliam, J. J., and Smith, J. L. (1964). Long term studies on individual hibernating animals. *Ann. Acad. Sci. Fenn., Ser. A4* **71**, 399-414.

Strumwasser, F., Schlechte, F. R., and Streeter, J. (1967). The internal rhythms of hibernators. *In* "Mammalian Hibernation III" (K. C. Fisher, A. R. Dawe, C. P. Lyman, E. Schönbaum, and F. E. South, eds.), pp. 110-139. Oliver & Boyd, Edinburgh.

Tucker, V. A. (1962). Diurnal torpidity in the California pocket mouse. *Science* **136**, 380-381.

Twente, J. W., and Twente, J. A. (1965a). Effects of core temperature upon duration of hibernation of *Citellus lateralis. J. Appl. Physiol.* **20**, 411-416.

Twente, J. W., and Twente, J. A. (1965b). Regulation of hibernating periods by temperature. *Proc. Natl. Acad. Sci. U.S.A.* **54**, 1058-1061.

Twente, J. W., and Twente, J. A. (1967). Seasonal variation in the hibernating behavior of *Citellus lateralis*. *In* "Mammalian Hibernation III" (K. C. Fisher, A. R. Dawe, C. P. Lyman, E. Schönbaum, and F. E. South, eds.), pp. 47-63. Oliver & Boyd, Edinburgh.

Twente, J. W., and Twente, J.A. (1968a). Progressive irritability of hibernating *Citellus lateralis*. *Comp. Biochem. Physiol.* **25**, 467-474.

Twente, J. W., and Twente, J. A. (1968b). Effects of epinephrine upon progressive irritability of hibernating *Citellus lateralis*. *Comp. Biochem. Physiol.* **25**, 475-483.

Twente, J. W., and Twente, J. (1970). Arousing effects of trophic hormones in hibernating *Citellus lateralis*. *Comp. Gen. Pharmacol.* **1**, 431-436.

Twente, J. W., and Twente, J. (1978). Autonomic regulation of hibernation by *Citellus* and *Eptesicus*. *In* "Strategies in Cold: Natural Torpidity and Thermogenesis" (L. C. H. Wang and J. W. Hudson, eds.), pp. 327-373. Academic Press, New York.

Twente, J. W., Twente, J., and Giorgio, N. (1970). Arousing effects of adenosine and adenine neucleotides in hibernating *Citellus lateralis*. *Comp. Gen. Pharmacol.* **1**, 485-491.

Wang, L. C. H. (1973). Radiotelemetric study of hibernation under natural and laboratory conditions. *Am. J. Physiol.* **224**, 673-677.

Willis, J. S., Goldman, S. S., and Foster, R. F. (1971). Tissue K concentration in relation to the role of the kidney in hibernation and the cause of periodic arousal. *Comp. Biochem. Physiol. A* **39A**, 437-445.

7

Mechanisms of Arousal

As mentioned in Chapter 5, there is very little information concerning the physiological changes that take place when an animal arouses spontaneously from the hibernating state, and few studies have emphasized this aspect of the cycle of hibernation. It has generally been assumed that evoked and spontaneous arousals are identical, but the evidence for this assumption is not very well founded. Actually, Sexton *et al.* (1976) report that spontaneous arousals are slower than evoked arousals in the woodchuck, but a detailed report has not yet been published. Tähti and Soivio (1977) observed this slow spontaneous arousal in the hedgehog, and this is discussed briefly below. The fact that a spontaneous arousal can only be recorded by chance during long-term measurements of a hibernating bout, and the difficulty in creating and monitoring a stimulus-free environment are enough to discourage most investigators, but another factor has served to confuse the issue. Quite naturally, body temperature is most frequently monitored in studies of hibernation. However, evidence from our laboratory and from others clearly shows that factors such as heart rate, respiratory rate, and oxygen consumption may increase markedly without a measurable change in T_b

(Lyman, 1948). An investigator who is monitoring T_b alone during a spontaneous arousal may note a change only when that arousal is well under way. From that point forward, the arousal may appear identical to an evoked arousal, for the rate of increase in T_b may be the same. It is possible, however, that the first changes took place gradually, perhaps hours before the change in temperature manifested itself. Thus, the assumption that the beginning of an evoked and a spontaneous arousal are the same may not be valid. With this caveat, there follows a description of the evoked arousal with some speculations on the spontaneous arousal.

It appears to be characteristic of all hibernators that a burst of muscle action potentials occurs if an animal in hibernation receives a stimulus that exceeds the threshold value. This is always followed by an immediate increase in heart and respiratory rate and a rise in oxygen consumption. If arousal is to follow the stimulus, the oxygen consumption, respiratory rate and heart rate remain elevated and gradually increase. In the study by Tähti and Soivio (1977) of spontaneous arousal in the hedgehog, it was found that heart rate increased before the cessation of apnea or any other sign while, in induced arousal, rise in blood pressure and heart rate and cessation of apnea all occurred at the same time. Muscle action potentials were not recorded, but in induced arousals with rodents the potentials decrease in voltage after the initial burst and may disappear entirely for several minutes. Following this brief period, muscle action potentials reappear on the electromyogram and then continue to increase in frequency and voltage. As arousal continues, visible shivering occurs, which can be correlated with the augmentation of the muscle action potentials. With the onset of shivering, T_b increases more rapidly, and, during the most rapid increase in T_b, the shivering in the anterior part of the body is violent.

The Syrian hamster can serve as a useful example in the study of the physiology of arousal from hibernation because there is little variation from arousal to arousal or from animal to animal. Changes in oxygen consumption and in temperature in various parts of the body are dramatic. At the start of arousal at a T_a of 5°C, oxygen consumption is 60–80 ml/kg/hr, but within about 170 min this rises to more than 8000 ml/kg/hr (Lyman, 1948), a level which is comparable to that of violent exercise (Benedict and Lee, 1938). At the start of arousal, cheek pouch temperature, which is actually the temperature of the thorax, is virtually the same as rectal temperature. As arousal progresses, the cheek pouch temperature rises more rapidly than the rectal temperature, so that there is a difference of more than 20°C between the two areas about 160 min after the start of the arousal process. A short time later, oxygen consumption reaches its peak and declines rapidly, whereas cheek pouch, and then rectal temperature, attain the euthermic level, as shown in Fig. 7-1. In Fig. 7-1 the oxygen consumption after complete arousal is higher than that found in the normal, resting hamster. Most hibernators are very irritable after arousal, and the struggles of the animals as they tried to rid

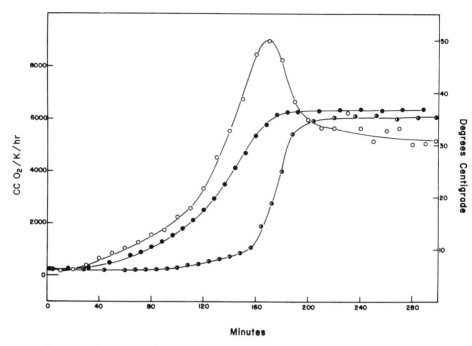

Fig. 7-1. The open circles represent the average oxygen consumption of three hamsters waking from hibernation at an environmental temperature of 4°-5°C. The closed circles represent the average cheek-pouch temperature, and the half-closed circles represent the rectal temperature of the same animals. Note that the rectal temperature rises more slowly than that of the cheek pouch. (After Lyman, 1948.)

themselves of the recording apparatus undoubtedly contributed to the high oxygen consumption.

The difference between anterior and posterior T_b during arousal has been observed in virtually all hibernators that have been examined, including bats. An exception to this temperature difference may occur in the Mohave ground squirrel (*Citellus mohavensis*) arousing from estivation. In this case it has been reported that the rectal temperature was usually no more than 0.5°C lower than the oral temperature during arousal (Bartholomew and Hudson, 1960). However, in one animal arousing from a T_a of 16°C, the rectal temperature was 25°C when the oral temperature reached 35°C, indicating that this species also possessed the ability to maintain a differential between anterior and posterior. The arousals were measured at high ambient and body temperatures and it is probable that one or both of these factors contributed to the lack of difference between the rostral and caudal parts of the body. The eastern chipmunk (*Tamias striatus*) and the hispid pocket mouse (*Perognathus hispidus*) do not maintain a large difference in temperature between anterior and posterior during arousal from low T_b (Wang and Hudson, 1970, 1971). However, as mentioned previously, neither of these species can be categorized as typical deep hibernators. Studier (1974), in his studies of the flight of bats, found that the chest cavities of *Eptesicus fuscus* and *Myotis sodalis* were warmer than the rectal temperatures during arousal, but the chest muscle temperature was not. It is generally agreed that respiratory rate, heart rate, and oxygen consumption increase virtually simultaneously at the start of arousal, followed by an increase in body temperature. However, Mokrasch *et al.* (1960) reported a rise in cheek pouch or rectal temperature prior to an increase in oxygen consumption in the arousing Syrian hamster, and they maintained that anaerobic processes took place at the beginning of arousal before aerobic metabolism got underway. Using the golden mantled ground squirrel, Hammel *et al.* (1968) presented contrary evidence and calculated that there was an uncorrected inherent lag of 15–20 min in the oxygen measurements due to low flow rate and the size of the metabolism chamber and the connectors.

It has long been known that the animal in hibernation undergoes profound circulatory adjustments. Marès (1892) and Dubois (1896) both reported that dyes injected into the circulation appeared first in the anterior part of the body, but it is not clear whether these investigators realized that this occurred during arousal rather than during the hibernating state. However, it is undoubtedly true that the process of injecting a dye into the circulation would have started the arousal. Since these early observations, the circulatory pattern of arousal has been studied in many laboratories. With a radio-opaque dye and X ray we were able to show that blood flow to the posterior was indeed restricted in Syrian hamsters arousing from hibernation (Lyman and Chatfield, 1950). Using the arctic ground squirrel (Johansen, 1961) and the thirteen-lined ground squirrel (Bullard and Funkhauser, 1962) and ^{86}Rb, it was demonstrated that the blood flow was proportionately higher in organs and tissues in the anterior part of the body. Bullard and Funkhauser calculated the blood flow to various organs and found that flow to forelimbs, heart, diaphragm, and thorax increased during the initial stages of arousal, whereas it remained almost unchanged in the abdominal viscera and the rest of the posterior portion generally. Both investigations revealed that the brown fat received a disproportionate amount of blood, and Bullard at the annual meeting of the Federation of American Societies for Experimental Biology in 1961 suggested that this so-called "hibernating gland" was in fact a source of heat during arousal and should be called the "dehibernating gland." The research on brown fat as a source of heat has grown and multiplied since, and some aspects of this study will be considered later (pp. 113–119).

Early investigations of blood pressure in hibernating animals were made acutely by Dubois (1896) on marmots and by Chao and Yeh (1951) on hedgehogs. The surgery that was necessary to obtain arterial pressure inevitably started the process of arousal, but the sequential changes in blood pressure during a complete arousal were not recorded. In the acutely cannulated Syrian hamster arousing from hibernation at a T_a of 5°C, the heart rate is reported to peak at over 500 beats a minute, but the blood pressure reaches its maximum of slightly more than 100 mg Hg when the heart rate is only 200 beats a minute (Chatfield and Lyman, 1950). Eliassen (1960) and later Kirkebö (1968) made acute measurements in hedgehogs arousing from hibernation and both concluded that maximum blood pressure occurred well before body temperature attained its euthermic level.

Measurements of changes in blood pressure during arousal without acute surgical preparation have been recorded using thirteen-lined ground squirrels (Lyman and O'Brien, 1960). Here peripheral resistance, as measured by the decrease in diastolic runoff time taken from the same systolic pressure, appears to lessen at the start of arousal. We have suggested that this indicates a vasodilation or passive distension of the previously mildly constricted blood vessels in the anterior parts of the body. In these experiments, the rise in blood pressure was not as rapid as that reported for hamster or hedgehog, for it reached its peak at about the same time as did the heart rate and heart temperature (Fig. 7-2). The reason for this difference is unknown. There have been no studies comparing the two methods and the thirteen-lined and golden mantled ground squirrels appear to be the only hibernating mammals in which the blood pressure has been studied in detail using chronically implanted vascular cannulae. It is tempting, however, to assign the difference to methodology, for Kirkebö (1968) found that forcefully moving a hibernating animal resulted in changes in circulatory pattern, and it is to be expected that surgery would produce similar effects.

In his detailed studies of the hedgehog, Kirkebö (1968) reported that peripheral resistance and viscosity of the blood both decreased during the warming process and ascribed the former to the latter. Even so, the blood pressure rose and remained high due to the increasingly rapid beating of the heart and a gradual rise in stroke volume.

Thus, it appears to be established that, in three phylogenetically distinct species, the blood pressure rises rapidly during arousal while the heart rate is also accelerating. This must mean that the rapidly beating heart is working against a high head of pressure. As such, the heart may be an inefficient pump, but a good source of heat as the animal warms from the cold temperatures of deep hibernation.

In the thirteen-lined ground squirrel, systolic, diastolic, and mean blood pressure are apt to decline just as the rectal temperature begins its rapid increase (Fig. 7-2). At the same time, peripheral resistance, as indicated by a more rapid diastolic runoff, appears to decrease. We have suggested that this indicates a sudden vasodilation in the previously constricted areas of the body, thus greatly increasing the blood flow in the whole vascular bed and resulting in a lowered blood pressure (Lyman and O'Brien, 1960).

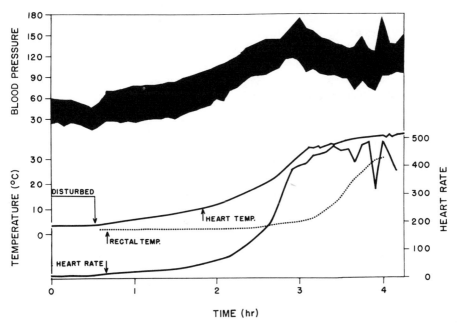

Fig. 7-2. Ground squirrel arousing from hibernation at a T_a of 3°C. Blood pressure in *dark area* is highest systole and lowest diastole recorded every 4 min for a 1-min period. (After Lyman, 1965.)

The importance of vasoconstriction during arousal can be demonstrated using the sympatholytic agent β-TM10 {[2-(2,6-dimethylphenoxy)propyl]-trimethylammonium chloride hydrate}. The drug may be introduced into the bloodstream of a hibernating ground squirrel via an aortic cannula. This results in a drop in peripheral resistance and blood pressure and, in some cases, the animal may die if not warmed artificially. Other animals remain in hibernation but, if stimulated, their arousal process is very slow and the posterior part of the body warms at the same rate as the anterior. Apparently vasoconstriction does not occur after sympathetic blockade, blood flow is not restricted to the area of most active thermogenesis, and cutaneous heat loss must be high during the warming process. These factors surely contribute to the long period required for arousal (Lyman and O'Brien, 1963). It seems reasonable to postulate that arousal is initiated and driven by a mass discharge of the sympatheticoadrenal system (Chatfield and Lyman, 1950). If this is so, the amount of catecholamines circulating in the blood should be high during arousal. A difference in threshold to these mediators might exist so that the areas with the lower threshold would be vasoconstricted during arousal. However, this appears not to be the case, for in arousing ground squirrels previously treated with β-TM10 there is no retardation of warming of the posterior if the animals are then infused with norepinephrine. β-TM10 does not inhibit the effect of norepinephrine at the effector cell, thus one must conclude that there is no difference in threshold to the vasoconstrictive action of norepinephrine throughout the animal during arousal.

Other methods can be used to illustrate the effect and control of vasoconstriction in the arousing, cannulated ground squirrel. When warming is well under way and there is a large difference in temperature between anterior and posterior parts of the body, infusion of ACh causes a rise in rectal temperature and a drop in the temperature near the heart. In this case, ACh appears to have caused a vasodilation of the posterior part of the animal. As a result, the warm blood from the anterior part of the body causes the temperature of the posterior to rise, whereas the cold blood from the posterior causes a drop in the temperature near the heart. The vasodilation cannot be maintained by continuous infusion of ACh, but periodic infusions will cause a series of steps in the warming process (Fig. 7-3).

When the temperature near the heart is approaching the euthermic level and the rectal temperature is increasing rapidly, this rise can be temporarily halted and reduced by intra-arterial infusion of norepinephrine. The temperature near the heart is unaffected by the infusion, and continuous infusion will not maintain the rectal temperature at one level, though each periodic infusion will temporarily reduce the rectal temperature (Fig. 7-4). In this case it is apparent that norepine-

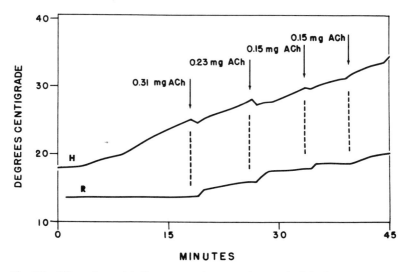

Fig. 7-3. Effect of acetylcholine on arousing ground squirrel while thoracic temperature is rising and rectal temperature is not. H, temperature in area of heart; R, rectal temperature. (After Lyman and O'Brien, 1963.)

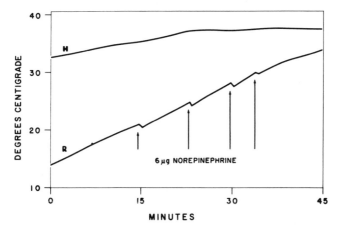

Fig. 7-4. Effect of norepinephrine (0.006 mg at arrows) on ground squirrel in last stages of arousal when rectal temperature is rising rapidly. H, heart area; R, rectal temperature. (After Lyman and O'Brien, 1963.)

phrine briefly reimposes vasoconstriction in the posterior part of the body. Under these circumstances one would expect the rectal temperature to stop rising and remain level rather than to decline. The decline may be due to the chilling effect of the ambient temperature which was 15°–23°C colder than the rectal temperature during the experiments.

The evidence presented here strongly suggests that there is a differential vasoconstriction in the posterior part of the body and in the periphery as the animal warms from hibernation, and that this constriction is virtually essential for an efficient arousal. However, students of the autonomic nervous system in euthermic mammals are generally agreed that this system does not have a discrete regional distribution of function, and an alternate explanation may be necessary to satisfy observed results in the arousing hibernator. The vasodilation of activity is believed to be prepotent over sympathetic activity, since sympathetic storms during exercise do not reduce muscle blood flow but actually augment it. One may suggest that the somatic reflexes in the anterior part of the body have the lowest threshold for producing skeletal muscle activity. Once this activity has begun, heat production rises and local, chemically produced vasodilation takes place. The drop in peripheral resistance at the start of arousal fits in well with this concept. When the warmed blood from the forepart of the body is perfused through the brain and spinal cord, activity may be extended to more posterior reflex centers in the somatic nervous system, followed by activity of the muscles and a rise of the temperature in the posterior part of the body. If this concept is valid, the vasoconstriction at the start of arousal must be profound, for the difference between the temperature of the anterior and posterior parts of the body during arousal is striking. Furthermore, the movement of warmed blood down the cord must take place quite quickly for the rise in rectal temperature, when it occurs, is very rapid, particularly in the Syrian hamster (Fig. 7-1). A determination of the time of onset of electrical signals of muscular activity in relation to the time of the occurrence of vasodilation should clarify whether activity, rather than regional differences in sympathetic tone, was the cause of vasodilation, but this has not been reported. However, it should be mentioned that the anterior-posterior difference in temperature is not as great if the animal arouses from hibernation in a warm room. This suggests that local cold, as might be expected, has a vasoconstrictor effect on the vascular bed.

The sources of heat that warm the arousing hibernator have been the object of study for more than a century, and even now there is not complete agreement on all aspects of the problem. The biochemical aspects of this question will be covered in Chapter 8, but reports on experiments with the whole animal are described here.

As early as 1896, Dubois reported his experiments on arousal in the hibernating European marmot after surgical intervention of various types, including destruction of parts of the brain. He found that tying the vena cava above the liver or tying the portal vein slowed the warming process. Furthermore, when liver and diaphragm were insulated from each other during arousal, the liver warmed faster than the mouth and slightly faster than the diaphragm. He concluded that the liver was a primary source of heat during arousal. In view of later work with nonhibernating mammals (Bollman and Mann, 1936), it appears likely that the impaired warming in the ligation experiments was due to the debilitating effects of acute portal stasis. The observation that the liver warmed faster than the mouth is difficult to interpret, as Dubois had previously shown that the esophagus warmed faster than the liver.

More than 50 years later, we reported similar experiments with hibernating Syrian hamsters (Lyman and Chatfield, 1950). Eviscerated animals with the blood supply to the liver and kidneys ligated could still warm at a T_a of 5°C. The arousal was not complete, however, for though the rise in cheek pouch temperature was as rapid as in the normal animal, it reached only about 32°C, and the rectal temperature rose very slowly and did not attain 15°C even 30 min after the cheek pouch temperature had reached its peak. Warming ceased soon thereafter and began an irreversible decline. These experiments indicated that there were sources of heat other than the viscera which could warm the anterior part of the animal to nearly normal levels.

In order to test the contribution of skeletal muscle to the arousal process, the rate of warming from hibernation was measured in a series of curarized hamsters supplied with artificial respiration. In this situation, warming of the cheek pouch was slower than normal and the temperature of 37°C was never attained. Rectal temperature rose very slowly and did not go above 20°C. Two curarized, eviscerated animals were able to attain cheek pouch temperatures of 25°C and 27°C, which indicated that some warming could occur in spite of this double insult.

The early appearance of muscle action potentials, the violent shivering, and the slower warming after curarization all emphasize the important role that skeletal muscle must play during arousal from hibernation. However, the warming that took place after curarization indicated that there must be other sources of heat than muscular activity. As mentioned previously, the demonstration of increased blood flow to brown fat during arousal had given rise to the suggestion that this tissue might be an important source of heat. Two reports, published almost simultaneously, sparked interest in this concept. Smith and Hock (1963) reported that the area of interscapular brown adipose tissue was warmer than other parts of the body, not including the heart, in the western marmot (*Marmota flaviventris*) warming from about 9°C to a T_b of 26°C, and Smalley and Dryer (1963) found that the same tissue in the hibernating big brown bat (*Eptesicus fuscus*) warmed more rapidly than other areas and organs, including the heart, as the bats aroused from hibernation at a T_a of 4° ± 2°C. Since we had previously stated that the region of the heart was the warmest area of the body in rodents during arousal (Lyman and Chatfield, 1950), we re-examined the question using woodchucks (*Marmota monax*), thirteen-lined and golden mantled ground squirrels, Syrian hamsters, and edible dormice (*Glis glis*). In the woodchucks and ground squirrels, heart temperature was measured with a thermocouple chronically implanted in the aortic arch via a carotid artery, whereas in the hamsters and dormice an end-to-end thermocouple was sewed in the thorax so that the sensing element was next to the heart. Temperature of the brown adipose tissue was measured by a thermocouple placed in the middle of a suitable brown fat pad. In every species the temperature of the heart was warmer than the brown fat during arousal from hibernation at a cold T_a (Lyman and Taylor, 1964). Probably because it was reported only in abstract form, these findings are seldom cited, and it is generally accepted that brown fat is warmer than the heart during arousal in all hibernators. We emphasize, however, that there is no published evidence to indicate that brown fat is warmer than the heart in any rodent during the arousal process.

In a later report, temperatures of arousing big brown bats (16–20 g) and golden mantled ground squirrels weighing ten times as much were compared using thermocouples which were chronically implanted in the aortic arch and a brown fat pad (Hayward *et al.*, 1965). The contrast in the arousal pattern was striking. The bats warmed from a T_b of about 8°C to 37°C in 25 min, whereas the arousal time in the ground squirrels was more than three times as long. More importantly, the area of the brown fat in the bat was always warmer than the heart during its rapid arousal, whereas the opposite was true in the ground squirrel. Further comparative studies showed that completely curarized bats with artificial respiration could arouse from hibernation as rapidly as normal animals, whereas curarization of dormice and Syrian hamsters greatly delayed the warming process, with the dormice failing to warm beyond 30°C (Hayward and Lyman, 1967). It was also demonstrated that the temperature of the brown fat was lower than that of the heart as the bat entered hibernation at a T_a of 6°–8°C. However, if the animal was disturbed by a stimulus, the temperature of the brown fat rose quickly above the temperature of the heart. Using thermography, which records the intensity of temperature-dependent infrared radiation, it was shown that the area of the interscapular brown fat pads in the arousing brown bat was the warmest portion of the body (Fig. 7-5).

There is not as much information on the distribution of heat during arousal in the order Insectivora. Edwards and Munday (1969) measured the temperature of brown fat, heart, rectum, and lumbar muscles of European hedgehogs arousing from a T_a of 22.5°C. In three animals that warmed rapidly the interscapular brown fat was warmer than the area near the heart, but in three animals that warmed more slowly there was no significant difference between the two temperatures. In these experiments the heart temperature was recorded from a thermistor placed in the esophagus behind the heart, and this might not reflect the temperature of the heart itself. Arousal took place at room temperature so that body temperatures are not strictly comparable with those of animals arousing at the low temperatures of hibernation.

Comparative temperatures from various areas of the body do not necessarily delineate the areas that are contributing the most heat. For example, a region with a high metabolic rate may be cooled by incoming blood and appear to be relatively inactive. However, the *in vivo* evidence that brown fat makes a large contribution to arousal in bats is quite compelling. The high rate of blood flow, the high temperature of the interscapular pads, and the extraordinary ability to warm rapidly when curarized all implicate the brown adipose tissue. The precise amount of heat contributed by brown fat remains a puzzle. Hayward and Ball (1966) measured the total oxygen consumption of big brown bats arousing from hibernation and calculated the amount of energy required for a complete arousal. They determined the *in vitro* metabolic rate of brown fat measured at 37°C with epinephrine or norepinephrine added to maximize the oxygen consumption. The mean metabolic rate was 137 μl O_2/100 mg/hr, which was as high or higher than that reported for brown fat in other species. The total amount of brown fat in the bats averaged only 3.3% of the total body weight. According to their calculations from these data, brown adipose tissue utilized but 5.7% of the total oxygen consumed during an arousal, and this figure would be even lower if the *in vitro* measurements had been made at the temperatures of the intact animal during arousal. Among other possible reasons for this great discrepancy, Hayward and Ball suggest that the *in vitro* conditions fail to match those found in the intact animal. Using the ^{86}Rb technique, Rausch (1973) determined the blood flow of various tissues and organs during arousal from a T_a of 5°C in the same species of

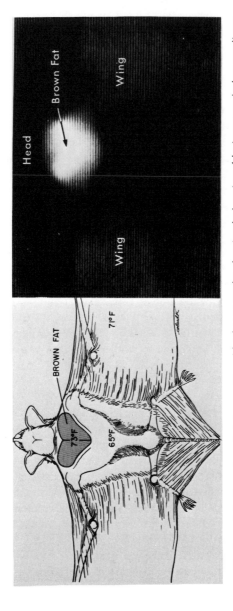

Fig. 7-5. Left: Illustration of a bat as positioned for thermography, showing the location of the interscapular brown adipose tissue and the major temperature prevailing at the commencement of thermographic scanning. Right: Thermogram of the dorsal surface of a bat during its arousal from hibernation. The higher the temperature and intensity of infrared radiation from the skin surface, the brighter is the image. (After Hayward and Lyman, 1967.)

bat. She calculated that about 16% of the total cardiac output flowed through the brown fat during arousal and suggested that this fraction represented the amount of heat contributed by the brown fat. If it is accepted that the blood flow through a tissue and its metabolic rate are synonymous, which in many cases they are not, then the heat contributed by brown fat is almost three times that calculated by Hayward and Ball. This is still not enough to warm the curarized bat as rapidly as the normal animal unless brown fat contributes more heat when shivering is repressed.

Recent work of Foster and Frydman (1978) indicate brown fat may contribute a large portion of the total heat in a homeotherm. Using cold-adapted rats infused with norepinephrine, they reported that the cardiac output to brown fat was 2.6% of the total output at rest and rose to a remarkable 33.5% during maximum calorigenesis. They measured the oxygen in arterial blood supplying the interscapular brown fat and in the venous blood leaving this tissue. Knowing the total brown fat in the animal, they calculated that this tissue accounted for at least 60% of the calorigenic response. They emphasize that studies employing the ^{86}Rb technique may grossly underestimate the contribution of brown adipose tissue to nonshivering thermogenesis in rats. They extrapolate these data to arousing hibernators, and, if the brown fat of the anesthetized cold-adapted rat and the hibernating brown bat react in the same way, they may be right, but experiments on the oxygen extraction of brown fat during arousal from hibernation have yet to be reported.

The contribution of brown fat to arousal from hibernation in rodents is less convincing. The reports of Smith and co-workers (Smith and Hock, 1963; Smith, 1964) have been misinterpreted or misquoted repeatedly. In their measurements, using needle thermocouples, it was shown that some sites of brown fat showed a decrease in temperature when the arousing marmot was exposed to extreme cold, whereas the interscapular fat pads and some other regions of brown fat showed an increase in temperature. This implied a differential control of blood flow to the concentrations of brown fat. There was no systematic exploration of the body in search of areas or organs that might be warmer than the brown fat. A countercurrent system via Sulzer's vein was demonstrated in the laboratory rat which may act as a heat trap between the interscapular fat pads and the anterior thorax (Smith and Roberts, 1964). It has been often assumed that hibernators have a similar vascular arrangement, but this has not been reported. Actually, many rodent hibernators, including the golden mantled and thirteen-lined ground squirrels, have no interscapular brown fat pads, yet the temperature of this area in *C. lateralis* has been reported as the temperature of the brown fat.

Joel (1965) presented further evidence to demonstrate the importance of brown fat during arousal from hibernation, but the data are open to criticism. In order to determine the role of brown fat during arousal, he killed a group of six *Citellus tridecemlineatus* in deep hibernation and six animals after arousal. All the available brown fat was dissected and weighed and the percent lipid per 100 g and the total lipid were determined. In the animals which had aroused from hibernation, the percent decrease of all three determinations was statistically significant when compared to the hibernating controls. In order to determine whether the brown fat was reconstituted after arousal, he sacrificed three animals in hibernation, three animals that had just completed arousal, and three animals 15–46 hr after arousal that had not been permitted to reenter hibernation. The same determinations were made and it was found that the lipid of the brown fat recovered after the depletion of arousal. The total lipid in the brown fat at the end of arousal was significantly lower than in the hibernating animals or in the ground squirrels sacrificed during the recovery period ($p = 0.03$ and 0.02, respectively). According to our calculations, the differences found in total weight of brown fat and in lipid per 100 g were not significant.

As Spencer *et al.* (1966) point out, these results are difficult to interpret. In the first experiment, $45 \pm 7\%$ or a total of 0.9 g of lipid had disappeared upon completion of arousal, but the total weight of brown fat decreased a total 1.2 g. This implies that the brown fat was losing something more than lipid. In the second experiment, the total weight of brown fat in the hibernating animals was 4.65 g or 15% less than the 5.47 g of the previous experiment. In the first experiment, about 0.9 g or 45% of total lipid was lost at the end of arousal, whereas in the second experiment only 0.49 g or 30% was lost.

The basic difficulty in these experiments lies in the great individual variation in the amount and condition of the brown fat in a normal colony of animals that hibernate, and this difficulty is compounded by a variation with the seasons. Using *C. lateralis,* Grodums *et al.* (1966) found that the individual variation in brown fat was too great to produce repeatable results, even when the animals were sacrificed at the same time of the year. Because of this, they were forced to use a biopsy at the start of arousal and compare this with the rest of the brown fat when arousal was complete and the animal killed. Although we have not kept precise records of the amount of brown fat found during dissections and autopsies, we have noted gross differences in the amount of brown fat among individual thirteen-lined and golden mantled ground squirrels and in Syrian and Turkish hamsters. In view of this, groups of three animals each, as used in Joel's (1965) experiments, should not be expected to yield significant results. For example, using a minimum of five animals in each group, Burlington *et al.* (1969) found that axillary brown fat decreased in weight during arousal in thirteen-lined ground squirrels, but the difference was not significant. Furthermore, in Joel's (1965) study of restoration of brown fat after arousal, it is stated that the animals were not allowed to reenter hibernation, but the method of preventing hibernation is not described. Thirteen-lined ground squirrels often arouse from hibernation and return to that state within a few hours, and this pattern is sequentially repeated several times. If the total lipid is depleted 45%

with each arousal, repletion of the lipid during the time the animal is active must be rapid indeed to supply energy to the brown fat for repeated arousals.

In their series of three papers, Grodums, Spencer, and Dempster examined these problems using the golden mantled ground squirrel as the experimental animal. In the first paper (Grodums *et al.*, 1966), they found no correlation between the size of the axillary fat pads and the time of year or the length of the hibernating period. On arousal they found more blood in the capillaries of the brown fat and noted small cells in which neutral fat globules had either disappeared or had been reduced in size. They point out that similar changes occurred in the liver and suggest that the opening of the capillary bed in other peripheral organs may be just as sudden and thorough as in the axillary brown fat pad. Spencer *et al.* (1966) carried out biochemical studies and found that both brown and white fat lost glyceride during arousal, but more was lost from the brown fat. In their third paper, Dempster *et al.* (1966) tested the effect of experimental infection of *C. lateralis* with coxsackie B3 virus. This virus differentially attacks the brown adipose tissue and it was their hope that this would modify the hibernating cycle. The expected modifications did not occur, but of particular interest was animal number 1585. This ground squirrel totally lacked axillary brown fat, which is the largest brown fat pad in this species, and there were only necrotic elements of brown fat cells in the strands of the cervical fat pads. In spite of this, animal number 1585 was able to arouse from hibernation, though the rate of arousal was not measured. The conservative conclusion appears to be that the precise contribution of brown fat to normal arousal in the order Rodentia and in the hedgehog of the order Insectivora is yet to be documented. The brown fat pads are reported to be somewhat depleted during arousal in most rodent species that have been studied, but not in the hamster (Feist and Quay, 1969). There have been numerous reports of changes in the level of plasma lipids and glycerides in various species of hibernators, and these are considered in Chapter 8. Obviously, these changes can not be attributed to the brown fat alone, but the concept of brown adipose tissue as a source of heat is so attractive that most of the research has concentrated on this tissue.

The lack of definitive data on this subject is due in large measure to methodological difficulties, Not only is brown fat located in different places in the various species, but also some deposits can not be surgically removed without causing the death of the animal. For example, the anterior part of the axillary brown fat in some ground squirrels inextricably surrounds the brachial plexus. Another mass of brown fat is found at the apex of the heart and a strip surrounds the aorta. The only known method of removing all the brown fat is with the coxsackie virus, and only one ground squirrel in experiments involving a great number of animals attained a virtually complete absence of brown fat.

A pharmacological approach is also beset with difficulties. It is generally agreed that the mobilization of brown fat is under the control of the sympathetic nervous system (Horowitz, 1972), but no drug is known that will block the response of brown fat alone. The process of arousal from hibernation is a delicately balanced system and one cannot be assured that an unsuccessful or delayed arousal is due to the lack of the thermogenic effect of brown fat or whether it is due to an imbalance in the physiology of the whole system. For example, the experiments of Hsieh *et al.* (1957) using the ganglionic blocking agent hexamethonium are often quoted as evidence for the sympathetic control of brown fat. With cold-adapted, curarized rats, these investigators were able to show that an intravenous infusion of 75 mg/kg hexamethonium was followed by a decline in metabolic rate and T_b. Similar doses infused intra-arterially into hibernating ground squirrels caused a drastic lowering of blood pressure followed by death unless the animals were treated with norepinephrine and warmed artificially (Lyman and O'Brien, 1964). Feist (1970) has shown that inhibition of norepinephrine synthesis blocks arousal in the Syrian hamster, but again the failure to arouse after such drastic treatment cannot be attributed only to the absence of the thermogenic effect of brown fat.

A reliable method of measuring blood flow and oxygen extraction in the brown fat during hibernation and arousal is sorely needed. It may be possible to chronically implant cannulae in an accessible fat pad of a large hibernator and measure arteriovenous differences, but to our knowledge this has not been attempted. The ^{86}Rb method and the use of radioactive microspheres produce similar, but not completely comparable, results in determining blood flow to various organs in the rat (see, e.g., Mendell and Hollenberg, 1971). The use of microspheres is further complicated by the necessity of infusion into the left atrium or left ventricle, and the difficulty of insuring complete mixing. There is the further possibility that functional arteriovenous shunts could influence results with either of these methods. However, Dr. Geoffrey Molyneux of the University of Queensland examined brown fat from hibernators during his recent stay in this country and reports that there is no evidence of shunts in this tissue.

The thermogenic action of brown fat has proved to be a fascinating biochemical study. The cold-adapted rat or the neonatal rabbit are better experimental models than the arousing hibernator and have been examined far more extensively. The danger lies in applying the results from these rather special conditions in rat and rabbit to the also specialized, but different, condition of the arousing hibernator.

Other possible sources of heat for arousal from hibernation have not been quantitated. The heart rate may increase 100-fold during arousal, and, as mentioned earlier, it is pumping against a high pressure during most of the arousal period. In both the normal and curarized animal it must contribute some unknown fraction of heat to the warming process. The amount of nonshivering thermogenesis from muscle is also unknown. In rodents and probably in the hedgehog the main source of heat in a normal arousal is undoubtedly generated by the shivering of skeletal muscle, but in bats this is apparently of lesser importance. Once well under way, the arousal almost always goes to completion in the normal, healthy hibernator. The arousal appears to involve all possible sources of heat, though the proportion of involvement of each source may differ from species to species. If one source of heat is experimentally denied, some compensation may take place and the animal may nevertheless achieve a successful arousal. Thus, the fact that the curarized big brown bat can arouse from hibernation as quickly as the control means that this animal can warm rapidly without shivering. It does not mean that no shivering occurs during a normal arousal.

As far as the basic driving force for arousal from hibernation is concerned, the sympathetic nervous system must certainly be heavily involved. In Chapter 4 evidence was presented that the parasympathetic system was not effectively functional during hibernation, and that it could only be effective some minutes after arousal had begun. As arousal progresses, it appears to regain its effectiveness, for if the arousing hibernator is physically disturbed, apnea and bradycardia occur for a brief period before polypnea and tachycardia are resumed. We have noted this in ground squirrels and hamsters, and assume it occurs in other hibernators, although it does not appear to have been reported.

Heller and Hammel (1972), using the golden mantled ground squirrel, conclude that at least part of the ultimate control of the arousal process originates in the POAH area. They were able to initiate the arousal process by chilling this area bilaterally with two thermodes to about 1°C. If the POAH area was then warmed to 11.5°-12.5°C the arousal process could be reversed provided that the warming occurred a few minutes after the beginning of arousal. If arousal had progressed for 15-30 min, it was necessary to heat the POAH area to euthermic levels before the rate of heat production could be decreased. When the rectal temperature reached 30°C during arousal, the oxygen consumption and rectal temperature dropped suddenly if the POAH area was heated to 39°C. With the POAH temperature held at 38°-40°C for 30 min the oxygen consumption and rectal temperature continued to decline, and only increased after the POAH area was relieved of its thermal clamp. In an earlier paper, Hammel et al. (1968) reported an experiment in golden mantled ground squirrels in which arousal was initiated by slowly reducing T_a to 2.6°C. Reversion to the deeply hibernating state occurred when T_a was raised to 2.8°C. We have observed similar results with hamsters. Perhaps this reaction is related to the responses achieved by heating the POAH area as reported by Hammel and his group.

Although details are not given, South et al. (1975) were not able to reverse arousal by heating the POAH area of the yellow-bellied marmot. These experiments differed from those of Heller and Hammel in that a single thermode was used on a different species with a much larger brain. There appears to be no information on the effect of heating the POAH area in animals that have begun arousal as a result of physical disturbance.

REFERENCES

Bartholomew, G. A., and Hudson, J. W. (1960). Aestivation in the Mohave ground squirrel, *Citellus mohavensis*. *Bull. Mus. Comp. Zool.* **124**, 193–208.

Benedict, F. G., and Lee, R. C. (1938). Hibernation and marmot physiology. *Carnegie Inst. Washington Publ.* No. 497.

Bollman, J. L., and Mann, F. C. (1936). The physiology of the impaired liver. *Ergeb. Physiol., Biol. Chem. Exp. Pharmakol.* **38**, 445–492.

Bullard, R. W., and Funkhauser, G. E. (1962). Estimated regional blood flow by rubidium[86] distribution during arousal from hibernation. *Am. J. Physiol.* **203**, 266–270.

Burlington, R. F., Therriault, D. G., and Hubbard, R. W. (1969). Lipid changes in isolated brown fat cells from hibernating and aroused thirteen-lined ground squirrels (*Citellus tridecemlineatus*). *Comp. Biochem. Physiol.* **29**, 431–437.

Chao, I., and Yeh, C. J. (1951). Hibernation of the hedgehog. III. Cardiovascular changes. *Chin. J. Physiol.* **18**, 1–16.

Chatfield, P. O., and Lyman, C. P. (1950). Circulatory changes during process of arousal in the hibernating hamster. *Am. J. Physiol.* **163**, 566–574.

Dempster, G., Grodums, E. I., and Spencer, W. A. (1966). Experimental Coxsackie B-3 virus infection in *Citellus lateralis*. *J. Cell. Physiol.* **67**, 443–454.

Dubois, R. (1896). "Physiologie comparée de la Marmotte." Masson, Paris.

Edwards, B. A., and Munday, K. A. (1969). The function of brown fat in the hedgehog (*Erinaceus europaeus*). *Comp. Biochem. Physiol.* **30**, 1029–1036.

Eliassen, E. (1960). Cardiovascular pressures in the hibernating hedgehog, with special regard to the pressure changes during arousal. *Arbok Univ. Bergen, Mat.-Naturvitensk. Ser.* No. 6, 1–27.

Feist, D. D. (1970). Blockade of arousal from hibernation by inhibition of norepinephrine synthesis in the golden hamster. *Life Sci.* **9**, Part I, 1117–1125.

Feist, D. D., and Quay, W. B. (1969). Effects of cold acclimation and arousal from hibernation on brown fat lipid and protein in the golden hamster (*Mesocricetus auratus*). *Comp. Biochem. Physiol.* **31**, 111–119.

Foster, D. O., and Frydman, M. L. (1978). Brown adipose tissue: the dominant site of nonshivering thermogenesis in the rat. *Experientia, Suppl.* No. 32, 147–151.

Grodums, E. I., Spencer, W. A., and Dempster, G. (1966). The hibernation cycle and related changes in the brown fat tissue of *Citellus lateralis*. *J. Cell. Physiol.* **67**, 421–430.

Hammel, H. T., Dawson, T. J., Abrams, R. M., and Andersen, H. T. (1968). Total calorimetric measurements on Citellus lateralis in hibernation. *Physiol. Zool.* **41**, 341–357.

Hayward, J. S., and Ball, E. G. (1966). Quantitative aspects of brown adipose tissue thermogenesis during arousal from hibernation. *Biol. Bull. (Woods Hole, Mass.)* **131**, 94–103.

Hayward, J. S., and Lyman, C. P. (1967). Non-shivering heat production during arousal from hibernation and evidence for the contribution of brown fat. *In* "Mammalian Hibernation III" (K. C. Fisher, A. R. Dawe, C. P. Lyman, E. Schönbaum, and F. E. South, eds.), pp. 346–355. Oliver & Boyd, Edinburgh.

Hayward, J. S., Lyman, C. P., and Taylor, C. R. (1965). The possible role of brown fat as a source of heat during arousal from hibernation. *Ann. N.Y. Acad. Sci.* **131**, 441–446.

Heller, H. C., and Hammel, H. T. (1972). CNS control of body temperature during hibernation. *Comp. Biochem. Physiol. A* **41A,** 349–359.

Horowitz, J. M. (1972). Neural control of thermogenesis in brown adipose tissue. *Proc. Int. Symp. Environ. Physiol.: Bioenerg. Temp. Regul.* (R. E. Smith, J.L. Shields, J. P. Hannon, and B. A. Horwitz, eds.), pp. 115–121. Fed. Am. Soc. Exp. Biol., Bethesda, Maryland.

Hsieh, A. C. L., Carlson, L. D., and Gray, G. (1957). Role of the sympathetic nervous system in the control of chemical regulation of heat production. *Am. J. Physiol.* **190,** 247–251.

Joel, C. D. (1965). The physiological role of brown adipose tissue. *Hand. Physiol., Sect. 5: Adipose Tissue* pp. 59–85.

Johansen, K. (1961). Distribution of blood in the arousing hibernator. *Acta Physiol. Scand.* **52,** 379–386.

Kirkebö, A. (1968). Cardiovascular investigations on hedgehogs during arousal from the hibernating state. *Acta Physiol. Scand.* **73,** 394–406.

Lyman, C. P. (1948). The oxygen consumption and temperature regulation of hibernating hamsters. *J. Exp. Zool.* **109,** 55–78.

Lyman, C. P. (1965). Circulation in mammalian hibernation. *Handb. Physiol., Sect. 2: Circ.* **3,** 1967–1989.

Lyman, C. P., and Chatfield, P. O. (1950). Mechanisms of arousal in the hibernating hamster. *J. Exp. Zool.* **114,** 491–516.

Lyman, C. P., and O'Brien, R. C. (1960). Circulatory changes in the thirteen-lined ground squirrel during the hibernating cycle. *Bull. Mus. Comp. Zool.* **124,** 353–372.

Lyman, C. P., and O'Brien, R. C. (1963). Autonomic control of circulation during the hibernating cycle in ground squirrels. *J. Physiol. (London)* **168,** 477–499.

Lyman, C. P., and O'Brien, R. C. (1964). The effect of some autonomic drugs on *Citellus tridecemlineatus* during the hibernating cycle. *Ann. Acad. Sci. Fenn., Ser. A4* **71,** 311–324.

Lyman, C. P., and Taylor, C. R. (1964). Temperature of heart and brown fat during arousal from hibernation. *Fed. Proc., Fed. Am. Soc. Exp. Biol.* **23,** 310.

Marès, M. F. (1892). Expériences sur l'hibernation des mammifères. *C. R. Seances Soc. Biol. Ses Fil.* **4,** 313–320.

Mendell, P. L., and Hollenberg, N. K. (1971). Cardiac output distribution in the rat: comparison of rubidium and microsphere methods. *Am. J. Physiol.* **221,** 1617–1620.

Mokrasch, L. C., Grady, H. J., and Grisolia, S. (1960). Thermogenic and adaptive mechanisms in hibernation and arousal from hibernation. *Am. J. Physiol.* **199,** 945–949.

Rausch, J. (1973). Sequential changes in regional distribution of blood in *Eptesicus fuscus* (big brown bat) during arousal from hibernation. *Can. J. Zool.* **51,** 973–981.

Sexton, J. N., Albert, T. F., Ingling, A. L., and Douglass, L. W. (1976). Comparison of spontaneous and evoked arousals from hibernation in the wooodchuck, *Marmota monax. Physiologist* **19,** 362.

Smalley, R. L., and Dryer, R. L. (1963). Brown fat: thermogenic effect during arousal from hibernation in the bat. *Science* **140,** 1333–1334.

Smith, R. E. (1964). Thermoregulatory and adaptive behavior of brown adipose tissue. *Science* **146,** 1686–1689.

Smith, R. E., and Hock, R. J. (1963). Brown fat: thermogenic effector of arousal in hibernators. *Science* **140,** 199–200.

Smith, R. E., and Roberts, J. C. (1964). Thermogenesis of brown adipose tissue in cold-acclimated rats. *Am. J. Physiol.* **206,** 143–148.

South, F. E., Hartner, W. C., and Luecke, R. C. (1975). Responses to preoptic temperature manipulation in the awake and hibernating marmot. *Am. J. Physiol.* **229,** 150–160.

Spencer, W. A., Grodums, E. I., and Dempster, G. (1966). The glyceride fatty acid composition and lipid content of brown and white adipose tissue of the hibernator *Citellus lateralis. J. Cell. Physiol.* **67,** 431-442.

Studier, E. H. (1974). Differential in rectal and chest muscle temperature during arousal in *Eptesicus fuscus* and *Myotis sodalis* (Chiroptera: Vespertilionidae). *Comp. Biochem. Physiol. A* **47A,** 799-802.

Tähti, H., and Soivio, A. (1977). Respiratory and circulatory differences between induced and spontaneous arousals in hibernating hedgehogs (Erinaceus europaeus L.). *Ann. Zool. Fenn.* **14,** 198-203.

Wang, L. C.-H., and Hudson, J. W. (1970). Some physiological aspects of temperature regulation in the normothermic and torpid hispid pocket mouse, *Perognathus hispidus. Comp. Biochem. Physiol.* **32,** 275-293.

Wang, L. C.-H., and Hudson, J. W. (1971). Temperature regulation in normothermic and hibernating eastern chipmunk, *Tamias striatus. Comp. Biochem. Physiol. A* **38A,** 59-90.

Intermediary Metabolism in Hibernation

Cellular metabolism must fulfill two fundamental needs of hibernators that experience profound body cooling: it must yield sufficient energy to support whatever activities are continuing at the low temperature, and it must produce the heat for rewarming the organism during arousals. Superimposed on these heterothermal requirements is the added challenge for many hibernators of surviving seasonal starvation. While these three issues provide the focus for past investigation and our own present discussion of intermediary metabolism in hibernation, they are not entirely separable. Thus, metabolic adequacy at low temperature is obviously a prerequisite for increase in the dissipation of energy as heat during the initial stages of arousal. Then, too, the metabolic arrangements for circumventing inanition may place constraints upon which pathways can or will be operating at low body temperature.

TEMPORAL CHANGES IN INTERMEDIARY METABOLISM

Seasonal Alterations

As with any fasting animal, including man, two themes pervade the metabolism of seasonal hibernation: fat management and gluconeogenesis. Fat, of course, because of its high yield of energy per unit weight and its nonhydrated (i.e., compact) character, is the chief form of energy storage in animals, and the main issue in hibernation is the timing of and changes in capacity for its deposition and utilization. Some tissues, notably brain, renal medulla, and erythrocytes, rely only or primarily on glucose metabolism. Gluconeogenesis (the resynthesis, mainly in the liver and kidney, of glucose from amino acids, lactate, or glycerol) is the only means of regeneration of depleted carbohydrate reserves. Again, the issues in hibernation concern the timing and extent to which this activity may be augmented.

Fat Metabolism

It is obvious from the many studies of cycles of weight change and body composition that seasonal hibernators, such as ground squirrels, hedgehogs, and dormice, lay on heavy reserves of fat in late summer and autumn and during the period of hibernation elect to fast and utilize these reserves (Lyman and Chatfield, 1955; Pengelley and Fisher, 1963; Mrosovsky, 1966).

Details of the latter, utilizational, aspect of this pattern are almost totally lacking. Seasonal changes in mechanisms of or capacity for lipolysis and cellular fatty acid degradation and serum transport have all received scant attention. Since lipid utilization for energy is an oxidative process, it is of interest that the size and the number of mitochondria of the heart of ground squirrels increase during the hibernating season (Moreland, 1962; Zimny and Moreland, 1968), and that the number of fat droplets near the mitochondria is also doubled (Burlington et al., 1972). Such changes have not been seen, however, in hedgehogs (Olsson, 1972b) or dormice (Poche, 1959), although in the latter case the numbers of mitochondrial christae were greater during the hibernating season.

Ambid and Agid (1974) estimated lipolytic activity of white adipose tissue of dormice by measuring glycerol release during incubation of tissue slices *in vitro* and found no changes during the hibernation season (November to April), but they unfortunately did not compare these results with activity during the summer.

Since one product of lipolysis is glycerol, the activities of the enzymes glycerol-phosphate dehydrogenase (GPDH) are of interest. The cytoplasmic form of the enzyme permits the entry of glycerol phosphate derived from lipid into the glycolytic path, and the mitochondrial form permits its direct oxidation. Olsson (1972a) found that the activity of total GPDH of hearts from hedgehogs was lower in the summer than during the rest of the year, but it did not vary much in liver and adipose tissues. His estimations were based on whole homogenates made in distilled water and it is not clear what the relative contributions from the cytoplasmic and mitochondrial form of the enzyme were. Chaffee *et al.* (1964, 1966) found that oxygen consumption with glycerol phosphate as the substrate is increased during hibernation in liver mitochondria from hamsters, but decreased in those from golden mantled ground squirrels.

There appears so far to have been only a single study of fatty acid oxidation as related to season in a hibernator. Entenman *et al.* (1975) found with incubated liver slices from four hibernating thirteen-lined ground squirrels that the production of CO_2 from radioactively labeled palmitate was one-third that in preparations from two contemporarily active animals (i.e., squirrels kept in a warm room and killed in December). One possible explanation of this result might have been a shift in balance between oxidation and esterification into di- or triglycerides or phospholipids, but these activities were also reduced in the livers from hibernating squirrels.

Fatty acids are transported in the blood in association with albumin, and in ground squirrels plasma albumin concentration is four times greater in the winter than it is after its sudden decline in the late spring (Galster and Morrison, 1966). In cold-exposed hamsters, however, a much more modest rise in albumin concentration occurs only when the animals are actually hibernating (South and Jeffay, 1958).

In comparison with utilization, the seasonal cycle of synthesis has been better explored. Thus, in a Colorado population of thirteen-lined ground squirrels, Whitten and Klain (1969) found that the activities of three liver dehydrogenase enzymes, which yield biosynthetic equivalents (NADPH), and one other enzyme, citrate cleavage enzyme, involved in the biosynthesis of fats, were about twice as great in June as in September. Slices of liver from June animals also were more active in incorporating ^{14}C from glucose into fatty acids than were those of the September animals. In livers of winter ground squirrels (both actually hibernating and those which were euthermic following a provoked arousal), the activities of the same enzymes were about 5% of those in June, and incorporation of [^{14}C]glucose and [^{14}C]acetate into fatty acids in slices of livers of winter animals was even more reduced. One of the enzymes investigated by Whitten and Klain (1969), glucose-6-phosphate dehydrogenase, which catalyzes the first step in the hexosemonophosphate shunt and reduces NADP, does not change so much in livers of other hibernators. In arctic ground squirrels it was only 50% lower in hibernation (Hannon and Vaughan, 1961), and in hedgehogs there was no difference with season (Olsson, 1972a).

A shutdown of potential for fatty acid synthesis with onset of hibernation was, however, also observed in livers of Syrian hamsters by Denyes and colleagues (Denyes and Carter, 1961; Denyes and Baumber, 1964), but in this species the situation is more complex than it is in thirteen-lined ground squirrels. Unlike ground squirrels, Syrian hamsters must first be exposed to cold for several weeks before they will hibernate. Denyes and Carter (1961) found in hamsters, as Masoro *et al.* (1954) had previously observed in rats, that introduction to cold caused an immediate reduction (i.e., within 48 hr) in capacity of liver to synthesize lipids from acetate *in vitro* at 37°C and that over a longer interval of cold exposure (up to 8 weeks) this capacity returned to about 75% of normal. The decrease that occurred when the hamsters began to hibernate after 8 weeks in the cold, therefore, represented a second cycle of inhibition.

Lipogenesis is closely tied to carbohydrate metabolism (Chaikoff, 1953) and Denyes and Carter (1961) also found that there was a correspondence between the glycogen content of the liver of cold-exposed and hibernating hamsters and its lipogenic capacity, as had been seen also in cold-exposed rats (Masoro *et al.,* 1954). In livers taken from hamsters that had been fully aroused from hibernation for 5 or 6 hr, the capacity to synthesize lipids was somewhat greater than in livers from hibernating hamsters, and in the livers of three aroused hamsters with relatively elevated glycogen content the incorporation of acetate was as high as that of hamsters cold-exposed for 8 weeks without hibernation.

The rapidity of the return of the inhibition of hepatic fatty acid synthesis with hibernation in hamsters, the tendency for the inhibition to disappear after arousal, and its apparent connection with glycogen depletion all suggest that at least a part of the inhibition in livers of hamsters is related to the direct effects of cold during a short bout of hibernation, rather than to a long-term seasonal effect like that in livers of ground squirrels.

The corresponding changes in lipogenic activity of white adipose tissue of Syrian hamsters have been found to be very different from those of liver. When hamsters were put into the cold there was a four- to sixfold increase in the capacity of white adipose tissue to synthesize lipid *in vitro* from acetate (Baumber and Denyes, 1963). During hibernation, this capacity did not significantly decrease. Acetate incorporation was somewhat less in adipose tissue taken from hamsters following stimulated arousal than in cold-exposed or hibernating hamsters, but still at least three times greater than in the non-cold-exposed controls.

Since white adipose tissue is a major, perhaps the principal, site of fatty acid synthesis (Farvager and Gerlach, 1955), the question arises as to the likely significance of shutdown only in the liver. A principal role of the liver in lipid metabolism may be to reprocess unutilized fatty acids into triglycerides and thus to "buffer" fatty acid concentration in the blood. Thus, a block in fatty acid synthesis in the liver may be viewed as sparing hepatic glycogen reserves. The adipose tissue, however, can also synthesize fatty acids from glucose, so one must still ask why this potential sink should not be plugged. Actually, the results of Denyes and Baumber do not show that it is not blocked, since they only traced incorporation of ^{14}C from acetate and not from glucose. It is unlikely, however, that acetate incorporation would have been increased without an increase in glucose incorporation as well. In cold-exposed rats, where a similar increase of acetate incorporation into lipids of white adipose tissue was observed (Patkin and Masoro, 1961), exogenous glucose stimulated the uptake still further and there were indications in the work of Baumber and Denyes that lipid synthesis in the hamsters was also strongly dependent upon the presence of glucose. Such stimulation is believed to involve the utilization of glucose in the synthesis (Fritz, 1961).

Clearly, then, there are gaps in the picture. The species for which we have information about adipose lipid synthesis, the Syrian hamster, is one which does not fast in the hibernating season. For ground squirrels and other "fasters" we have no comparable information, and can only cite two relevant bits of information that suggest the situation for adipose tissue is different in such other species. In *intact* ground squirrels hibernating with a body temperature of 5°C, the incorporation of ^{14}C from *glucose* into total lipids of liver was much less reduced than was incorporation into white fat (Tashima *et al.*, 1970). (Aside from the obvious, and italicized, differences in technique of this study from those with *in vitro* incorporation of acetate at high temperature, the controls used were actually winter animals.) In white adipose tissue of hedgehogs the NADP-linked glucose-6-phosphate dehydrogenase is greatly reduced during the winter hibernating period, even though that of the liver was not (Olsson, 1972a).

The question thus arises (see also Entenman *et al.*, 1969) of how the seasonal changes in hibernators which do, or can, fast during the hibernating period differ from those that might occur simply as a result of fasting at any time of the year. In the studies described here the controls were not fasted. The same question also arises with gluconeogenesis (see below).

Aside from the quantitative aspects of lipid synthesis, there has also been interest in seasonal changes in the character of fatty acids and phospholipids of hibernators which might reflect either qualitative changes in synthesis or else selective degradation. More than 25 years ago, Fawcett and Lyman (1954) described changes in the lipid melting point and iodine number (a crude measure of relative unsaturation of fatty acids) in hamsters and ground squirrels. They found that in hamsters the iodine number of depot lipids rose during the period of cold exposure prior to hibernation. While the relative unsaturation could be biased upward with a peanut oil diet or downward with a beef tallow diet, the prehibernation shift still occurred. No difference was found in the saturation of the lipids of hibernating and cold-exposed, euthermic hamsters maintained on a normal diet.

Other investigators have determined the relative abundance of specific fatty acids in other tissues of hibernating and active individuals of hibernating species and have found changes in relative composition, but either no change in overall saturation or relatively small changes [heart of ground squirrels (Aloia *et al.,* 1974), red cells of hamsters and mitochondria of liver (Chaffee *et al.,* 1968), kidney of hamsters (Wilkinson, 1976), brains of hamsters (Goldman, 1975).] Goldman (1975), however, did find a striking decrease in saturation of the fatty aldehydes (plasmalogens) of the microsomal fraction of hamster brains with hibernation.

Phospholipids have been investigated in the hearts of golden mantled ground squirrels and the kidneys of hamsters, and in both cases surprisingly high levels of lysophosphatides (i.e., those with the glycerol bearing only a single esterified fatty acid) were found in the hibernating individuals (Aloia *et al.,* 1974; Wilkinson, 1976). "Lyso" compounds are normally toxic to cells because they disrupt cell membranes, but at low temperature they may also serve to make them more fluid. These findings, along with those of Goldman (1975), may, therefore, be of relevance for cold adaptation of function.

How these changes are produced has so far not been investigated, except that with regard to unsaturation of fatty acids the effects are sometimes subject to modification by diet (iodine number in adipose tissue of hamster, specific composition in lipids of kidney) and in other cases they are not (iodine number in ground squirrels, total unsaturation in kidney of hamsters). Paulsrud and Dryer (1968) found that in brown adipose tissue of bats the concentration of oleic acid (18 carbons:1 double bond) was low in the summer and increased during the winter, whereas palmitic (18:0) and linoleic (18:2) acid concentrations were high in the summer and declined in the winter. By measuring $^{14}CO_2$ production from labeled fatty acids in homogenates of brown fat, they observed that palmitic acid was used to a greater extent than oleic at temperatures below 20°C. Thus, it seems likely that the seasonal shift in this case was due to selective utilization during the more nearly continuously low body temperature of the hibernating season.

Gluconeogenesis

Burlington and Klain (1967) found that, at incubation temperatures of 40° and 6°C, the capacity of kidney slices to synthesize glucose from several appropriate organic acids and from glycerol was doubled in kidneys from winter thirteen-lined ground squirrels (both actually hibernating and aroused) as compared with kidneys from active ground squirrels killed in the spring.

A key issue in gluconeogenesis is the natural source of precursor. Since acetate cannot yield net synthesis of glucose (Lehninger, 1970), fatty acids cannot be that source. Lipolysis can, however, yield a precursor in the form of glycerol. Burlington and Klain (1967) noted that even in slices of kidney taken from summer ground squirrels, the utilization of glycerol for gluconeogenesis at 40°C was already twice as high as in rats. Assuming that oxygen consumption in winter arctic ground squirrels was totally devoted to oxidation of fatty acids,

Galster and Morrison (1975) estimated that the glycerol corresponding to the fatty acids released from triglycerides would be sufficient to restore as much as two-thirds of the carbohydrate reserve depleted during a single bout of hibernation and arousal. From the small amount of urea excreted during a bout of hibernation, they also computed that the corresponding contribution from protein could not be more than 20%. There was, however, no indication in these studies of whether these proportions would be changed with season.

Other studies suggest that the role of proteins or amino acids is important and does change with seasonal onset of hibernation. Thus, $^{14}CO_2$ fixation in glucose and glycogen was found to increase by two- to threefold in incubated slices of liver at 37°C (three- to fivefold at 6°C) from hibernating, as compared with nonhibernating, ground squirrels (Klain and Whitten, 1968a). This observation not only confirmed the earlier results of Burlington and Klain (1967), but suggested that the pathways favoring amino acid (or at any rate pyruvate) as precursor were enhanced in the winter period. Whitten and Klain (1968) also found that ^{14}C incorporation into glycogen from alanine was increased 20-fold in liver slices from hibernating, as compared with active, ground squirrels, a much greater increase than for the other correlates of gluconeogenesis mentioned above. In man, alanine contributes disproportionately to gluconeogenesis during fasting and appears to play a special role (Felig et al., 1970). The bulk of the amino acids are released from muscle protein, and there is no reason to suppose that alanine-rich proteins are used preferentially. Rather, alanine appears to be produced by transamination of pyruvate in muscle. Thus, alanine could perhaps be viewed as a common transport form for delivery of gluconeogenic substrate to liver and kidney (Felig et al., 1970).

One may suppose that, since proteins play many roles in cell metabolism, this gluconeogenic reserve is severely limited, and indeed it is generally considered that an organism cannot survive more than a 50% depletion during a fast (Masoro, 1973). Tashima et al. (1970) did find, however, that the protein content of skeletal muscle of ground squirrels after a period of hibernation was 39% less than that of active, nonfasting ground squirrels. (There was a 12% reduction in heart and a 22% reduction in liver.)

Conservation of protein during fasting undergoes marked seasonal changes in black bears (Nelson, 1978). During the summer, fasting leads, as in man, to increased formation of urea and, if deprived of water, bears become dehydrated and blood urea rises (Nelson et al., 1975). As noted in Chapter 2, hibernation in bears consists of a long winter sleep in a den with natural fasting and no formation of urine and only slight fall in body temperature. During this natural fast, there is very little net urea formation (Nelson et al., 1975; Nelson, 1978). The little urea that is excreted at a slow rate into the urine is reabsorbed through the bladder wall (Nelson et al., 1975). When urea is injected into the blood of a hibernating bear, urine formation does occur as normally (Nelson et al., 1973).

In fasting man there is a decline in gluconeogenesis over a long period of fasting, as the brain appears to become adapted to using ketone bodies released directly as end products of degradation of fatty acids (Owen *et al.*, 1967; Masoro, 1973). This possibility has not been studied in hibernators, but there is a suggestion that a reduction in gluconeogenesis may occur near the end of the hibernating season in the studies of Behrisch (1973) on fructose diphosphatase (FDPase). [FDPase is a regulatory enzyme which catalyzes a step (dephosphorylation of fructose 1,6-diphosphate) that is both unique to the reverse-glycolytic portion of the gluconeogenic pathway and is also rate-limiting for gluconeogenesis.] Behrisch found that the kinetic properties of this enzyme differed in livers of hibernating arctic ground squirrels from those of summer active ground squirrels in such a way as to favor gluconeogenesis at low temperature (Chapter 9). The kinetic properties of this enzyme change back to the summer form several weeks before seasonal emergence.

Protein Synthesis

Since protein is an apparently important, if indirect, source of energy during the hibernation fast, one must ask, as with lipids, whether the resynthesis of proteins competes both for energy and for available amino acids. Whitten and Klain (1968) found that the incorporation of ^{14}C from methionine into the protein of liver slices was reduced by about two-thirds in tissue from thirteen-lined ground squirrels in winter (both hibernating and aroused) as compared with summer active individuals. Whitten and co-workers pursued this finding by showing that the polyribosome levels were greatly diminished in liver tissue from winter ground squirrels and that, while cell sap of liver of winter ground squirrels was capable of supporting protein synthesis with ribosomes from summer active animals, the converse was not the case (Whitten *et al.*, 1970a,b). Thus, the deficiency lies in the disaggregation of ribosomes and not in the activity of the synthetic enzymes.

Satisfying as these results are, one must ask, as with lipids, whether liver is necessarily the most appropriate tissue with which to resolve this question. The results of Tashima *et al.* (1970) with *in vivo* [^{14}C]glucose incorporation into various tissues and fractions thereof in ground squirrels show that while utilization of glucose and the absolute rate of incorporation into any constituent were greatly reduced in hypothermic, hibernating ground squirrels, the proportion of incorporation of label into protein, as a fraction of the glucose that was utilized, was increased fivefold. Furthermore, in hibernation the net incorporation into skeletal muscle was less reduced than it was into liver, heart, or adipose tissue. These results do not necessarily conflict with those of Whitten *et al.* (1970a), since the ^{14}C of glucose could have been incorporated into the polysaccharide component of glycoproteins as well as into amino acids. Nevertheless, they serve to keep the general question open.

Except in bears, there has been no attempt to determine the effect of hibernation on the turnover of proteins, despite well-developed techniques for doing this (Goldberg and Dice, 1974) and their previous application to other similar problems of comparative physiology and thermal adaptation (Sidell, 1977). In bears during hibernation, when, as noted previously, no net catabolism of protein is occurring, there is actually a three- to fivefold increase in rate of protein turnover as judged by half-time of disappearance of radiolabeled albumin and by rate of incorporation of radiolabeled leucine (Lundberg et al., 1976). From these and later results, Nelson (1978) has proposed that in hibernation the nitrogen end product is recaptured by ammonia fixation with α-ketoglutaric acid to yield glutamic acid, which is then transaminated with pyruvate to yield alanine and to give back α-ketoglutaric acid. The pyruvate thus utilized arises as a product of the glycerol being released from lipid metabolism. Thus, the lack of net production of protein end products described previously appears not to reflect any absence of catabolism, but rather an increased recapture of nitrogen and resynthesis of peptides, coupled with enhanced protein breakdown. Nelson (1978) suggests that this "futile cycle" may serve as an accessory source of heat production in the hibernating bear.

TEMPORAL RELATIONS OF METABOLISM WITHIN THE SHORT CYCLE OF HIBERNATION

Our discussion of intermediary metabolism has so far been concerned with seasonal adjustments of major pathways. The question that links this subject with others more directly connected with the temperature demands of hibernation, (namely, mechanisms of cold adaptation and of heat production) is whether there is a shift in the relationships among these pathways in the various tissues on a shorter time scale as the organism cycles into and out of a bout of hibernation within the winter period.

Numerous venerable studies have shown that during deep hibernation the respiratory quotient (RQ), that is, moles of CO_2 respired per moles of O_2 consumed, is near 0.7, a value characteristic of lipid metabolism. From this and the changes in morphology of fat pads during hibernation, the conclusion has arisen and has been universally accepted without further comment that fat is the principal energy source (for review, see South and House, 1967).

Evidence of utilization or nonutilization of alternative sources rests mainly on inference from changes in plasma or urine concentrations during the hypothermic phase of a bout of deep hibernation. Thus, there is no progressive accumulation of urea in the urine or plasma of hedgehogs, marmots, or ground squirrels, a finding which implies little or no net oxidative degradation of amino acids (Kristoffersson, 1965; Pengelley et al., 1971; Galster and Morrison, 1975; Kastner *et al.*, 1977). Plasma glucose concentration is lower in hibernation in most species than in the inter-bout, aroused period (Lyman and Chatfield, 1955; Twente and Twente, 1967), and in several species (dormice, hedgehogs, arctic ground squirrels, but not golden mantled ground squirrels) there is a progressive decline in glucose from the beginning of a single bout of hibernation until that bout is terminated by arousal (Sarajas, 1967; Twente and Twente, 1967; Agid and Ambid, 1969; Galster and Morrison, 1970).

This result merely implies faster utilization than production or release of glucose. In arctic ground squirrels, the activity of the enzyme phosphorylase, which catalyzes the release of glucose 6-phosphate from glycogen, completely disappears, a result which would appear to mean that liver glycogen is eliminated as a source of glucose during the bout of hibernation (Hannon and Vaughan, 1961). Activity was restored in the liver upon arousal, and in the myocardial muscle phosphorylase was not diminished during hibernation. In hibernating bats there is a similar, but less complete, suppression of liver phosphorylase (Leonard and Wimsatt, 1959). Galster and Morrison (1975) found, however, that liver glycogen in arctic ground squirrels was depleted very considerably during the hypothermic phase of a bout of hibernation in arctic ground squirrels.

Lactate does not accumulate in the blood and tissues of hibernators during the steady-state period of hypothermia (Twente and Twente, 1968; Agid and Ambid, 1969; Galster and Morrison, 1975), so that the overall utilization of glucose must either be as synthesis of other substances or as oxidative degradation. Tashima *et al.* (1970) found that in thirteen-lined ground squirrels specific activity of ^{14}C from intra-aortically injected glucose appearing as CO_2 was only 16% of that in active individuals (contemporary winter animals kept in a warm animal room). Unfortunately for this comparison these investigators did not study ground squirrels fully rewarmed from a bout of hibernation, but in partially aroused squirrels (rectal temperature 15°C) the specific activity was already 2.8 times higher than in the hibernating squirrels (i.e., specific activity was 44% of that in active animals). One may also see in their data that, of the glucose label that was utilized for any purpose, only 38% went into CO_2 in the hibernating ground squirrels, whereas the fraction going into CO_2 in the arousing animals was higher (i.e., 49%; 76% in active animals).

A difficulty with this study is that the kinetics of distribution of ^{14}C label were not determined and that different intervals of incorporation were used for the active and arousing squirrels (1 hr) from those for the hibernating squirrels (24 hr). It is possible, therefore, that the results could have been influenced by different relative degrees of equilibration with utilizable metabolic pools as well as by differences in the size of those pools. On the other hand, Forssberg and Sarajas (1955) achieved similar results from measurements of expired CO_2 and $^{14}CO_2$ from [^{14}C]glucose injected subcutaneously in hedgehogs, hibernating and recently aroused. They found that while CO_2 production in hibernating hedgehogs was reduced to about 6% of that in aroused animals, $^{14}CO_2$ production was reduced to about 0.3%. Of course, with these studies a problem not present in those of Tashima *et al.* (1970) is that the diffusion from the subcutaneous space to the plasma might have caused a significant delay in equilibration and might, therefore, have accounted for part of the slower appearance of $^{14}CO_2$ at low temperature. Nevertheless, despite the difficulties, the results of both studies favor the conclusion that there is markedly reduced relative utilization of glucose during hibernation.

During arousal, the RQ is generally thought to change to about 1.0, a value compatible with carbohydrate (glycogen) as the primary energy source. In keeping with this, Lyman and Leduc (1953) found profound depletion of liver and muscle glycogen during the latter period of arousal in hamsters with maintenance of constant blood glucose levels, but when the liver was ligated, arousing hamsters became hypoglycemic. A considerable confusion of results regarding glycogen levels in various species of hibernators suggests that these results with hamsters are probably not representative of other species, especially those that fast in the hibernating season. It has been suggested that, even during arousal, lipid degradation may be the chief source of oxidative energy and that the high RQ results from the fact that during hibernation the blood has a higher total CO_2 content, the excess of which is released with warming (Galster and Morrison, 1975). [This explanation would not apply to hamsters, which experience little change in total CO_2 content in hibernation (Lyman and Hastings, 1951).]

So far, there have been only two relatively complete studies of changes in metabolic reserves through the complete short cycle of hibernation with detailed investigation of the arousal phase, that of Agid and co-workers in dormice (Agid and Ambid, 1969) and that of Galster and Morrison (1975) in arctic ground squirrels. In both of these cases plasma glucose concentration was found to rise during arousal (in dormice, after a short initial drop). Liver glycogen reserves of the dormouse fell, but only late in the arousal phase. In the arctic ground squirrels the liver glycogen generally increased throughout the arousal phase, although there was a slight dip late in arousal. Muscle glycogen of ground squirrels also decreased late in arousal, after having initially increased.

With regard to our prime issue of timing, it is apparent from the early rise in carbohydrate reserves in the arctic ground squirrel that gluconeogenesis increases early in arousal and continues through most of the active period. The inference is that it does not keep pace with depletion during hibernation, when carbohydrate reserves decline by approximately 50% (Galster and Morrison, 1975). The relationships with the dormouse are different: glycogen reserves do not begin to be restored until well into the active period and apparently are not depleted during the hibernation phase (Agid and Ambid, 1969).

In other species, while the timing of gluconeogenesis has not been so adequately described, there is fragmentary evidence of mobilization of both lipids and amino acids during arousal. Thus, plasma concentrations of amino acids increase during arousal in both thirteen-lined ground squirrels and hedgehogs (Klain and Whitten, 1968b; Kristoffersson and Broberg, 1968). (In the latter, they are depressed in hibernation and their increase only brings them to normal levels.) Whitten and Klain (1968) found that arginase activity of livers from arousing ground squirrels was more than twice as great as in livers from hibernating ground squirrels, a change which presumably reflects an increase in potential for amino acid catabolism.

Plasma levels of fatty acids also increase dramatically, not only in the dormice and arctic ground squirrels in the two studies discussed above, but also in bats and, reportedly, in hedgehogs (Suomalainen and Saarikoski, 1967; Esher *et al.*, 1973). On the last, however, there is some disagreement between published reports, which also serves to illustrate the pitfall of making judgements about utilization simply from concentrations in the blood. In contrast to Suomalainen and Saarikoski (1967), Konttinen *et al.* (1964) found a decrease of free fatty acids in the plasma of arousing hedgehogs below a level already reduced. Nevertheless, both groups of authors concluded that lipids were being utilized for arousal, Konttinen *et al.* (1964) referring to their "combustion in excess of release" and Suomalainen and Saarikoski to the "intensive lipolysis."

Accumulation in the plasma, of course, might as easily represent blocked or inadequate oxidation as lipolysis, and more persuasive evidence of mobilization would be a concomitant decrease in stored reserves. This was observed in bats by Esher *et al.* (1973) who found not only a threefold elevation in fatty acids during arousal but a decrease in all lipid components of liver (90% for fatty acids, 73% for triglycerides), but such a report has not appeared for any of the rodent hibernators or for hedgehogs. In this context, it is important to note that bats rely far more heavily than other hibernators, perhaps even exclusively, on heat produced by the metabolism of brown fat (Chapter 7).

To summarize this discussion of intermediary metabolism, until 1940 science had succeeded in establishing what might have been guessed by an intelligent savage: that since many hibernators get fat in the autumn and thin by spring, they are mainly utilizing fat reserves during the winter. Modern biochemistry has added little to this conclusion. There is no information about whether the enzymatic basis of fat catabolism is qualitatively or quantitatively altered with season. As might be expected, there is evidence that enzymes of synthesis and synthetic pathways in cells are most active during the summer and virtually inactive in the winter in ground squirrels, which are fasting animals, but this conclusion is in need of confirmation for this group and seems decidely shaky for hamsters, which do not fast. Gluconeogenesis is most active in the winter and synthesis of proteins, at least in liver, is strongly inhibited.

During a single bout of hibernation, hibernating mammals use predominantly lipid metabolism. There is indirect evidence that in dormice and arctic ground squirrels, lipid metabolism is also the major source of energy during arousal, as it certainly is in bats, which rely heavily upon heat produced by brown fat. Evidence for lipid as a major immediate energy source for arousal in other species is more questionable. Mobilization of glycogen contributes to the latter part of the arousal process (largely through glycolysis) in arctic ground squirrels and dormice, and in hamsters it is perhaps the principal source during this period.

The utilization of carbohydrates during the normothermic and hypothermic phases of the short cycle of hibernation appears to differ with species. In dormice (and probably in hamsters) glycogen reserves are first depleted then restored during the active phase and are not depleted again until the next arousal, whereas in arctic ground squirrels they are restored throughout the arousing and active phases and depleted throughout the hypothermic (hibernating) phase. In one species, congeneric with the arctic ground squirrel, however, there is no progressive depletion of glucose during a hibernation bout and in another there is evidence for a restriction of entry of glucose into the Krebs cycle during hibernation.

REFERENCES

Agid, R., and Ambid, L. (1969). Effects of corporeal temperature on glucose metabolism in a homeotherm, the rat, and a hibernator, the garden dormouse. *In* "Depressed Metabolism" (X. J. Musacchia and J. F. Saunders, eds.), pp. 119–157. Am. Elsevier, New York.

Aloia, R. C., Pengelley, E. T., Bolen, J. L., and Rouser, G. (1974). Changes in phospholipid composition in hibernating ground squirrel, *Citellus lateralis,* and their relationship to membrane function at reduced temperature. *Lipid* **9,** 993–999.

Ambid, L., and Agid, R. (1974). Activité lipolytique de cellules isolées du tissu adipeux blanc du lérot pendant l'hibernation, influences hormonales. *C. R. Seances Soc. Biol. Ses Fil.* **168,** 610–614.

Baumber, J., and Denyes, A. (1963). Acetate-1-C^{14} metabolism of white fat from hamsters in cold exposure and hibernation. *Am. J. Physiol.* **205,** 905–908.

Behrisch, H. W. (1973). Molecular mechanisms of temperature adaptation in arctic ectotherms and heterotherms. *In* "Effects of Temperature on Ectothermic Organisms" (W. Wieser, ed.), pp. 123–137. Springer-Verlag, Berlin and New York.

Burlington, R. F., and Klain, G. J. (1967). Gluconeogenesis during hibernation and arousal from hibernation. *Comp. Biochem. Physiol.* **22,** 701–708.

Burlington, R. F., Bowers, W. D., Jr., Damm, R. C., and Ashbaugh, P. (1972). Ultrastructural changes in heart tissue during hibernation. *Cryobiology* **9,** 224–228.

Chaffee, R. R. J., Allen, J. R., Cassuto, Y., and Smith, R. E.(1964). Biochemistry of brown fat and liver of cold-acclimated hamsters. *Am. J. Physiol.* **207,** 1211–1214.

Chaffee, R. R. J., Pengelley, E. T., Allen, J. R., and Smith, R. E. (1966). Biochemistry of brown fat and liver of hibernating golden mantled ground squirrels (*Citellus lateralis*) *Can. J. Physiol. Pharmacol.* **49,** 217–223.

Chaffee, R. R. J., Platner, W. S., Patton, J., and Jenny, C. (1968). Fatty acids of red blood cell ghosts, liver mitochondria and microsomes of cold-acclimated hamsters. *Proc. Soc. Exp. Biol. Med.* **127,** 102–106.

Chaikoff, I. L.(1953). Metabolic blocks in carbohydrate metabolism in diabetes. *Harvey Lect.* **47,** 99-125.

Denyes, A., and Baumber, J. (1964). Lipogenesis of cold-exposed and hibernating golden hamsters. *Ann. Acad. Sci. Fenn., Ser. A4* **71,** 129-140.

Denyes, A., and Carter, J. D. (1961). Utilization of acetate-1-C¹⁴ by hepatic tissue from cold exposed and hibernating hamsters. *Am. J. Physiol.* **200,** 1043-1046.

Entenman, C., Hillyard, L. A., Holloway, R. J., Albright, M. L., and Leong, G. F. (1969). Intermediary metabolism in hypothermic rat livers. *In* "Depressed Metabolism" (X. J. Musachia and J. F. Saunders, eds.), pp. 159-177. Am. Elsevier, New York.

Entenman, C., Ackerman, P. D., Walsh, J., and Musacchia, X. J. (1975). Effect of incubation temperature on hepatic palmitate metabolism in rats, hamsters and ground squirrels. *Comp. Biochem. Physiol. B* **50B,** 51-54.

Esher, R. J., Fleischman, A. I., and Lenz, P. H. (1973). Blood and liver lipids in torpid and aroused little brown bats, *Myotis lucifugus. Comp. Biochem. Physiol. A* **45A,** 933-938.

Farvager, P., and Gerlach, J. (1955). Recherches sur la synthese des graises a partir d'acetate ou de glucose. II. Les roles respectifs du foie, du tissu adipeux et de certaines autres tissus dans la lipogenese chez la souris. *Helv. Physiol. Acta* **13,** 96-105.

Fawcett, D. W., and Lyman, C. P. (1954). The effect of low environmental temperature on the composition of depot fat in relation to hibernation. *J. Physiol. (London)* **126,** 235-247.

Felig, P., Pozefsky, T., Marliss, E., and Cahill, G. (1970). Alanine: key role in gluconeogenesis. *Science* **167,** 1003-1004.

Forssberg, A., and Sarajas, H. S. S. (1955). Studies on the metabolism of ¹⁴C-labelled glucose in awake and hibernating hedgehogs. *Ann. Acad. Sci. Fenn., Ser. A4* **28,** 1-8.

Fritz, I. B. (1961). Factors influencing the rates of long-chain fatty acid oxidation and synthesis in mammalian systems. *Physiol. Rev.* **41,** 52-129.

Galster, W. A., and Morrison, P. (1966). Seasonal changes in serum lipids and proteins in the 13-lined ground squirrel. *Comp. Biochem. Physiol.* **18,** 489-501.

Galster, W. A., and Morrison, P. R. (1970). Cyclic changes in carbohydrate concentrations during hibernation in the arctic ground squirrel. *Am. J. Physiol.* **218,** 1228-1232.

Galster, W., and Morrison, P. R. (1975). Gluconeogenesis in arctic ground squirrels between periods of hibernation. *Am. J. Physiol.* **228,** 325-330.

Goldberg, A. L., and Dice, J. F. (1974). Intracellular protein degradation in mammalian and bacterial cells. *Annu. Rev. Biochem.* **43,** 835-869.

Goldman, S. S. (1975). Cold resistance of the brain during hibernation. III. Evidence of a lipid adaptation. *Am. J. Physiol.* **228,** 834-838.

Hannon, J.P., and Vaughan, D. A. (1961). Initial stages of intermediary glucose catabolism in the hibernator and non-hibernator. *Am. J. Physiol.* **201,** 217-223.

Kastner, P. R., Zatzman, M. L., South, F. E., and Johnson, J. A. (1977). The renin-angiotensin aldosterone system of the hibernating marmot. *Am. J. Physiol.* **233,** R37-R43.

Klain, G. J., and Whitten, B. K. (1968a). Carbon dioxide fixation during hibernation and arousal from hibernation. *Comp. Biochem. Physiol.* **25,** 363-366.

Klain, G. J., and Whitten, B. K. (1968b). Plasma free amino acids in hibernation and arousal. *Comp. Biochem. Physiol.* **27,** 617-619.

Konttinen, A., Rajasalmi, M., and Sarajas, H. S. S. (1964). Fat metabolism of the hedgehog during the hibernating cycle. *Am. J. Physiol.* **207,** 845-848.

Kristoffersson, R. (1965). Hibernation in the hedgehog (*Erinaceus europaeus*). Blood urea levels after known lengths of continuous hypothermia and in certain phases of spontaneous arousals and entries into hypothermia. *Ann. Acad. Sci. Fenn., Ser. A4* **96,** 1-8.

Kristoffersson, R., and Broberg, S. (1968). Free amino acids in blood serum of hedgehogs in deep hypothermia and after spontaneous arousal. *Experientia* **24,** 148-150.

Lehninger, A. L. (1970). "Biochemistry." Worth, New York.

Leonard, S. L., and Wimsatt, W. A. (1959). Phosphorylase and glycogen levels in skeletal muscle and liver of hibernating and nonhibernating bats. *Am. J. Physiol.* **197,** 1059-1062.

Lundberg, D. A., Nelson, R. A., Wahner, H. W., and Jones, J. D. (1976). Protein metabolism in the black bear before and during hibernation. *Mayo Clin. Proc.* **51,** 716-722.

Lyman, C. P., and Chatfield, P. O. (1955). Physiology of hibernation in mammals. *Physiol. Rev.* **35,** 403-425.

Lyman, C. P., and Hastings, A. B. (1951). Total CO_2, plasma pH and pCO_2 of hamsters and ground squirrels during hibernation. *Am. J. Physiol.* **167,** 633-637.

Lyman, C. P., and Leduc, E. H. (1953). Changes in blood sugar and tissue glycogen in the hamster during arousal from hibernation. *J. Cell. Comp. Physiol.* **41,** 471-492.

Masoro, E. J. (1973). Factors influencing intermediary metabolism. *In* "Physiology and Biophysics: Digestion, Metabolism, Endocrine Function and Reproduction" (T. C. Ruch and H. D. Patton, eds.), pp. 136-162. Saunders, Philadelphia, Pennsylvania.

Masoro, E. J., Cohen, A. I., and Panagos, S. S. (1954). Effect of exposure to cold on some aspects of hepatic acetate utilization. *Am. J. Physiol.* **179,** 451-456.

Moreland, J. E. (1962). Electron microscopic studies of mitochondria in cardiac and skeletal muscle from hibernated ground squirrels. *Anat. Rec.* **142,** 155-167.

Mrosovsky, N. (1966). Acceleration of annual cycle to 6 weeks in captive dormice. *Can. J. Zool.* **44,** 903-910.

Nelson, R. A. (1978). Urea metabolism in the hibernating black bear. *Kidney Int.* **13,** Suppl. 8, S177-S179.

Nelson, R. A., Wahner, H. W., Jones, J. D., Ellefson, R. D., and Zollman, P. E. (1973). Metabolism of bears before, during, and after winter sleep. *Am. J. Physiol.* **224,** 491-496.

Nelson, R. A., Jones, J. D., Wahner, H. W., McGill, D. B., and Code, C. F. (1975). Nitrogen metabolism in bears: urea metabolism in summer starvation and in winter sleep and role of urinary bladder in water and nitrogen conservation. *Mayo Clin. Proc.* **50,** 141-146.

Olsson, S.-O. R. (1972a). Dehydrogenases (LDH, MDH, G-6-PDH, and α-GPDH) in the heart, liver, white and brown fat. *Acta Physiol. Scand., Suppl.* No. 380, 62-95.

Olsson, S.-O. R. (1972b). Ultrastructure of brown adipose tissue and myocardium. *Acta Physiol. Scand., Suppl.* No. 380, 117-130.

Owen, O. E., Morgan, A. P., Kemp, H. G., Sullivan, J. G., Herrera, M. G., and Cahill, C. F., Jr. (1967). Brain metabolism during fasting. *J. Clin. Invest.* **46,** 1589-1595.

Patkin, J. K., and Masoro, E. J. (1961). Effects of cold acclimation on lipid metabolism in adipose tissue. *Am. J. Physiol.* **200,** 847-850.

Paulsrud, J. R., and Dryer, R. L. (1968). Circum-annual changes in triglyceride fatty acids of bat brown adipose tissue. *Lipids* **3,** 340-345.

Pengelley, E. T., and Fisher, K. C. (1963). The effect of temperature and photoperiod on the yearly hibernating behavior of captive golden-mantled ground squirrels, *Citellus lateralis tescorum.* *Can. J. Zool.* **41,** 1103-1120.

Pengelley, E. T., Asmundson, S. J., and Uhlman, C. (1971). Homeostasis during hibernation in the golden-mantled ground squirrel, *Citellus lateralis.* *Comp. Biochem. Physiol. A* **38,** 645-653.

Poche, R. (1959). Elektronmikroskopische Untersuchungen zur Morphologie des Herzmuskels vom Siebenschläfer während des aktiven und des letargischen Zustandes. *Z. Zellforsch. Mikrosk. Anat.* **50,** 332-360.

Sarajas, H. S. S. (1967). Blood glucose studies in permanently cannulated hedgehogs during a bout of hibernation. *Ann. Acad. Sci. Fenn., Ser. A4* **120-** 1-11.

Sidell, B. D. (1977). Turnover of cytochrome c in skeletal muscle of green sunfish (*Lepomis cyanellus,* R.) during thermal acclimation. *J. Exp. Zool.* **199,** 233-250.

South, F. E., and House, W. A. (1967). Energy metabolism in hibernation. *In* "Mammalian Hibernation III" (K. C. Fisher, A. R. Dawe, C. P. Lyman, E. Schönbaum, and F. E. South, eds.), pp. 305–324. Oliver & Boyd, Edinburgh.

South, F. E., II, and Jeffay, H. (1958). Alterations in serum proteins of hibernating hamsters. *Proc. Soc. Exp. Biol. Med.* **98,** 885–887.

Suomalainen, P., and Saarikoski, P.-L. (1967). The content of non-esterified fatty acids and glycerol in the blood of the hedgehog during the hibernation period. *Experientia* **23,** 457–458.

Tashima, L. S., Adelstein, S. J., and Lyman, C. P. (1970). Radioglucose utilization by active, hibernating and arousing ground squirrels. *Am. J.Physiol.* **218,** 303–309.

Twente, J. W., and Twente, J. A. (1967). Concentrations of d-glucose in the blood of *Citellus lateralis* after known intervals of hibernating periods. *J. Mammal.* **48,** 318–386.

Twente, J. W., and Twente, J. A. (1968). Concentrations of 1-lactate in the tissues of *Citellus lateralis* after known intervals of hibernating periods. *J. Mammal.* **49,** 541–544.

Whitten, B. K., and Klain, G. J. (1968). Protein metabolism in hepatic tissues of hibernating and arousing ground squirrels. *Am. J. Physiol.* **214,** 1360–1362.

Whitten, B. K., and Klain, G. J. (1969). NADP-specific dehydrogenases and hepatic lipogenesis in the hibernator. *Comp. Biochem. Physiol.* **29,** 1099–1104.

Whitten, B. K., Posviata, M. A., and Bowers, W. D. (1970a). Seasonal changes in hepatic ribosome aggregation and protein synthesis in the hibernator. *Physiologist* **13,** 339.

Whitten, B. K., Schrader, L. E., Huston, R. L., and Honold, R. G. (1970b). Hepatic polyribosomes and protein synthesis: seasonal changes in a hibernator. *Int. J. Biochem.* **1,** 406–408.

Wilkinson, H. L. (1976). Lipid involvement in ion transport of the hamster kidney: its role in cold adaptation. Ph.D. Thesis, Univ. of Illinois, Urbana.

Zimny, M. L., and Moreland, J. E. (1968). Mitochondrial populations and succinic dehydrogenase in the heart of a hibernator. *Can. J. Physiol. Pharmacol.* **46,** 911–913.

9

Is There Cold Adaptation of Metabolism in Hibernators?

Survival of extreme low body temperature and ability to rewarm from that low temperature are two unique qualities of deeply hibernating small mammals that would seem to require special adaptations of the metabolism of cells. The question which will begin, end, and pervade this discussion is: are adaptations in fact necessary? At first glance the question of what one should expect seems simple. Cooling slows metabolism and production of such energy-rich compounds as adenosine triphosphate (ATP) exponentially. A vast literature on poikilothermic forms shows that often when such organisms are moved from a warm to a cold environment, enzymatic components of their exergonic metabolism are altered through a variety of mechanisms to compensate for the inhibitory effect of cooling and to allow for greater metabolic rate (Precht, 1958). Surely, in tissues of hibernating mammals that experience such a profound decrease in temperature for extended periods one ought to expect similar differences in the temperature sensitivity of their metabolism in comparison with that of the normally stenothermic and warm-adapted tissues of ordinary mammals.

The matter, however, is not so simple. To begin with, the case of the hibernator differs in two important respects from that of the model poikilotherm stably adapted to one temperature or another. First, the tissues of the deeply hibernating mammal must be able to function adequately at both of the two temperatures at which it regulates. The separation between these two (as much as 35°C) represents a thermal challenge for carrying on normal physiological function as great as, or greater than, that faced by any vertebrate poikilotherm. Furthermore, the change from one to the other of these two temperatures takes only a few hours, a time too short for most compensatory acclimations observed in poikilotherms.

Second, the kind of changes referred to in poikilotherms are "capacitative;" that is, the organism not only survives the given change of temperature, but continues to operate in its environment. The problem faced by the hibernator requires something closer to what is called "resistance" adaptation, in which the organism merely survives, with minimal activity, the period of duress, and is effectively withdrawn from the environment. In this case, the problem is one of deciding what is limiting for survival: does the nonhibernator that dies from hypothermia do so because its energy metabolism is inadequate?

Even stated in this way the answer may seem obvious. If rate of production of ATP (or other convertibly energy-rich compound) is profoundly decreased, is that not in itself limiting? Not necessarily. The matter is one of balance between energy release and energy utilization, of balance between ATP production and consumption.

In Chapter 8 we saw evidence of reduced synthetic activity at low temperature or during the hibernating season in general (hepatic lipogenesis and protein synthesis; gluconeogenic activity enhanced mainly during arousal and activity). If this were a general phenomenon, the energy load on hibernator tissues might be greatly reduced. Of course, there is a limit to the possibility of reducing need for available energy. Some kinds of cellular activity must continue at the low temperature, notably muscle contraction (heart, diaphragm), and some level of ion transport and other energy-requiring membrane activity, at least in functioning, excitable tissues.

A final complication is the one implicit within the first sentence of this chapter: the hibernator faces two problems, surviving and getting warm again. So even when we find an adaptation for low temperature we may be left with the difficulty of deciding whether it is essential for hypothermic survival or only for the increased energy demand of producing heat during the earliest stages of arousal.

MANAGEMENT OF ATP AT LOW TEMPERATURE IN MAMMALIAN TISSUES

The most readily available and utilized source of energy in animal cells is that which is released in the hydrolysis (i.e., splitting off) of the terminal phosphate of ATP to yield ADP, and inorganic phosphate (P_i). Accordingly, the concentrations of ATP, ADP, and P_i are usually tightly regulated within cells and the three molecules are powerful "effectors" or determinants of activity of key steps in the pathways of glycolysis and respiration, which in turn, of course, release energy from carbon-chain molecules and recover it largely in the form of ATP (Figs. 9-1 and 9-2) (for reviews, see Hochachka and Somero, 1973; Neely and Morgan, 1974).

To begin, therefore, to determine whether release of energy is limiting at low temperature in a specific tissue, perhaps the most useful single criterion would be cellular ATP concentration. A fall in ATP concentration at low temperature, one would think, must surely indicate that metabolism is not keeping pace with utilization.

This is not, however, an infallible test. In the first place, the important issue is whether there is sufficient ATP to support essential cellular activities. In a particular cell the concentration of ATP at high temperature (or for that matter, even a reduced concentration at low temperature) could be saturating for given processes. Van Rossum (1972), for example, demonstrated that in liver slices of rats, ATP concentration at 37°C had to be lowered by more than 75% before ion

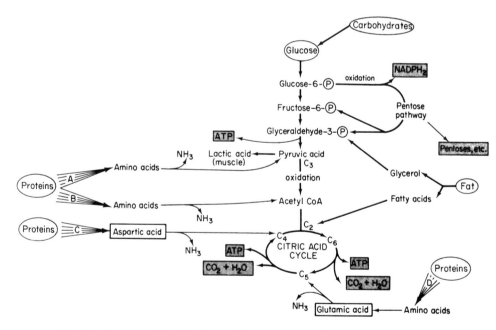

Fig. 9-1. Summary of principal metabolic pathways. (After W. S. Hoar, "General and Comparative Physiology," 2nd ed., © 1975, p. 230. Reprinted by permission of Prentice-Hall, Inc., Englewood Cliffs, N.J.)

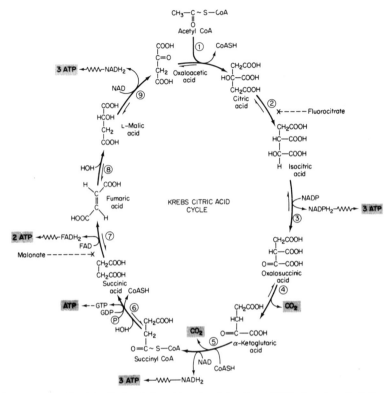

Fig. 9-2. Details of the Krebs citric acid cycle. (After Hoar, 1975. Based on T. P. Bennett and E. Frieden, "Modern Topics of Biochemistry," © 1966 by Macmillan Publishing Co., Inc., New York.)

transport was affected. Hence, in this case the energy utilizing element (Na^+,K^+-ATPase) had a high affinity for ATP, and given the prevailing ATP concentration, a wide safety margin.

It would also be possible, in principle at least, for a cell to have a high ATP concentration at low temperature, but for the ATP to be so compartmentalized that it could not reach the sites of utilization. (Imagine, for example, a high temperature sensitivity of the transfer system in mitochondria that shuttles ATP and ADP. At low temperature, ATP would be trapped within the mitochondria. The cell would be starved for ATP even though the apparent total concentration would be high.)

Finally, a fall in concentration of ATP might be a result not of direct inhibition by low temperature of ATP production, but a secondary consequence of some other cellular damage.

Notwithstanding these limitations, an inquiry into the status of ATP and other energy reserves of a tissue at low temperature still seems a reasonable first step in establishing its metabolic adequacy. The first and third objections would not even arise unless ATP were found to fall, and the second would be comparatively easy to test.

Determination of cellular ATP concentration is tedious, but not otherwise difficult, and it is surprising that it has so seldom been employed in studies of mammalian tissues at low temperature. Surgical hypothermia and storage of organs and tissues are two research areas of active clinical interest in which the metabolic status of cells at low temperature is important and in the literature of which may be found a few estimates of ATP. Unfortunately, these are rarely suitable to answer the present need, because in both operations limitation of metabolites is so often a direct cause of loss of ATP. Indeed, in both fields, cold is often regarded as having a sparing rather than a deleterious effect.

Thus, in his detailed comparison of the differences in hypothermic resistance between hibernating and nonhibernating adult and infant mammals, Andjus (1969) has clearly demonstrated that what limits survival in the whole organism in forced hypothermia is hypoxic tolerance. He showed that the time course and character of decline in ATP of brain and changes in other intermediates in brain of rats or ground squirrels dying of hypothermia were similar to those of changes occurring from respiratory occlusion at comparable temperatures.

Brain

Mendler and associates (1972) attempted to separate the effects of hypoxia and cold in brain by comparing the changes of electrolytes, water content, and metabolites in perfused brains of rats made hypothermic by extracorporeal circulation with those made hypothermic by surface cooling with only respiratory support. The governing hypothesis of this group was that irreversible brain injury in hypothermia is caused by the structural damage arising from brain swelling, which in normothermia is prevented by electrolyte regulating mechanisms dependent upon metabolism. They found that in brains of the surface-cooled rats with a body temperature of $0°-4°C$, ATP concentration was one-fourth to one-eighth that of the initial, normothermic value and, as was also observed by Andjus, this drop was accompanied by comparable decrease in glycogen and glucose reserves and by brain swelling (i.e., water uptake following uptake of Na). When, however, cooling was achieved by extracorporeal circulation which maintained oxygenation close to normal, ATP concentration and ATP/ADP ratio stayed at normal levels even though brain swelling and loss of electrolyte regulation were unabated. Thus, they concluded, as Andjus had, that primary metabolic deterioration in ordinary hypothermia (even with respiratory support) is due to hypoxia, not to a direct effect of cold.

The concentration of ATP in brain of hibernating hedgehogs was found by

Kristoffersson (1961) to be actually higher than in active individuals. The body temperature of the hibernating animals in this study, however, was not very low, ranging between 7° and 17°C.

Heart and Other Organs

Studies of the status of energy reserves of heart in deep hypothermia or hypothermic storage have been less detailed than those in brain and do not permit a choice between limitation of oxygen (or other substrate) and inhibition by cold. The results of Covino and Hannon (1959) have frequently been cited as demonstrating the superior cold resistance of metabolism of heart of a hibernator, but they suffer from this same uncertainty (Hannon et al., 1972). They found that when arctic ground squirrels and rabbits were made hypothermic to a rectal temperature of 15°C by surface cooling (and no respiratory support), the ATP concentration in rabbit hearts declined by 38%, whereas that in the ground squirrel was unchanged. It seems likely that both circulation and respiration would have been failing in the rabbits at 15°C and that the fall of ATP was at least accelerated, if not caused entirely, by hypoxia.

Zimny and Taylor (1965) obtained similar results with hypothermic rats and thirteen-lined ground squirrels, except in their case ATP and creatine phosphate concentrations did decline by about 60% in the hypothermic ground squirrels (rectal temperature, "7°-17°C"). In the hypothermic rats (19°-11°C) a similar decline was marked by evidence of hypoxia (increase of lactate, changes in ECG record), but not in the ground squirrels. Zimny and Gregory (1958) had previously shown that the levels of ATP (+ADP) and of creatine phosphate in hearts of ground squirrels is not diminished in 3-5 days of natural hibernation. In a later paper (Zimny and Gregory, 1959), however, the same authors claimed that over a longer period of "uninterrupted hibernation" (30 days) both "ATP" (=ATP + ADP) and creatine phosphate did fall by 50-60%. (It is unfortunate that the authors did not describe their method of monitoring hibernation, because a 30-day bout without at least partial arousal is very unlikely in this species.) ATP was seen to fall by about 30% in cardiac muscle of hibernating hedgehogs (Kristoffersson, 1961), and by a similar amount in skeletal muscle, even though the hibernation in these animals resulted in body temperatures no lower than 7°C and usually considerably higher. The decrease in ATP in hibernation was also about the same as that produced by forced hypothermia in the same species.

More recently Burlington et al. (1976) determined concentrations of ATP, ADP, and P_i in perfused hearts of rats and thirteen-lined ground squirrels. As the hearts of rats were cooled below 20°C (at which temperature they ceased to beat) the concentration remained unchanged, but the ratio ATP/ADP + P_i, sometimes called the "energy change" and considered by some to be more indicative of available energy (see Hochachka and Somero, 1973), did decrease. In the ground squirrel hearts beating persisted below 6°C, and at that temperature ATP concentration was diminished by 25% from the value at 35°C. The energy charge, however, was unchanged.

When surgically excised hearts of nonhibernating species are "stored" at 4°C (i.e., suspended in medium without perfusion at the low temperature), ATP levels decline, but much less rapidly than under the same conditions at higher temperature (Fedelešová et al., 1969; Siska et al., 1969).

In their studies of hypothermically perfused dog hearts, Fedelešová and colleagues found that exogenous ATP injected into the coronary vein not only leads to less decline in ATP, but also to improved cardiac performance and forestalled decline in glycogen. With doubly labeled [^{14}C,^{32}P]ATP, Fedelešová et al. obtained indirect evidence that some of the ATP was actually entering the heart intact, although they concluded that much of it was split by extracellular ATPases. Calman (1974a,b) subsequently found that exogenous ATP also improved the poststorage "survival" (i.e., ability to start up again at high temperature when perfused with blood) of stored rat hearts. Aside from the practical implications, these results are important in suggesting that under conditions of hypoxia or hypothermia the plasma cell membrane, normally quite impermeable to such a large and charged molecule as ATP, becomes permeable. Thus, ATP could conceivably leak out as well as in, and inhibition of metabolism is not the only potential cause of decline in ATP concentration.

There is little information about ATP levels in kidney at low temperature. Collste et al. (1971) found that in cortex of kidneys of dogs perfused at 8°-10°C after brief circulatory arrest, the ATP level actually increased from the depressed level caused by ischemia and remained elevated for up to 48 hr. At 0°-4°C, however, in a somewhat different perfusion solution, the ATP concentration fell to nil in 24 hr. It is not clear whether the difference was due to the composition of the medium or to its temperature, and there was no proper control for any higher temperature with either of the media. In rabbit kidneys perfused at 4°C ATP level declined from 1.5 μmoles/g to 0.2, over 6 hr, but again, there was no control for maintenance of ATP at a higher temperature under the perfusion conditions (Schmidt-Mende and Brendel, 1967).

The ATP concentration in kidney of hedgehogs and in liver of bats, hedgehogs, and ground squirrels increases during hibernation, in the liver sometimes as much as severalfold (Zimny and Gregory, 1958; Kirstoffersson, 1961; Dryer and Paulsrud, 1966).

Red Blood Cells

Although function of erythrocytes is not so crucial to survival of hibernation or hypothermia as the brain or the heart, they nevertheless provide a more suitable model for exploring this question of ATP management at low temperature for several reasons. There is a long history of storing erythrocytes at low temperature for later transfusion, and there is, therefore, more information specifically about ATP retention under these conditions than for the other tissues. Further, because the cell is structurally simple and has but few avenues of ATP utilization (namely, transport of Na, K, of Ca, retention of shape, phosphorylation of glucose and fructose in glycolysis and synthesis of glutathione) it is possible to evaluate the demands for ATP in relation to the supply.

Many studies of human red blood cells mixed with the standard preservative ACD (i.e., acid—pH 5; citrate, to chelate Ca and prevent clotting, also to provide a non-penetrating solute to reduce hemolysis; and dextrose, or glucose, as a substrate), and stored at 4°C have shown that ATP concentration declines gradually over a period of weeks. Furthermore, it is well accepted that the concentration of ATP retained during storage is directly correlated with the survival of the cells upon subsequent transfusion (Nakao *et al.*, 1962; Valeri, 1974). [In the only study of erythrocytes of a hibernating species stored at 4°C in ACD, Brock (1960) found that "easily hydrolyzable phosphate" (= ATP + ADP) declined by about 54% in 3 days. Other organic phosphates increased and inorganic phosphate declined slightly. The last result, however, is at variance with her later experiments on stored blood and with her findings discussed below on changes *in vivo* during hibernation with inorganic phosphate increased.]

Several factors must be recognized before concluding that the results with human erythrocytes stored in ACD necessarily imply a failure of glycolysis to keep pace with ATP utilization because of inhibition by cold. First, the decrease in ATP is often very slow; indeed, in some studies it was virtually negligible in the first 2 weeks of storage (Jorgenson, 1957; Bartlett and Barnet, 1960).

On the other hand, there are other reserves of high energy phosphate in the cell. ADP, by virtue of the adenylate kinase reaction ($2ADP \rightarrow ATP + AMP$) is one, but consideration of its changes do not alter the picture significantly. In general, its concentration is either relatively constant or else it declines with ATP. Of greater significance is 2,3-diphosphoglycerate (DPG), which is produced from 1,3-diphosphoglycerate. DPG can be converted to 3-phosphoglycerate and thence to pyruvate yielding an ATP. Although the net effect of 1,3-diphosphoglycerate going to pyruvate via DPG is to lose production of one ATP (because a step in the main glycolytic path which produces an ATP is bypassed), DPG is usually far more abundant in the cell than ATP and, therefore, is the most important reserve form of phosphate bond energy. Part of the reason that ATP declines so slowly in ACD is that it is maintained by an initially more rapid depletion of DPG. After 2 weeks, when DPG is virtually exhausted, ATP declines steeply (Prankerd, 1956; Bartlett and Barnet, 1960; Valeri, 1974).

A third consideration is whether the common storage conditions would allow a steady state at a higher temperature, and the evidence is they would not. Bishop (1961) showed that at 37°C, time for 50% exhaustion of ATP in erythrocytes stored in ACD was about 38 hr compared with 28 days for cells at 4°C. Similarly, Valeri (1974) found that DPG decreased by 75% at 22°C in 24 hr, whereas at 4°C there was almost no decrease. (In this case, storage was in a medium with a higher pH than usual in ACD; see below.)

Thus, it appears clear that the gradual decrease in high energy phosphate reserves is due in large measure to the conditions of storage rather than to the direct inhibition by cold. (Indeed, low temperature retards these effects.) Workers in this field often refer to this subject as the "lesion of storage." Two candidates for the cause of the "lesion" are glucose depletion or other progressive change caused by the ongoing metabolism (pH, lactic acid accumulation) of the cells and the effects of ACD itself. Although glucose is always provided in the additive storage medium, it is utilized during storage and lactic acid does increase (Pappius *et al.*, 1954; deVerdier and Killander, 1962) and pH declines (Maizels, 1944).

With regard to the effects of ACD, two obvious choices are the chelation of Ca^{2+} and the (even initially) low pH. The latter has been extensively investigated. Low pH inhibits glycolysis (deVerdier *et al.*, 1964; Beutler and Duron, 1965; Chanutin, 1967). Because of this, another storage medium buffered with phosphate to a pH nearer neutral has been introduced to replace ACD (CPD, citrate-phosphate-dextrose). With CPD, glycolysis occurs at a greater rate during low temperature storage and DPG is better preserved, but ATP according to some investigators is rapidly depleted, so that on balance the maintenance of total available phosphate bond energy is not as well maintained (Chanutin, 1967; Valeri, 1974).

Thus, after all of this effort in blood banking, and the prodigious amount of blood-letting that it represents, we are left without a firm impression of what would be the effect directly of cold on phosphate energy reserves, all other factors being optimal. Perhaps the best indication can be obtained—not in the world of blood storage, but rather, and quite incidentally, from that of ion transport in red cells—in the work of Whittam (1958) who washed human erythrocytes and placed them in a physiological saline medium with glucose, but no citrate, and at an initial pH of 7.4. In this medium at 37°C the cells were able to maintain a constant ATP concentration for at least 5 hr, although DPG did decline slightly. After storage, the same medium also permitted a restoration of ATP. When the cells were stored for 5 days in this medium at 4°C their ATP concentration fell by 60% and ADP rose by an amount equal to the fall in ATP; DPG was unchanged.

No comparable studies have been done of *in vitro* preservation of erythrocytes of a hibernating species under conditions closer to physiological than those provided by ACD. There have, however, been several determinations of changes in organic phosphates of erythrocytes *in vivo* during low body temperature of hibernation. ATP was not appreciably lower in erythrocytes of hibernating woodchucks although DPG fell by about 20% (Harkness *et al.*, 1974). In all other hibernators studied, including hedgehogs (Kristoffersson, 1961), hamsters (Brock, 1967; Tempel and Musacchia, 1975), golden mantled, Mexican, and thirteen-lined ground squirrels, ATP and DPG fell by 40–60% (Burlington and Whitten, 1971; Larkin, 1973). Unfortunately, most of these reports do not state the elapsed time of the hibernating bout within which the blood sample was collected. In hamsters, however, Brock (1967) combined and reported the results of hamsters hibernating for 3–4 days. She found a drop in ATP of 33% (compared with her value of 43% for fall in easily hydrolyzable phosphate of stored hamster blood) and in DPG of 26% [compared with 39% for 48 hr observed by Tempel and Musacchia (1975)]. She also observed that inorganic phosphates rose by three to eightfold, but that fructose diphosphate was unchanged, a result suggesting that there is no deficit in the early stages of glycolysis. Interestingly, in view of the many studies showing improved survival of human cells stored with nucleosides (Prankerd, 1956; deVerdier *et al.*, 1964; Beutler and Duron, 1965; Chanutin, 1967), such as adenosine, she also observed that pentose phosphates increased in erythrocytes of cold-exposed hamsters and then decreased by 80% in 3 or 4 days of hibernation. (Adenosine can be utilized in glycolysis by conversion to hypoxanthine and pentose phosphate.)

ATP Utilization

One of the main sources of depletion of ATP (aside from phosphorylation of glycolytic intermediates) is the pumping of ions. Na and K transport for 20–45% of the ATP utilization at steady-state in human erythrocytes at 37°C (Whittam *et al.*, 1964). When red cells are stored the cells lose K and gain Na, a process that leads to the swelling of the cell and its eventual hemolysis. Wood and Beutler (1967) found that the loss of ion regulation in cold-stored human red cells showed no correlation with loss of ATP and occurred even in cells with well-maintained ATP reserves. They concluded from this and parallel studies of Na^+,K^+-ATPase that it was not lack of ATP that was inhibiting ion regulation, but the direct effect of cold itself. They achieved a spectrum of ATP concentrations, however, by using a variety of storage media (not described in detail), and it is possible that the components of some of these media (low pH, citrate) might also have exerted an inhibitory influence on ion pumping or else increased the leak of ions.

Summary

This inquiry into the ability of mammalian tissues to maintain ATP in the cold may be summarized as follows. When oxygen and metabolites are provided at low temperature, brains of at least some nonhibernators appear to be able to maintain normal, elevated concentrations of ATP, as do hibernators during hibernation. On the other hand, the number of relevant published reports is quite skimpy and for those that exist the periods of hypothermia (hours) are quite short compared with the period of a single bout of hibernation.

Low temperature storage and hypothermic studies of hearts of nonhibernators both show decline of ATP, but in neither case has the problem of concurrent hypoxia been solved. Phosphate energy reserves do fall in hearts of hibernators, too, both in hypothermia and in hibernation, although very slowly in the latter case. In other organs such as kidney and liver there is little useful information for nonhibernating mammals. In hibernation the levels of ATP in these organs increase.

Only with erythrocytes does there seem to be enough information to answer the original question in its own terms. (1) Under the usual conditions of storage, ATP in human and rabbit erythrocytes declines very slowly at 4°C. (2) Under more physiologically moderate conditions (higher pH, less citrate) the decline of ATP is more rapid, although the total high energy phosphate reserve still declines slowly. (3) The loss of ATP is not the limiting factor in erythrocytes of nonhibernators for one major avenue of expenditure (Na, K pumping). (4) The loss of ATP in erythrocytes of hibernators, either stored or *in vivo* during hibernation, is not less severe than that occurring in red cells of nonhibernators under comparable conditions.

Thus on the whole, the capacity of ordinary mammalian tissues to maintain ATP concentrations at low temperature is not bad and that of tissues of hibernating mammals has not yet been shown to be consistently or significantly better. If, therefore, metabolic adaptations to cold are to be found among hibernators, it is likely that rather than being generally prevalent and easily demonstrable, they will exist in specific tissues to provide either a margin of safety or else to serve some particularly acute needs and play specific roles.

EFFECT OF TEMPERATURE ON METABOLIC RATE

A more common approach to describing cold adaptation or sensitivity of intact mammals or isolated tissues therefrom has been to measure and to compare the effects of temperature on their metabolic rates. As with measurements of ATP and other high energy reserves, this approach has yielded mixed results and has the further disadvantage of being even less interpretable.

Oxidation in Intact Cells

Fuhrman and Fuhrman (1963, 1964) found that in hypothermic rats, blood glucose was not utilized, although in observations of glucose uptake and utilization of tissues *in vitro* no unusually temperature-sensitive step could be found.* Failure of organ systems in the whole organism may intervene to a greater extent than inhibition of cellular oxidation.

With slices of kidney cortex, Kayser (1954) found that the temperature characteristic (i.e., slope of the Arrhenius curve) of respiration was less in those of European hamsters than in those of rat. South (1958) obtained similar results for cardiac ventricular slices from Syrian hamsters compared with those of rat and slices of bat ventricle had an even lower temperature sensitivity than those of hamsters. With the same three species, however, he found no difference in Arrhenius slope either of respiration of brain slices or of glycolysis in that tissue. Burlington and Wiebers (1967), on the other hand, did find that glycolysis in brains of ground squirrels was much less inhibited at 5°C (7–8% of glycolysis at 38°C) than in brains of rats (0.6%).

Horwitz (1964) did not find any difference in temperature sensitivity of respiration in slices of kidney or of liver of bats as compared with those of rat, nor did Willis (1968) find any difference in respiration of kidney of Syrian hamster or thirteen-lined ground squirrels compared with guinea pigs and rats. (Respiratory rates in the rat tissues of Horwitz at 37°C, however, were about a one-fourth or one-third of those observed by others. It seems likely, therefore, that the slices were oxygen-limited because of their thickness or the air atmosphere used, and that the slope of the Arrhenius curve may have been artificially decreased.)

Several studies have shown that the temperature sensitivity of respiration of isolated tissues taken from hibernating individuals is no less than that of tissues from active individuals of the same species. An upward translation on the Arrhenius plot (i.e., higher respiration at every temperature) has, however, been reported (Kayser, 1954; South, 1958). Meyer and Morrison (1960) found instances of increased temperature sensitivity in some tissues of hibernating ground squirrels, but there was only a single estimate in each case. Horwitz (1964) reported the same in bat liver and kidney, but again the number of cases was small.

Although *in vitro* studies with intact tissues have given useful information about relative activity of specific pathways at different temperatures, which will be discussed more fully below, such preparations cannot be expected to provide an unequivocal, nor even a very useful, test for cold resistance or sensitivity of energy metabolism. There are two reasons for this. In intact cells it is energy-utilizing (i.e., ATP-consuming) activities that often set the pace of metabolism by producing ADP and inorganic phosphate, which are the limiting factors for mitochondrial respiration. Thus, the relatively greater respiration, seen by, say,

---

*In much of what follows it will be assumed that a rate function should be expected to decrease by about an eightfold factor between 37° and 7°C, i.e., a Q_{10} of 2. This would be an "ordinary" temperature sensitivity. Departures both above and below this expectation have been observed.

Kayser (1954) in kidney and by South (1958) in cardiac muscle of the hibernating species, might only have indicated greater activity of ATPase splitting by ion pumps or by muscle contractile mechanisms (South, 1960a; Willis, 1967). There have been only two attempts to test this point, and unfortunately, both were rather inconclusive. Saarikoski (1967) used the method of stimulation of brain slices of rats and hedgehogs to produce an "active" or "suprabasal" respiration, and Willis (1968) used the method of ouabain inhibition or Na deprivation to produce a "basal" respiration (i.e., one supposedly not supporting active transport) in kidney slices prepared from individuals of two hibernating and three nonhibernating species. In neither case was a systematic difference in temperature sensitivities observed between species, but in both cases the "active" respiration was found to be somewhat more steeply inhibited by cooling than the "basal" respiration (except in hamster kidney).

The second reason why whole-cell respiration may be a misleading test of cold adaptation is that if low temperature had a directly deleterious effect on cellular integrity, increased respiration could be a response to this. As an example, if regulation of cytoplasmic Ca were to fail, thus causing increased ionic Ca concentration in the cell, respiration of mitochondria would be stimulated by their "uncoupled" uptake of Ca.

Thus, in neither the organism nor the cell can simple metabolic rate be used as the criterion of cold adaptation because with either level of organization both regulatory feedback loops between energy demand and energy release and changes at the reaction sites due to failure of the system must be considered. It has, therefore, been found necessary to make determinations in "open loop" situations or even purified enzymes. Such an approach is of course well tried in other contexts and also subject to its own well-recognized hazards. Among these are the fallacy that the temperature optimum or characteristic of an isolated enzyme, separated from many impinging cytoplasmic factors, will necessarily be the same as that *in situ* and the (very frequent!) conclusion that a greater or lesser rate of a reaction at a given temperature in a preparation from a hibernator (as compared with some other species) reflects an adaptation for hibernation. Of course, in either instance one must erect a framework, in the context of which the isolated observation may achieve relevance, hoping meanwhile that the structure thus erected does not turn out to be a hanging scaffold.

Respiration in Cell-Free Preparations

Because the citric acid cycle (Fig. 9-2) and the associated oxidative phosphorylation is the greatest available source of ATP, more attention has been paid to these pathways than to any other. Both respiration and respiratory efficiency (judged by moles of ATP produced per mole oxygen consumed, the P/O ratio) have been used as criteria of mitochondrial activity.

The effects of temperature on homogenates or purified mitochondria of two tissues, heart and liver, have been investigated extensively. Continued function of the heart is, of course, central to survival in hibernation and, happily, it provides suitable mitochondrial preparations. Liver mitochondria are of course the "darlings" of biochemists and have not been neglected, even though the effects of temperature on respiration of liver cells have been.

The main interest for our purpose is a comparison between species. It is important for this purpose, however, to know also for each species of hibernator and for each tissue whether a change occurs with the onset of hibernation. For this reason we shall digress at this point to consider what changes have been reported to be associated with the hibernating state.

In heart mitochondria, South (1960b) found an upward displacement of the Arrhenius curve of respiration in the sarcosomes from hibernating hamsters, similar to, but less pronounced than, that seen for respiration of slices of cardiac muscle. (There was no change in P/O ratio.) Roberts and Chaffee (1973) saw a similar shift for the succinoxidase activity of heart homogenates of cold-exposed but not of hibernating hamsters. Woodard and Zimny (1973) described the reverse change in succinate dehydrogenase of heart mitochondria of hibernating ground squirrels, a downward vertical displacement of the Arrhenius curve, but the error was very large.

The respiration of liver mitochondria of hamsters and ground squirrels has generally been found to be lower in preparations made from hibernating individuals (Daudova, 1968; Liu et al., 1969; Roberts et al., 1972). Daudova (1968), however, found that this difference disappeared when the mitochondria were washed. Her procedure, however, also differed in other details (she used glutamate as substrate and was apparently measuring basal, so-called state 4 respiration).

The cytochrome oxidase activity of liver mitochondria of ground squirrels has also been reported to be 50% lower in those of hibernating individuals, but Shug et al. (1971) concluded that this was not apparently what was limiting respiration. When mitochondria of hibernating ground squirrels were uncoupled with salicylamide they exhibited a greater maximal respiratory rate than those from active animals, even though their maximal respiration with ADP had previously been only one-half that of the active animals' mitochondria. These results favor the interpretation that it is decreased nucleotide permeation that may be depressing respiratory rate during hibernation.

The P/O ratio of liver mitochondria of hamsters, ground squirrels, and bats is not generally altered with hibernation. Roberts et al. (1972) did report, however, a higher P/O at low incubation temperatures with liver mitochondria of hibernating hamsters, but this was not observed in the similar studies of Liu et al. (1969).

With respect to the effect of hibernation on the temperature sensitivity of respiration, especially of succinate, Roberts et al. (1972) found a straighter and less steep Arrhenius slope in the mitochondria from livers of hibernating hamsters. Raison and Lyons (1971) reported much the same result in liver mitochondria of golden mantled ground squirrels. In the latter case a downward "break" at 20°C in the Arrhenius curve for succinate oxidase of mitochondria of active individuals was not present in the curves based on two hibernating squirrels, so that the overall slope for the mitochondria of hibernating squirrels was less than that of the lower limb of the curve for mitochondria of active animals. Using much the same methods, however, but averaging results of several animals, Liu *et al.* (1969) did not observe such differences in liver mitochondria either of hamsters or of thirteen-lined ground squirrels. In hibernation, the "break" persisted and the slopes of the Arrhenius curves at lower temperatures were the same or even a little steeper than those of active animals.

Thus, while hibernation is not without effects on the quality of respiration in heart and liver mitochondria, it does not appear to alter the temperature sensitivity in heart mitochondria, and it may or may not do so in liver mitochondria.

Returning, then, to the main point of comparing cold adaptation between species, South (1960b) found the slope of the Arrhenius curve (or energy of activation, E_a) was smaller in preparations made from hearts of hamsters than it was in those made from hearts of rats. Thus, the lower temperature sensitivity of the hamster hearts, which this investigator had previously reported for cellular respiration, seemed to be present in the mitochondria as well.

Roberts and Chaffee (1973), however, measuring oxygen consumption polarigraphically in homogenates of hearts of hamsters and rats, with succinate as substrate, failed to find any difference in temperature sensitivity between preparations from the two species. There were several differences in technique between these two studies, but one which would seem to us most likely to generate a difference in apparent temperature sensitivity was the choice of substrate.

With pyruvate as a substrate (South, 1960b), the entire Krebs cycle would necessarily be involved. In particular, two highly complex reactions before succinate oxidation would seem likely candidates for the observed discrepancy. The pyruvate dehydrogenase complex, which catalyzes the five-step oxidation of pyruvate to acetyl-CoA, is a 4 million dalton entity comprising many molecules of individual pyruvate dehydrogenase enzyme and two other enzymes that are lipoproteins. The system is negatively inhibited by ATP. Isocitrate dehydrogenase catalyzes the oxidation of isocitrate to α-ketoglutarate, a prime rate-limiting step in the citric acid cycle, and it is also a polymeric regulatory structure, strongly stimulated by ADP and strongly inhibited by ATP and NADH. Such complex regulatory enzymes may be particularly vulnerable to the structural effects of changing temperature, especially when the structure contains a lipid moiety.

Hannon (1958) had previously observed in rat heart homogenates that the

Arrhenius curve for respiration with succinate showed two linear segments with different slopes, the steeper one lying between 20° and 5°C with a break at 20°C. (The state of the mitochondria in these studies is a little uncertain since, according to the description, no ADP was added as a phosphate acceptor, and they might, therefore, have been in "state 4," minimal oxygen consumption. However, 2 mM ATP was present, and since it was a homogenate, presumably containing ATPases, there was probably enough ADP present to keep them in "state 3," maximal respiration, as were the other cases discussed here.) With pyruvate and malate, the curve below 20°C was curvilinear or broken again at 10°C. Thus, in the range of 10°-0°C the E_a was more than 36,000 cal/mole ($=Q_{10}$ of 10), much greater than that for succinate in the same range ($E_a =$ 20,000; $Q_{10} = 3.3$).

South (1960b), who used pyruvate and malate, chose to characterize the Arrhenius slope of respiring rat heart mitochondria with a single regression line, but a discerning and determined eye may perceive the breaks, or the curvilinearity, at lower temperatures described by Hannon for pyruvate–malate fueled respiration. It is at these lower temperatures, of course, that the values for rats depart from those for hibernating hamsters, because the latter do not exhibit the downward curvilinearity. In the only study comparing the mitochondria of hearts of another hibernator, Covino and Hannon (1959) found no difference in temperature sensitivity between arctic ground squirrels and rabbit, but they only investigated two temperatures, 37° and 15°C. (At 15°C, the intact hearts or rabbits in their study were not even blocked.)

In South's (1960b) study with hamster and rat hearts there were no significant changes with temperature nor differences between species in the phosphorylating efficiency of mitochondria (i.e., moles ADP phosphorylated per mole oxygen consumed or P/O ratio).

In liver mitochondria Chaffee et al. (1961) reported that with two temperatures (37° and 7°C) there was no difference in cold sensitivity of "state 3" mitochondria of hamsters compared with those of rat. This was confirmed in the very extensive studies of Liu et al. (1969), not only for hamsters compared with rat, but also for ground squirrel, and not only for succinate, but also for β-hydroxybutyrate, which enters the Krebs cycle through acetyl-CoA. Roberts et al. (1972) cited a steeper Arrhenius slope of "succinoxidase" at low temperature (6°–20°C) in rat liver, than in hamster liver mitochondria. Close inspection reveals, however, that this apparent discrepancy is largely due to an upward break of the Arrhenius curve between 20° and 15°C and then a steeper slope below 15°C in the rat mitochondria. Between 37° and 6°C or between 20° and 6°C there was no overall difference between the non-cold-exposed individuals of the two species in the decrease of respiration of their liver mitochondria.

With hibernation, however, the Q_{10} between 20° and 6°C went down slightly in hamsters and with cold exposure it went up in rats (Roberts et al., 1972) so that the comparison between these two groups showed a more attestable difference. Liu *et al.* (1969), however, with much the same techniques, observed no change with cold exposure in rats. [Roberts and colleagues (1972, 1973), incidentally, found that with both heart and liver mitochondria the succinoxidase of squirrel monkey mitochondria was more temperature-sensitive than that of either rat or hamsters. This finding is a reminder that "the rat," the common control species, is perhaps actually fairly cold tolerant for a nonhibernating mammal.]

There is no consensus regarding the effect of temperature on P/O ratios in liver mitochondria. In rat liver mitochondria with succinate, it has been reported to go up, to go down, and not to change at 6°C relative to 37° or 20°C, and to reach a minimum at 13°C, and then to go up again at 5°C (Horwitz, 1964; Liu *et al.,* 1969). The effect of temperature in rats on P/O ratio has also been reported to be (Horwitz, 1964) and not to be (Liu *et al.,* 1969) dependent upon substrate. In hamsters, it is agreed that the P/O ratio of liver mitochondria is not different at 6°C from that at 37°C, but the findings between these two temperatures are as conflicting as they are in rats. In liver mitochondria of ground squirrel there was no decrease in P/O with temperature (Daudova, 1968; Liu *et al.,* 1969), but Horwitz (1964) found in those of bat the same minimum at 13°C that they described in those of rat.

Thus, while there is evidence that mitochondria of both heart and liver of hibernating species are less temperature-sensitive than those of nonhibernators, either in terms of respiratory rate or efficiency or both, and that, at least in liver this improves still further with hibernation, there is also contradictory evidence in each instance. With heart mitochondria the conditions of the separate investigations were sufficiently different to allow the possibility that the adaptation may be present when the full oxidative pathway is in operation, but not when it is "short-circuited" by use of succinate as starting substrate. In liver mitochondria the methods of the several investigators were more closely similar, involving polarigraphy as the measuring system and succinate as substrate, so that the divergence in results precludes a general conclusion.

Lipid Metabolism

Perhaps a majority of all papers on metabolism of hibernation begin with the catechism that the principal source of energy in hibernation is fat; it is little short of astounding, therefore, that there is not a single published account of any attempt to examine, comparatively with a hibernator, the temperature sensitivity of any enzyme or (cell-free) reaction, specifically and exclusively concerned with lipid catabolism.

Only very recently, indeed, have there even been attempts to examine the temperature characteristics of lipid metabolism in whole organ and cellularly intact tissue preparations. While interpretation of rates of consumption of fatty acids in such preparations may be subject to criticisms outlined previously, they nevertheless can provide useful initial information about hibernators as compared with nonhibernators. The value of this information is enhanced by the simultaneous consideration of two other questions, the effect of temperature on catabolism as opposed to synthesis of fatty acids (or their esterification into triglycerides) and the relative effect of temperature on lipid catabolism as compared with glucose catabolism.

Taking this last point first, Shipp et al. (1965) found that in perfused hearts of rats depleted of glycogen, the utilization of [1-^{14}C]palmitate as a source of $^{14}CO_2$ was reduced between 37° and 17°C by about half, a much smaller reduction than that for $^{14}CO_2$ from glucose (75%) or for lactate production (84%). The uptake of both glucose and palmitate into the heart was diminished under these conditions by only about 45%.

In liver slices and fat pads of rats, however, palmitate oxidation was not found to be unusually cold resistant, the reduction between 37°C and 17°C or 7°C in fat pads being a little greater than for glucose oxidation and in liver slices a little less (Hillyard and Entenman, 1973). Comparing palmitate oxidation to CO_2 in livers of hamsters and rats, Entenman et al. (1969, 1975) found that the rate at 7° relative to 37°C was the same in the two species (11%). In active ground squirrels, it was very low (2% based, however, on only one estimate at 7°C), and in hibernating ground squirrels it was higher (14%). In keeping with the depression of respiration in liver mitochondria, the rate of oxidation of palmitate in the liver slices of hibernating ground squirrels at 37°C was only one-third that of slices from livers of active animals.

In fat pads of rats and the livers of both rats and hamsters the overall incorporation of palmitate carbon into triglycerides was somewhat less temperature-sensitive than its oxidation to CO_2, and in the livers of both species the formation of diglycerides was virtually unaffected by cooling. In the ground squirrel livers, however, diglyceride formation was conventionally temperature-sensitive, a factor that could have spared fatty acids for oxidation.

If one considers synthesis of fatty acids or cholesterol as opposed to mere esterification of fatty acids, however, then such "sparing" is very evident even in rat tissues. In perfused livers of rats, Entenman et al. (1969) found no incorporation of [^{14}C]acetate into cholesterol, phospholipids or triglycerides even at 17°C, and in slices of liver it was only 1 and 6% fatty acids and cholesterol, respectively, at that temperature and nil at 7°C. In rat fat pads the inhibition of incorporation of label was also much more severe at low temperature than was inhibition of oxidation of acetate, or even of palmitate (Hillyard and Entenman, 1973). Comparable in vitro studies on synthesis have not been reported for tissues of a hibernator, but for hamsters in forced hypothermia with a body temperature of 7°C the same group of investigators has observed a very much greater relative incorporation of acetate into lipids of various tissues (Entenman et al., 1971).

Finally, it should be noted that the uptake of either fatty acids or triglycerides from albumin complexes in the milieu surrounding cell membranes is a process that is affected very little by temperature (only about 50% between 37° and 0°C (Markscheid and Shafrir, 1965; Reshef and Shapiro, 1965).

It thus appears that even in nonhibernators, two of the processes required for lipid utilization for energy (permeation, oxidation) are not particularly temperature-sensitive, so that there is little room for a species-related cold adaptation of this overall process in tissues of hibernators. And indeed when such adaptation is specifically sought it is not found.

On the other hand, it must be remembered that such determinations of oxidations in whole cells are uncontrolled for phosphate acceptor and other regulatory factors. While it is true that a difference in tissue oxidation between a hibernator and a nonhibernator does not prove an adaptation, it is equally so that absence of a difference does not eliminate that possibility. Furthermore, in the studies discussed, there was no report of pool size for fatty acids, specific activity of $^{14}CO_2$, or time course of incorporation of fatty acid label into lipid fractions. Hence it is possible, for example, that mobilization of fatty acid at low temperature in tissues of hibernators could have reduced the apparent use of palmitate, and that recovery of labels in lipids sometimes reflected steady-state concentrations of products rather than rates of syntheses.

Still to be investigated, even in an exploratory way, are the comparative effects of temperature on fatty acid oxidation in isolated mitochondria and on the activity of lipases and other mechanisms of mobilization. So, while it seems on the face of it that there can be little expectation that species-related cold adaptation will be found for lipid catabolism, it is too early to foreclose that possibility.

Glycolysis

Animals of hibernating species possess unusual hypoxic tolerance (Hiestand *et al.*, 1950; Bullard *et al.*, 1960; Andjus *et al.*, 1964). Because of this fact, the initial, anaerobic stages of glucose catabolism have received more attention than lipid catabolism, although much of it is tangential to the issue of cold adaptation. Attempts to account for hypoxic tolerance have included demonstrations that normally ''aerobic'' organs (heart and brain) are resistant in that they either possess or can be induced to possess higher glycolytic rates (Kayser and Malan, 1963; Andjus *et al.*, 1964; Burlington and Wiebers, 1966, 1967; Burlington *et al.*, 1970), and that with the onset of hibernation the isozymes of a key enzyme in glycolysis, lactic dehydrogenase, are altered. The last show an increase in the anaerobic, ''M-type'' isozyme or decrease in the aerobic, ''H-type'' LDH in various tissues including liver, heart, and brain (Burlington and Sampson, 1968).

Bullard *et al.* (1960) found that the ability of ground squirrels to undergo voluntary hypothermia enhanced their hypoxic survival. It is not clear, however, that the converse is the case, that is, that resistance to hypoxia in some way improves the ability of the organism to survive the hypothermia of hibernation. Viewing the array of information referred to above, several investigators (Bullard *et al.*, 1960; Kayser and Malan, 1963) have assumed, explicitly or otherwise, that hibernation does involve hypoxia, at least at the tissue level. Evidence for this, however, is seldom offered.

There have been few studies of oxygen levels in blood during hibernation which would be useful for dealing with this question. Musacchia and Volkert (1971) found in permanently cannulated ground squirrels that, although the P_{O_2} of arterial blood was not decreased, that of the venous blood (6 mmHg) was only 20% of the value in the venous blood of normothermic animals (28 mmHg). On the other hand, the hemoglobin at 6°C and pH 7.4 (the conditions prevailing in their hibernating ground squirrels) is 50% saturated at a P_{O_2} of 6 mmHg, as it is at 28 mmHg at 36°C (and pH 7.4). Thus, the actual oxygen content of the blood was probably not significantly different in the two states. For reference, Burlington *et al.* (1969) found that the P_{O_2} of actually hypoxic ground squirrels (in an atmosphere of 6% O_2, or 46 mmHg) was between 17 (arterial) and 14 (venous), under which conditions the hemoglobin saturation would have been well under 50% at the prevailing body temperature of 37° or 38°C.

Furthermore, the presumed changes in anaerobic capacity reflected in change of LDH isozymes has not been shown to be effective at low temperature. Burlington and Sampson (1968) who reported the increase of both total and M-type LDH activities, as assayed at 32°C, in livers, brains, and hearts of ground squirrels, with the onset of hibernation, did not find greater activities at an assay temperature of 15°C in hearts and brains. Finally, as noted in Chapter 8, there is seldom an accumulation of lactate in blood of deeply hibernating mammals, nor other indication of anaerobic function.

On the other hand, during the arousal process such an accumulation of lactate does sometimes occur and it may be that during this period of intense activity metabolic rate exceeds the prevailing capabilities to deliver oxygen and that a superior anaerobic capacity would play a role under such conditions.

Specifically with regard to the question of cold resistance of glycolysis, we have already dealt with comparisons at the cellular level between species and between competing pathways (Chapter 8). The conclusions of these comparisons were that glycolysis is not usually less temperature-sensitive in hibernating species, with the exception of ground squirrel hearts and brains (Burlington and Wiebers, 1966, 1967), and that glycolysis probably is generally more temperature-sensitive than lipid degradation in comparable tissues.

There have been no cell-free studies on the overall process of glycolysis (or alternate paths such as the pentose shunt pathway; see Fig. 9-1) but, perhaps because of the threefold interest in cold adaptation, gluconeogenesis, and hypoxic tolerance, there have been more explicit and detailed studies on enzymes involved with this than with any other pathway.

Comparing LDH of several tissues of guinea pigs, hedgehogs, and bats, Olsson (1974) found no lower E_a's (between 40° and 0°C) in tissues of the hibernators, either in comparison with summer animals of the same species or in comparison with guinea pigs, nor did Burlington and Sampson (1968) in tissues of ground squirrel and rat between 32° and 15°C.

Pyruvate kinase is the only enzyme in all of the catabolic pathways of metabolism so far to have been accorded a serious attempt at a deep and sophisticated analysis of effects of temperature on kinetics and effect of hibernation on molecular form, and it has received this attention from two laboratories dealing with two species of hibernators and two tissues: Behrisch, with purified enzyme from liver of arctic ground squirrels (Behrisch, 1974; Behrisch and Johnson, 1974); Moon and Borgmann (1976; Borgmann and Moon, 1976) with homogenates from liver and muscles of little brown bats. Pyruvate kinase catalyzes conversion of phosphoenol pyruvate (PEP) to pyruvate with transphosphorylation of ADP to ATP. This is an essentially irreversible step in glycolysis, and this property has three significant consequences. The thermal properties of this reaction could be one of the determinants of the temperature characteristic of flux through the entire path (Hearon, 1952); it is a candidate as a major regulatory step in glycolysis, and it is one of the steps that must be bypassed in gluconeogenesis (Llorente et al., 1970; Seubert and Schoner, 1971). Thus, as it exists in such gluconeogenic tissues as liver and kidney, pyruvate kinase is a "crossroads enzyme" (Llorente et al., 1970), and as such it is subject to the complex and regulatory interaction of several factors. First, it has two substrates, PEP and ADP, and of these, PEP exhibits cooperative (sigmoid) activation kinetics. Second, it is allosterically activated by fructose diphosphate (and also H^+ and K^+), which increases the affinity for PEP and abolishes the sigmoidicity of PEP activation. Third, it is negatively affected by ATP and alanine, both of which could play a role in switching the cell from glycolysis to gluconeogenesis.

The liver enzyme of summer active ground squirrels and summer, nonhibernating bats exhibits all these properties qualitatively as described in other forms. Quantitatively, however, the squirrel liver enzyme is relatively insensitive to the negative effects of ATP and of amino acids (Behrisch and Johnson, 1974).

The enzyme from rat muscle (and other nongluconeogenic tissues) does not exhibit allosteric properties, although ATP exerts a weak competitive inhibition on both the PEP and ADP sites (Llorente et al., 1970). Again, pyruvate kinase from flight muscle of summer bats behaves much the same way (Moon and Borgmann, 1976).

As described by Llorente et al. (1970), the effect of low temperature in rat liver pyruvate kinase is complex. Those investigators were particularly interested in a persistent (but reversible) desensitization of the enzyme to its allosteric modulators caused by preincubation at low temperature (0°C). They also reported, but did not extensively discuss, the effects of incubation at low temperature on activation by PEP and fructose diphosphate (FDP). Cooling to 0°C acted somewhat like the positive effector, FDP, in that K_m for PEP was decreased and sigmoidicity vanished. The maximal activity of the enzyme with saturating substrate (V_{max}) showed a normal, linear Arrhenius slope with a Q_{10} of 1.8.

Behrisch and Johnson (1974) examined the effects of cooling the enzyme from active ground squirrels in more detail and found that cooling to 5°C caused a decrease in affinity for PEP (as measured by the concentration required for half-maximal activity, the "K_m"), and also a decrease in cooperativity. These effects on PEP activation were relatively slight and, as was also true in rat, the K_m was not changed by cooling when FDP was also present. Of far more noticeable magnitude, however, was an increase in the affinity for ADP (indicated by a decrease in K_m from 0.18 to 0.03 mM in the presence of FDP). The affinity for ATP as a negative effector decreased somewhat. The net effect of these changes would be to create a low sensitivity to change in temperature; in the presence of a concentration of ADP near its K_m (and presumably no heterotropic effectors), Behrisch (1974) calculated the Q_{10} was 1.3.

In the liver of active bats, cooling caused a decrease in affinity for FDP as well as for PEP (the latter only between 20° and 0°C). In the muscle enzyme the affinity for PEP also increased with cooling and the K_i (i.e., concentration for half-maximal for inhibition) for ATP decreased markedly, so that, whereas ATP could not be a physiological inhibitor at high temperature (with a K_i near 30 mM) at 4°C it could be (K_i about 3 mM) (Moon and Borgmann, 1976). Effects of temperature on affinity for the other substrate for these enzymes was not discussed by these authors.

The changes that these authors found with hibernation in the pyruvate kinase of these two species with respect to all of the kinetic variables and their interactions are too complex to describe in detail here, but for our immediate purpose they can be summarized in two sentences. For the substrates, PEP and (in the ground squirrel) ADP, the effect of temperature on affinities is abolished, i.e., the affinity becomes temperature-independent (Behrisch, 1974; Borgmann and Moon, 1976). The enzyme from ground squirrel liver incubated with organic heterotropic allosteric effectors (ATP, FDP, alanine) becomes more temperature-sensitive (Q_{10} of 2.5 compared with 1.3) (Behrisch, 1974), and the pyruvate kinase from liver and muscle of bats incubated also without heterotropic effectors becomes (for the range 4°–16°C) less temperature-sensitive (Moon and Borgmann, 1976).

These results are not so contradictory as they may at first sound; in the ground squirrel liver, the decrease in ADP affinity with cooling presumably overrides the weaker increase in PEP affinity, and absence of these effects raises the Q_{10}. In the bat enzyme, on the other hand, where ADP affinities were apparently not independently measured, there may not have been a change in affinity with temperature for this substrate, and the change in PEP affinity (which was stronger than in ground squirrel) had the predominant and opposite effect on the Arrhenius slope (i.e., to make it steeper) as did its removal. Furthermore, the changes in thermal properties of the bat enzymes are more complex than is conveyed by the terse description above (Fig. 9-3). The change observed is more fully described as conversion from a broken (or curvilinear) curve with small slope at high temperatures and steep slope at low temperatures, to a straight curve (at least above 4°C) with an intermediate slope (Borgmann and Moon, 1976). Thus, *mirabile dictu,* we have at last found an apparently untrammeled instance of low temperature adaptation in a hibernator.

What are we to make, on the other hand, of the equally solid result that the ground squirrel perversely alters its liver pyruvate kinase from a relatively temperature-compensated form in the summer active state to a relatively temperature-sensitive one in the winter hibernating state (Behrisch, 1974)? One could resort to a dust kicking retreat by saying that with any enzyme, and especially one as complex as this, one cannot pretend to have adequately explored its behavior until all impinging factors (ions, pH, etc.) are under control and either correctly mimicking *in situ* conditions or all permutations thereof have been tried. As a particular example, Behrisch, like any good biochemist, took pains to maintain the same pH at all temperatures. The rules of good pH management with changing temperature are no longer simple, however, because attention has been drawn to the fact that the neutral pH of water (=pN) rises as temperature falls (Reeves *et al.,* 1977). Keeping a "constant pH," therefore, in effect produces a relative acidity (a higher H^+/OH^- ratio) at lower temperature (Chapter 13). Effects of temperature on enzyme kinetics are different with a regime of constant H^+/OH^- than with one of constant pH (Wilson, 1977). In pyruvate kinase, moreover, in which H^+ is a specific allosteric effector, such a consideration has added force. It is possible, for example, that the observed effects of temperature on PEP affinity were actually effects of the relative acidity at lower temperature with constant pH.

We cannot abide, however, until all hobgoblins have been banished before drawing conclusions about such results, and it is a mark of the technically advanced nature of these studies that such considerations even arise.

Behrisch (1974) offers two comments. The first is that this result would help to account for the apparent block of glucose carbon to the Krebs cycle in hibernating ground squirrels described by Tashima *et al.* (1970). Of course, all of the tissues were involved in those whole-animal studies and it is by no means to be expected that the pyruvate kinase of other tissues will necessarily behave like that of liver. In any case, the presumed adaptive value of that hypothetical block is obscure.

Second, he suggests that the high Q_{10} coupled to low inhibition by ATP would favor thermogenesis. (With hibernation the ATP affinity is constant with temper-

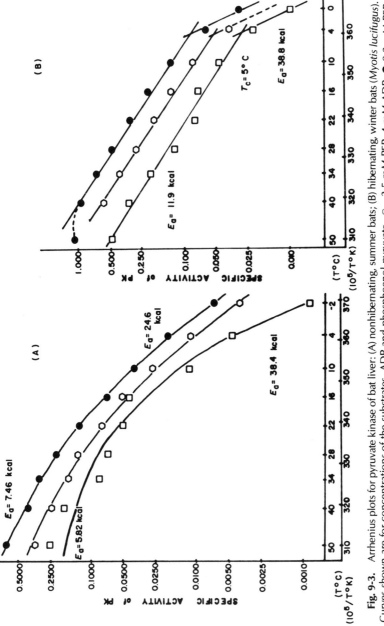

Fig. 9-3. Arrhenius plots for pyruvate kinase of bat liver: (A) nonhibernating, summer bats; (B) hibernating, winter bats (*Myotis lucifugus*). Curves shown are for concentrations of the substrates, ADP and phosphoenol pyruvate. ○ 2.5 mM PEP, 4 mM ADP; ● 0.2 mM PEP, 4 mM ADP; □ 2.5 mM PEP, 0.4 mM ADP. For the enzyme from nonhibernating bat liver the curves are concave with a steep slope at low temperatures, whereas for the enzyme from hibernating bats the curves are straight with an intermediate slope. (After Borgmann and Moon, 1976).

ature, but holds the low value seen in active squirrel liver enzyme only at low temperature.)

The need for thermogenesis at a low body temperature is a wonderfully flexible hypothesis for assigning an adaptive significance to a result, because it can make use of both temperature-sensitive and insensitive, both exergonic and endergonic, reactions. In the present case, the idea is that with a highly temperature-sensitive reaction, a small rise in temperature would provide a positive feedback, giving the explosive character needed for warming up. A temperature-insensitive reaction, however, would also be useful for insuring enough flux at low temperature. Not surprisingly, therefore, Moon and Borgmann (1976) also implicate the pyruvate kinase of liver of hibernating bats (which, it will be recalled, did have a reduced thermal sensitivity) in thermogenesis, but in their case the suggestion is based upon an elevated responsiveness of the enzyme to the positive allosteric effector, fructose diphosphate.

Regardless of the character of the change, it is apparent that in both of the studies of pyruvate kinase the enzyme is altered qualitatively with the onset of hibernation, and the next question is, what is changing? Pyruvate kinase is a multimeric enzyme with several isozymes, the immunological, electrophoretic, and kinetic properties of which do not tidily coincide (Seubert and Schoner, 1971). Borgmann and Moon (1976) found that with polyacrylamide gel electrophoresis the sets of bands reacting as pyruvate kinase in preparations made from hibernating bats were somewhat different from those from summer bats. Hence, the "hibernating enzymes" could have been composed of polypeptides different from those of the enzymes of summer bats.

This point is clearer in Behrisch's (1974) results with ground squirrel liver. His studies were based upon a population of arctic ground squirrels from an island off the coast of Alaska separated from the mainland for approximately 10,000 years. Perhaps because of the resultant inbreeding of the population (occasionally cloned still further by periodic volcanic eruptions!), the pyruvate kinase of active animals shows a single peak in electrofocusing experiments. So also does the pyruvate kinase of liver of hibernating ground squirrels, but it is separate and distinct from that of the active animals. In preparations from livers of animals taken a few weeks prior to the time for spring emergence, both peaks may be observed. It is clear that the enzymes have subunit structure, since ATP causes dissociation of two subunits as identified by gradient centrifugation, but it is not yet certain to what extent the two forms of the enzyme may share subunits.

Finally, fructose diphosphatase has also been investigated in livers of hibernating and active ground squirrels (Behrisch, 1974). This enzyme is involved in gluconeogenesis and is responsible for reversing and bypassing another irreversible step in glycolysis, the formation of FDP. FDPase is also regulatory and is inhibited by AMP and stimulated by ATP, as well as being cooperatively activated by its substrate, FDP. Behrisch (1973) found that, in the enzyme from

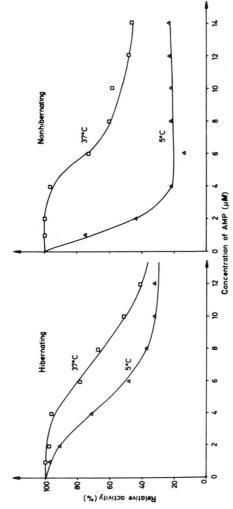

Fig. 9-4. Effect of temperature on the inhibition of fructose diphosphatase by adenosine monophosphate (AMP) at 37°C and at 5°C. In livers of hibernating arctic ground squirrels the inhibitor is less effective at low temperature. This change in kinetics favors an increase in activity at low temperature in the hibernator enzyme. (Derived from Behrisch, 1973.)

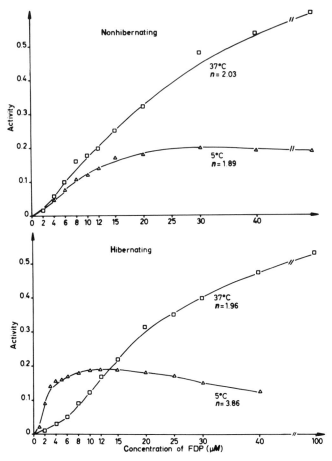

Fig. 9-5.   Effect of temperature on affinity and saturation of fructose diphosphatase for its substrate. In the enzyme of livers from winter (hibernating) arctic ground squirrels the affinity is increased at low temperature but not at high temperature. (Derived from Behrisch, 1973.)

livers of active ground squirrels, temperature acted as a ''positive modulator'' to increase the affinity for both substrate and inhibitor, although the effect on the inhibitor is rather stronger so that at low temperature, and even low concentrations of ATP, the enzyme would be virtually shut off. In the enzyme from the livers of hibernating squirrels, however, the effect of cooling on substrate affinity was much stronger and on inhibitor affinity it was much weaker than in those of awake squirrels, so that at low temperature the net effect would be to promote relatively greater gluconeogenic activity in the livers of hibernating ground squirrels (Figs. 9-4 and 9-5) at low temperature.

Summary and Conclusion

This survey of metabolism of cells has yielded two instances of apparent and as yet unrefuted cold adaptation of energy release in hibernators: mitochondrial respiration in the heart of hamsters and pyruvate kinase activity in liver and muscle of hibernating bats. Other cases have been cited, liver mitochondrial (succinate) respiration in hamsters and ground squirrels, kidney slice respiration in hamsters, ventricular slice respiration in bats, glycolysis in brain and hearts of ground squirrels, but none of these have gone without challenge. Even the "unchallenged" cases can not be extrapolated to other species; respiration in heart mitochondria of ground squirrels shows little likelihood of having low sensitivity to temperature and the pyruvate kinase of hibernating ground squirrels' liver certainly does not.

On the other hand, it is apparent that numerous changes do occur in metabolic machinery with the onset of hibernation (upward or downward displacements in respiration of whole cells and mitochondria, kinetic properties of pyruvate kinase and fructose diphosphatases in ground squirrel liver, levels of specific dehydrogenase isozymes and total activity in hedgehogs, bats, ground squirrels). The fact that these cannot be easily related to greater energy release at low temperature either means that "cold adaptation" involves more subtle interrelations than are as yet dreamt of in our philosophy or else that these observed changes are simply the, not necessarily adaptive, consequences of other adjustments for winter shutdown.

REFERENCES

Andjus, R. K. (1969). Some mechanisms of mammalian tolerance to low body temperatures. *Symp. Soc. Exp. Biol.* **23,** 351–394.

Andjus, R. K., Ćirković, T., Čupelović, N., Davidović, J., Marković-Uskoković, V., and Velimirović, T. (1964). Brain metabolism and resistance of a hibernator (*Citellus citellus*) and the rat to different anoxic conditions, including cardiac arrest in deep hypothermia. *Ann. Acad. Sci. Fenn., Ser. A4* **71,** 9–23.

Bartlett, G. R., and Barnet, H. N. (1960). Changes in the phosphate compounds of the human red blood cell during blood bank storage. *J. Clin. Invest.* **39,** 56–61.

Behrisch, H. W. (1973). Molecular mechanisms of temperature adaptation in arctic ectotherms and heterotherms. *In* "Effects of Temperature on Ectothermic Organisms" (W. Wieser, ed.), pp. 123–137. Springer-Verlag, Berlin and New York.

Behrisch, H. W. (1974). Temperature and the regulation of enzyme activity in the hibernator. Isoenzymes of liver pyruvate kinase from the hibernating and non-hibernating Arctic ground squirrel. *Can. J. Biochem.* **52,** 894–902.

Behrisch, H. W., and Johnson, C. E. (1974). Regulatory properties of pyruvate kinase from liver of the summer-active Arctic ground squirrel. *Can. J. Biochem.* **52,** 547–559.

Bennett, T. P., and Frieden, E. (1966). "Modern Topics in Biochemistry." Macmillan, New York.

Beutler, E., and Duron, O. (1965). Effect of pH on preservation of red cell ATP. *Transfusion* **5,** 17–24.

Bishop, C. (1961). Changes in the nucleotides of stored or incubated human blood. *Transfusion* **1**, 349-354.

Borgmann, A. I., and Moon, T. W. (1976). Enzymes of the normothermic and hibernating bat, *Myotis lucifugus:* temperature as a modulator of pyruvate kinase. *J. Comp. Physiol.* **B 107**, 185-200.

Brock, M. A. (1960). Hibernation and cold storage effects on phosphates in hamster blood. *Am. J. Physiol.* **199**, 195-197.

Brock, M. A. (1967). Erythrocyte glycolytic intermediates of control, cold-exposed and hibernating hamsters. *In* "Mammalian Hibernation III" (K. C. Fisher, A. R. Dawe, C. P. Lyman, E. Schönbaum, and F. E. South, eds.), pp. 409-420. Oliver & Boyd, Edinburgh.

Bullard, R. W., David, G., and Nichols, C. T. (1960). The mechanisms of hypoxic tolerance in hibernating and non-hibernating mammals. *Bull. Mus. Comp. Zool.* **124**, 321-336.

Burlington, R. F., and Sampson, J. H. (1968). Distribution and activity of lactic dehydrogenase isozymes in tissues from a hibernator and a non-hibernator. *Comp. Biochem. Physiol.* **25**, 185-192.

Burlington, R. F., and Whitten, B. K. (1971). Red cell 2,3-diphosphoglycerate in hibernating ground squirrels. *Comp. Biochem. Physiol. A* **38A**, 469-471.

Burlington, R. F., and Wiebers, J. E. (1966). Anaerobic glycolysis in cardiac tissue from a hibernator and non-hibernator as effected by temperature and hypoxia. *Comp. Biochem. Physiol.* **17**, 183-189.

Burlington, R. F., and Wiebers, J. E. (1967). The effect of temperature on glycolysis in brain and skeletal muscle from a hibernator and a non-hibernator. *Physiol. Zool.* **40**, 201-206.

Burlington, R. F., Maher, J. T., and Sidel, C. M. (1969). Effect of hypoxia on blood gases, acid-base balance and in vitro myocardial function in a hibernator and a nonhibernator. *Fed. Proc., Fed. Am. Soc. Exp. Biol.* **28**, 1042-1046.

Burlington, R. F., Whitten, B. K., Sidel, C. M., Posiviata, M. A., and Salkovitz, I. A. (1970). Effect of hypoxia on glycolysis in perfused hearts from rats and ground squirrels (*Citellus lateralis*). *Comp. Biochem. Physiol.* **35**, 403-414.

Burlington, R. F., Meininger, G. A., and Thurston, J. T. (1976). Effect of low temperatures on high energy phosphate compounds in isolated hearts from a hibernator and a non-hibernator. *Comp. Biochem. Physiol. B* **55B**, 403-407.

Calman, K. C. (1974a). The prediction of organ viability. I. An hypothesis. *Cryobiology* **11**, 1-6.

Calman, K. C. (1974b). The prediction of organ viability. II. Testing an hypothesis. *Cryobiology* **11**, 7-12.

Chaffee, R. R. J., Hoch, F. L., and Lyman, C. P. (1961). Mitochondrial oxidative enzymes and phosphorylations in cold exposure and hibernation. *Am. J. Physiol.* **201**, 29-32.

Chanutin, A. (1967). The effect of the addition of adenine and nucleosides at the beginning of storage on the concentrations of phosphates of human erythrocytes during storage in acid-citrate-dextrose and citrate-phosphate-dextrose. *Transfusion* **7**, 120-132.

Collste, H., Bergström, J., Hultman, E., and Melin, B. (1971). ATP in the cortex of canine kidneys undergoing hypothermic storage. *Life Sci.* **10**, 1201-1206.

Covino, B. G., and Hannon, J. P. (1959). Myocardial metabolic and electrical properties of rabbits and ground squirrels at low temperatures. *Am. J. Physiol.* **197**, 494-498.

Daudova, G. M. (1968). Oxidative phosphorylation in the liver and skeletal muscles of ground squirrels under various physiological conditions. *Biochemistry, Leningrad (Engl. Transl.)* **33**, 126-130.

deVerdier, C.-H., and Killander, J. (1962). Storage of human red blood cells. I. The effects of ACD solution on breakdown of glucose. *Acta Physiol. Scand.* **54**, 346-353.

deVerdier, C.-H., Högman, C., Garby, L., and Killander, J. (1964). Storage of human red blood cells. II. The effect of pH and of the addition of adenine. *Acta Physiol. Scand.* **60**, 141-149.

Dryer, R. L., and Paulsrud, J. R. (1966). Effect of arousal on ATP levels in bats. *Fed. Proc., Fed. Am. Soc. Exp. Biol.* **25,** 1293-1296.

Entenman, C., Hillyard, L. A., Holloway, R. J., Albright, M. L., and Leong, G. F. (1969). Intermediary metabolism in hypothermic rat liver. *In* "Depressed Metabolism" (X. J. Musacchia and J. F. Saunders, eds.), pp. 159-197. Am. Elsevier, New York.

Entenman, C., Ackerman, P., and Musacchia, X. J. (1971). Incorporation of intravenously injected acetate-2-^{14}C into tissue lipids of hypothermic hamsters. *Proc. Soc. Exp. Biol. Med.* **137,** 47-51.

Entenman, C., Ackerman, P. D., Walsh, J., and Musacchia, X. J. (1975). Effect of incubation temperature on hepatic palmitate metabolism in rats, hamsters and ground squirrels. *Comp. Biochem. Physiol. B* **50B,** 51-54.

Fedelešová, M., Ziegelhöffer, A., Krause, E.-G., and Wollenberger, A. (1969). Effect of exogenous adenosine triphosphate on the metabolic state of the excised hypothermic dog heart. *Circ. Res.* **24,** 617-627.

Fuhrman, G. J., and Fuhrman, F. A. (1963). Utilization of glucose by the hypothermic rat. *Am. J. Physiol.* **205,** 181-183.

Fuhrman, G. J., and Fuhrman, F. A. (1964). Effect of temperature on metabolism of glucose *in vitro. Am. J. Physiol.* **207,** 849-852.

Hannon, J. P. (1958). Effect of temperature on the heart rate, electrocardiogram and certain myocardial oxidations of the rat. *Circ. Res.* **6,** 771-778.

Hannon, J. P., Beyer, R. E., Burlington, R. F., Somero, G. N., and Bland, J. H. (1972). A guide for future studies of low temperature metabolic function. *In* "Hibernation and Hypothermia, Perspectives and Challenges" (F. E. South, J. P. Hannon, J. S. Willis, E. T. Pengelley, and N. R. Alpert, eds.), pp. 99-115. Elsevier, Amsterdam.

Harkness, D. R., Roth, S., and Goldman, P. (1974). Studies on the red blood cell oxygen affinity and 2,3-diphosphoglyceric acid in the hibernating woodchuck (*Marmota monax*). *Comp. Biochem. Physiol. A* **48A,** 591-599.

Hearon, J. Z. (1952). Rate behavior of metabolic systems. *Physiol. Rev.* **32,** 499-523.

Hiestand, W. A., Rockhold, W. T., Stemler, F. W., Stullken, D. E., and Wiebers, J. E. (1950). The comparative hypoxic resistance of hibernators and nonhibernators. *Physiol. Zool.* **23,** 264-268.

Hillyard, L. A., and Entenman, C. (1973). Effect of low temperature on metabolism of rat liver slices and epididymal fat pads. *Am. J. Physiol.* **224,** 148-153.

Hoar, W. S. (1975). "General and Comparative Physiology," 2nd ed. Prentice-Hall, Englewood Cliffs, New Jersey.

Hochachka, P. W., and Somero, G. N. (1973). "Strategies of Biochemical Adaptation." Saunders, Philadelphia, Pennsylvania.

Horwitz, B. A. (1964). Temperature effects on oxygen uptake of liver and kidney tissues of a hibernating and a non-hibernating mammal. *Physiol. Zool.* **37,** 231-239.

Jorgenson, S. (1957). Adenine nucleotides and oxypurines in stored donor blood. *Acta Pharmacol. Toxicol.* **13,** 102-106.

Kayser, C. (1954). L'incrément thermique critique de la respiration, *in vitro,* du tissu rénal de rat blanc et de hamster (*Cricetus cricetus*). *C. R. Hebd. Seances Acad. Sci.* **239,** 514-515.

Kayser, C., and Malan, A. (1963). Central nervous system and hibernation. *Experientia* **19,** 441-451.

Kristoffersson, R. (1961). Hibernation of the hedgehog (*Erinaceus europaeus*). The ATP and o-phosphate levels in blood and various tissues of hibernating and non-hibernating animals. *Ann. Acad. Sci. Fenn., Ser. A4* **50,** 1-45.

Larkin, E. C. (1973). The response of erythrocyte organic phosphate levels of active and hibernating ground squirrels (*Spermophilus mexicanus*) to isobaric hyperoxia. *Comp. Biochem. Physiol. A* **45A,** 1-6.

Liu, C.-C., Frehn, J. L., and LaPorta, A. D. (1969). Liver and brown fat mitochondrial response to cold in hibernators and nonhibernators. *J. Appl. Physiol.* **27**, 83–89.

Llorente, P., Marco, R., and Sols, A. (1970). Regulation of liver pyruvate kinase and the phosphoenolpyruvate crossroads. *Eur. J. Biochem.* **13**, 45–54.

Maizels, M. (1944). Phosphate, base and haemolysis in stored blood. *Q. J. Exp. Physiol. Cogn. Med. Sci.* **32**, 143–181.

Markscheid, L., and Shafrir, E. (1965). Incorporation of lipoprotein-borne triglycerides by adipose tissue *in vitro. J. Lipid Res.* **6**, 247–257.

Mendler, N., Reulen, H. J., and Brendel, W. (1972). Cold swelling and energy metabolism in the hypothermic brain of rats and dogs. *In* "Hibernation and Hypothermia, Perspectives and Challenges" (F. E. South, J. P. Hannon, J. S. Willis, E. T. Pengelley, and N. R. Alpert, eds.), pp. 167–190. Elsevier, Amsterdam.

Meyer, M. P., and Morrison, P. (1960). Tissue respiration and hibernation in the thirteen-lined ground squirrel, *Spermophilus tridecemlineatus. Bull. Mus. Comp. Zool.* **124**, 405–420.

Moon, T. W., and Borgmann, A. I. (1976). Enzymes of the normothermic and hibernating bat, *Myotis lucifugus:* Metabolites as modulators of pyruvate kinase. *J. Comp. Physiol. B* **107**, 201–210.

Musacchia, X. J., and Volkert, W. A. (1971). Blood gases in hibernating and active ground squirrels: HbO_2 affinity at 6 and 38°C. *Am. J. Physiol.* **221**, 128–130.

Nakao, K., Wada, T., Kamiyama, T., Nakao, M., and Nagano, K. (1962). A direct relationship between adenosine triphosphate-level and *in vivo* viability of erythrocytes. *Nature (London)* **194**, 877–878.

Neely, J. R., and Morgan, H. E. (1974). Relationship between carbohydrate and lipid metabolism and the energy balance of heart muscle. *Annu. Rev. Physiol.* **36**, 413–459.

Olsson, S.-O. R. (1974). Metabolic aspects of hibernation in the hedgehog (*Erinaceus europaeus*). Ph.D. Thesis, Univ. of Lund, Lund, Sweden.

Pappius, H. M., Andreae, S. R., Woodford, V. R., and Denstedt, O. F. (1954). Studies on the preservation of blood. I. Glycolytic behavior of blood during storage at 5°C in a medium containing an excess of glucose. *Can. J. Biochem. Physiol.* **32**, 271–281.

Prankerd, T. A. J. (1956). Chemical changes in stored blood, with observations on the effects of adenosine. *Biochem. J.* **64**, 209–213.

Precht, H. (1958). Concepts of the temperature adaptation of unchanging reaction systems of cold-blooded animals. *In* "Physiological Adaptation" (C. L. Prosser, ed.), pp. 50–78. Am. Physiol. Soc., Washington, D.C.

Raison, J. K., and Lyons, J. M. (1971). Hibernation: Alteration of mitochondrial membrane as a requisite for metabolism at low temperature. *Proc. Natl. Acad. Sci. U.S.A.* **68**, 2092–2094.

Reeves, R. B., Howell, B. J., and Rahn, H. (1977). "Protons—Proteins—Temperature," pp. III–IX. Dep. Physiol., State Univ. of New York, Buffalo.

Reshef, L., and Shapiro, B. (1965). Fatty acid adsorption by liver- and adipose-tissue particles. *Biochim. Biophys. Acta* **98**, 73–80.

Roberts, J. C., and Chaffee, R. R. J. (1973). Effects of cold acclimation, hibernation and temperature on succinoxidase activity of heart homogenates from hamster, rat and squirrel monkey. *Comp. Biochem. Physiol. B* **44B**, 137–144.

Roberts, J. C., Arine, R. M., Rochelle, R. H., and Chaffee, R. R. J. (1972). Effects of temperature on oxidative phosphorylation of liver mitochondria from hamster, rat and squirrel monkey. *Comp. Biochem. Physiol. B* **41B**, 127–135.

Saarikoski, J. (1967). Respiratory responses of cerebral cortex slices from a hibernating and a normal homeothermic mammal to electrical stimulation at 10–36°C. *Ann. Acad. Sci. Fenn., Ser. A4* **117**, 1–22.

Schmidt-Mende, M., and Brendel, W. (1967). Experimentelle Untersuchungen zum Energiestoffwechsel der Kühlkonservierten Niere. *Z. Gesamte Exp. Med. Einschl. Exp. Chir.* **143,** 250–255.

Seubert, W., and Schoner, W. (1971). The regulation of pyruvate kinase. *Curr. Top. Cell. Regul.* **3,** 237–267.

Shipp, J. C., Matos, O. E., and Crevasse, L. (1965). Effect of temperature on metabolism of glucose-U-C¹⁴ and palmitate-1-C¹⁴ in perfused rat heart. *Metab. Clin. Exp.* **14,** 639–646.

Shug, A. L., Ferguson, S., Shrago, E., and Burlington, R. F. (1971). Changes in respiratory control and cytochromes in liver mitochondria during hibernation. *Biochim. Biophys. Acta* **226,** 309–312.

Siska, K., Fedelesová, M., Stadelmann, G., Ziegelhöffer, A., and Holec, V. (1969). Protection of the myocardium during ischaemic asystole with intracoronary administration of ATP. *J. Cardiovasc. Surg.* **10,** 274–281.

South, F. E., Jr. (1958). Rates of oxygen consumption and glycolysis of ventricle and brain slices, obtained from hibernating and non-hibernating mammals, as function of temperature. *Physiol. Zool.* **31,** 6–15.

South, F. E. (1960a). Some metabolic specializations in tissues of hibernating mammals. *Bull. Mus. Comp. Zool.* **124,** 475–492.

South, F. E., Jr. (1960b). Hibernation, temperature and rates of oxidative phosphorylation by heart mitochondria. *Am. J. Physiol.* **198,** 463–466.

Tashima, L. S., Adelstein, S. J., and Lyman, C. P. (1970). Radioglucose utilization by active, hibernating, and arousing ground squirrels. *Am. J. Physiol.* **218,** 303–309.

Tempel, G. E., and Musacchia, X. J. (1975). Erythrocyte 2,3-diphosphoglycerate concentrations in hibernating, hypothermic, and rewarming hamsters. *Proc. Soc. Exp. Biol. Med.* **148,** 588–592.

Valeri, C. R. (1974). Liquid and freeze preservation of human red blood cells. *In* "The Red Blood Cell" (D. M. Surgenor, ed.), Vol. I, pp. 511–574.

van Rossum, G. D. V. (1972). The metabolic coupling of ion transport. *In* "Hibernation and Hypothermia, Perspectives and Challenges" (F. E. South, J. P. Hannon, J. S. Willis, E. T. Pengelley, and N. R. Alpert, eds.), pp. 191–218. Elsevier, Amsterdam.

Whittam, R. (1958). Potassium movements and ATP in human red cells. *J. Physiol. (London)* **140,** 479–497.

Whittam, R., Ager, M. E., and Wiley, J. S. (1964). Control of lactate production by membrane adenosine triphosphatase activity in human erythrocytes. *Nature (London)* **202,** 1111–1112.

Willis, J. S. (1967). Cold adaptation of activities of tissues of hibernating mammals. *In* "Mammalian Hibernation III" (K. C. Fisher, A. R. Dawe, C. P. Lyman, E. Schönbaum, and F. E. South, eds.), pp. 356–381. Oliver & Boyd, Edinburgh.

Willis, J. S. (1968). Cold resistance of kidney cells of mammalian hibernators: cation transport vs. respiration. *Am. J. Physiol.* **214,** 923–928.

Wilson, T. L. (1977). Interrelations between pH and temperature for the catalytic rate of the M_4 isozyme of lactate dehydrogenase from goldfish (*Carassius auratus* L.). *Arch. Biochem. Biophys.* **179,** 378–390.

Wood, L., and Beutler, E. (1967). Temperature dependence of sodium–potassium activated erythrocyte adenosine triphosphatase. *J. Lab. Clin. Med.* **70,** 287–294.

Woodard, C. B., and Zimny, M. L. (1973). Energy of activation of succinic dehydrogenase in a hibernator. *Am. J. Physiol.* **225,** 429–431.

Zimny, M. L., and Gregory, R. (1958). High energy phosphates during hibernation and arousal in the ground squirrel. *Am. J. Physiol.* **195,** 233–236.

Zimny, M. L., and Gregory, R. (1959). High-energy phosphates during long-term hibernation. *Science* **129,** 1363–1364.

Zimny, M. L., and Taylor, S. (1965). Cardiac metabolism in the hypothermic ground squirrel and rat. *Am. J. Physiol.* **208,** 1247–1252.

10

Hibernation: Some Intrinsic Factors

The state of deep hibernation in mammals is unique, and it is natural that many questions arise concerning the effect of the prolonged cold body temperature accompanied by the possible effect of hibernation itself. Some of the changes that may occur during hibernation must take place in every mammal that hibernates. They must be intrinsic to the hibernating state, and these will be considered in this chapter. In the chapter that follows, the response of the animal in hibernation to various extrinsic influences, such as neoplastic growth, radiation injury, and parasitism and symbionts will be discussed.

One paramount question concerns the problem of growth and cell replacement during the hibernating state. It is to be expected that growth and replacement would be reduced or stopped during deep hibernation, but it seems likely that some phases of the process would be more affected than others. For example, mitosis and DNA synthesis are very different processes, and their reaction to cold and hibernation are now known to be different.

Hock (1960) has emphasized that the active life of a typical hibernator is crammed into the summer months. During this time it must breed, reproduce, raise the young, and prepare for hibernation. Unless the animal can reconstitute its forces during hibernation, all these yearly processes including growth must take place during the period of activity, which may be as short as 4 months in Richardson's ground squirrel (Wang, 1979). If little or nothing takes place during hibernation, perhaps every process is more rapid in the active hibernator than it is in the nonhibernator which is active throughout the year.

In considering these problems it must be reemphasized that hibernation is not a continuous process during the hibernating season, and that it is always interrupted by periods of arousal. Unless their previous history has been monitored, animals collected in the field during hibernation can not supply data on the effect of uninterrupted hibernation, for it must be assumed that periodic arousals have occurred. Thus, any differences that appear between the experimental animals and the controls are due to a mixture of both hibernation and activity.

In studies of cell replacement and growth during hibernation, the experimenter faces further inherent difficulties. In order to establish a base line for comparative measurements, the hibernating animal must be disturbed, and this almost always causes arousal from hibernation. The return to the hibernating state is usually delayed some hours or even days, and during that time the growth or cell replacement that does occur can not be regarded as having taken place during hibernation. If a stimulus to cell replication such as injection of testosterone is applied, it usually is followed by a diminution of the hibernating periods, so that a clear picture of what occurs during actual hibernation is difficult to obtain.

Various species of ground squirrels, particularly the thirteen-lined ground squirrel, were used in early endocrinological investigations as models to study the sexual cycle of seasonally breeding animals. The emphasis of these studies was on the changes that occur annually in the reproductive system, and on the external and endogenous factors that controlled these changes. The effect of hibernation per se was not emphasized or seriously considered (Rasmussen, 1918; Zalesky, 1934, 1935; Wells and Moore, 1936). In the 1960s a study of the woodchuck (*Marmota monax*) was carried out by a series of investigators, with specific emphasis on reproduction, seasonal fattening, and hibernation. This constitutes one of the most complete life histories of a hibernating species, and includes laboratory studies as well as extensive work with the animals in the wild (Snyder and Christian, 1960; Snyder et al., 1961; Bailey, 1965; Bailey and Davis, 1965; Davis, 1967a,b). In both the ground squirrel and the woodchuck studies, the emphasis was on the life history of the animals, the changes that took place during the annual cycle, and the factors that controlled these changes. The hibernating season was considered as part of the life cycle, and statements such as "woodchucks, like ground squirrels, come into full sexual activity during hibernation" (Davis, 1976b) created the impression that morphological and physiological changes took place while the animals were actually in the hibernating state. In point of fact, no study has been undertaken in which the changes in preparation for the spring breeding season have been partitioned between the short periods of activity and the longer periods of hibernation.

It appears unlikely that notable changes do take place in the reproductive systems of animals hibernating at body temperature near 0°C. Lyman and Dempsey (1951) were able to show that the gain in weight of the seminal vesicles of castrated Syrian hamsters in response to injected testosterone varied inversely over a 10-day period with the time spent in deep hibernation. The seminal vesicles of injected animals that spent 7 or more days in hibernation were no heavier than uninjected controls, and the histological appearance of the tissues was comparable, indicating that regrowth had not taken place during hibernation even in the presence of adequate stimulus from the hormone. These experiments were carried out at an ambient temperature of $5° \pm 2°C$, so that T_b's of the hibernating animals were not more than 8°C. There appear to be no similar experiments in which growth was monitored in animals hibernating at higher body temperatures. It is not known whether the response of tissues under these conditions varies proportionally with the body temperature or whether there is a sharp break between relatively rapid growth and no growth at all. Although it seems highly probable that cell replacement is dependent on temperature alone, an experiment with hibernation at different temperatures should settle the matter. As it is, the possibility always remains that the state of hibernation itself contributes to the lack of cell replacement in some unexplained manner.

A detailed knowledge of cell growth and replacement during hibernation is certainly important in reference to the yearly sexual cycle as well as being of interest for its own sake. The problem has been around for some time, for Hansemann (1898) reported that there were no mitotic figures in tissues of hibernating ground squirrels and hedgehogs. Many years later Mayer and Bernick (1958) were unable to find cells in mitosis in the hibernating arctic ground squirrel, though the number of days of uninterrupted hibernation was not recorded. Actually, the presence of mitotic figures in the tissues of an animal in the hibernating state does not necessarily indicate that active cell division is taking place during hibernation, and their absence does not mean that all parts of the replacement cycle stop during hibernation. One may assume that cell division is going on in all euthermic hibernators just before they enter hibernation. Unless DNA synthesis stops and all mitosis is completed during the time the animal is entering hibernation, there must be mitotic figures present when the state of deep hibernation is attained. If the replacement cycle is totally halted at this time, then mitotic figures will be present and, presumably, will remain until the animal arouses. Thus, an unchanging number of mitoses during a bout of hibernation could be interpreted to indicate that there is no progression through the cell cycle. If mitotic figures are present when the animal attains deep hibernation, and if cell maturation continues and cell replacement stops, then the mitotic figures will disappear as the bout of hibernation progresses. Paradoxically, the total absence of mitotic figures could indicate that cell maturation takes place during deep hibernation.

There have been a few studies on the effect of stimuli on cell replacement during hibernation, but the results do not favor clear-cut conclusions. Hansemann (1898) found mitotic figures in the healing epidermis of a hibernating hedgehog, whereas none were seen in the normal tissue surrounding it. As outlined above, this result can be interpreted in several ways. Lyman *et al.* (1957) found that the erythropoietic response to massive bleeding was greatly reduced in hibernating hamsters, but the experimental design did not permit a precise correlation of the number of reticulocytes that appeared after bleeding, and the amount of hibernation. The hamsters were bled while in hibernation, but the cardiac puncture invariably caused arousal and hibernation was sporadic thereafter, with no animal returning immediately to the hibernating state. Thomson *et al.* (1962) studied regeneration of the liver in ground squirrels after partial hepatectomy and reported that regeneration was slowed but not stopped during hibernation. For the first 10 days after the operation, recovery was quite rapid, but little further gain in weight of the liver occurred in the next 26 days. The amount of hibernation that occurred was not monitored, but it seems virtually certain that the ground squirrels could not have remained in hibernation for the total duration of the experiment which lasted 36 days.

Sarnat and Hook (1942) studied the effect of hibernation on the growth of the incisor teeth in hibernating ground squirrels. They found that growth was retarded during hibernation, but the same problem of disturbing the animals to establish a base line prevented a precise correlation of the time in hibernation with the amount of growth.

In our present state of knowledge only chronic intubation of an artery or vein offers a technique where stimulants or drugs may be introduced into the hibernating animal and blood drawn for analysis and inspection without starting the process of arousal. In her studies of aging and red blood cells during hibernation, Brock (1960a) used this approach. Syrian hamsters were chronically intubated with small polyethylene aortic cannulae. When the animals hibernated, radioiron was introduced into the bloodstream via the cannula. After 24 hr of uninterrupted hibernation at a T_a of $8° ± 2°C$, blood was withdrawn without disturbing the animals and the amount of radioactivity in the red blood cells was determined. The concentration of radioiron found in the erythrocytes of the warm room control animals was 14.1 times greater than that found in the hibernators, but the amount of radioiron found in the plasma of the hibernators was only 4.6 times that found in the active controls. As iron is believed to enter the erythroid cell during the process of maturation, it was concluded that the maturation of these cells was greatly slowed during hibernation, but not stopped. Since the incorporation of radioiron was reduced 14-fold in the hibernators compared with the controls, the amount of radioiron remaining in the plasma of the hibernators should be 14 times greater than that found in the plasma of the controls. The lower figure of 4.6 was taken to indicate that the radioiron was trapped in the erythroid elements of the bone marrow. Brock (1960a) emphasized that the classic experiments of Huggins and Blocksom (1936) demonstrated that erythropoiesis was increased when the distal tail bones of rats were incorporated in the body cavity, and suggested that red cell production was directly proportional to temperature of the bone marrow, provided that the erythropoietic stimulus is constant. Brock's (1960a) work appears to be the first in which the state of hibernation was rigorously monitored and it supplies firm evidence that erthropoietic maturation takes place in a hibernator with a body temperature of about 8°C.

The techniques developed for the study of the kinetics of cellular proliferation using radioactive thymidine offered a new approach to the problem of cell division during continuous hibernation. Initial experiments with thymidine in active thirteen-lined ground squirrels resulted in negative radioautographs. Further examination, both *in vitro* and *in vivo,* showed that the amount of incorporation of thymidine varied with the species of rodent. Good labeling was found in the house mouse (*Mus musculus*), the Syrian hamster (*Mesocricetus auratus*), the edible dormouse (*Glis glis*), the garden dormouse (*Eliomys quercinus*), and the gray squirrel (*Sciurus carolinensis*). Inconstant labeling occurred in the red squirrel (*Tamiasciurus hudsonicus*) and the eastern chipmunk (*Tamias striatus*), and no labeling took place in the woodchuck (*Marmota monax*), the thirteen-lined ground squirrel (*Citellus tridecemlineatus*), and the golden mantled ground squirrel (*Citellus lateralis*) (Adelstein *et al.,* 1964). The animals tested were only a small sample of the many species in the order Rodentia, but it is clear that the variation in the ability to incorporate thymidine is not correlated with the ability to hibernate.

There was some correlation with the phylogenetic relationship, for chipmunks, ground squirrels, and woodchucks are all members of the tribe Marmotini. The two species of ground squirrels and the woodchuck showed the least ability to incorporate thymidine and are closely related, whereas their relationship to the chipmunk is more remote. The red squirrel and the gray squirrel are placed in two different tribes (Tamiasciurini and Sciurini, respectively). The other species in the experiment are members of a different suborder and differ from the squirrels both morphologically and phylogenetically (Adelstein *et al.,* 1964). *In vitro* comparisons of spleen cells from *Citellus lateralis* and *Mesocricetus auratus* led to the conclusion that the reduced incorporation in the former species was due to an unusually active catabolic pathway for thymidine and other pyrimidine nucleosides, including deoxyuridine, uridine, deoxycytidine, and cytidine (Adelstein and Lyman, 1968).

Since the dormouse (*Glis glis*) showed ready incorporation of injected tritiated thymidine and was known to undergo prolonged bouts of hibernation, it was chosen for the studies on cellular proliferation (Adelstein *et al.*, 1967). The hibernating animals were infused with the nucleoside via indwelling intraperitoneal or intracarotid cannulae while T_b was monitored to assure that arousal had not been precipitated. Only animals whose T_b remained below 9°C throughout the experiment were included in the analysis. The cell proliferation kinetics of active animals was established by counting the number of labeled mitoses in tongue, duodenum, and ileum in animals sacrificed hourly after injection of [³H]thymidine.

The cycle of cellular replacement and division is divided into four distinct periods: M, G_1, S, and G_2. M is the period of mitosis proper; G_1 is a postmitotic gap, sometimes called true interphase; S is the period of DNA synthesis, and G_2 is a premitotic gap period. When the time of injection of thymidine is known and the number of labeled mitotic figures is established, the duration of each of the periods can be calculated. In the investigation with *Glis*, no labeled mitotic figures were seen at the first postinfusion hour, after which the fraction of labeled figures rose rapidly (Fig. 10-1). This established G_2 as lasting a minimum of 1 hr and a $G_2 + \frac{1}{2}$M of 1.6 hr, setting the value of G_2 as between 1.1 and 1.3 hr depending on whether M is assumed to be 1 or 0.5 hr, respectively. From the 0.5

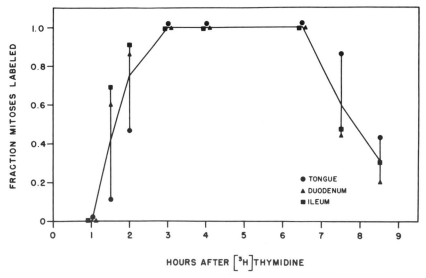

Fig. 10-1. Cell proliferation kinetics in normal dormice. Fraction mitoses labeled in tongue, duodenum, and ileum against time after [³H]thymidine administration. Each set of three symbols connected by a vertical line represents a single animal; point-to-point straight lines have been drawn from the mean values. From this graph $G_2 + \frac{1}{2}$M equals 1.6 hr and S, determined from the midpoint of the ascending and descending slopes, is 6.2 hr. (After Adelstein *et al.*, 1967.)

value on the ascending slope to the 0.5 value on the descending slope is 6.2 hr, which is the value of S.

If hibernating *Glis* were infused with tritiated thymidine and killed 4 hr later, it was found that the number of labeled nuclei was slightly greater than in the active animal. However, biochemical analysis of DNA synthesis revealed that the specific activity in the hibernating dormice was 24 times lower than in the active animals, giving a Q_{10} of 2.9. A reduction in the amount of synthesis in each cell rather than a reduction in the number of cells engaged in synthesis accounted for this marked decline. Spleen cell suspensions incubated at 37°C showed a specific activity which was 43 times greater than that of cells incubated at 10°C, yielding a Q_{10} of 3.9. Since the Q_{10}'s of the *in vivo* and the *in vitro* measurements were of the same order of magnitude, it was concluded that the reduction in synthesis during hibernation was due in large part, if not exclusively, to the low temperature at which the cells were functioning (Adelstein *et al.*, 1967).

Kunkel and Schubert (1959) reported that the incorporation of radiophosphate into DNA in hibernating dormice was about 46% of that of active controls. The injections were intraperitoneal and the measurements were made 4 hr later. The hibernating condition of the time of sacrifice is not mentioned and it seems likely that the animals were at least partially aroused. This would explain the relatively high rate of incorporation in their "hibernating" group. In either case, DNA synthesis is shown to continue in hibernation, and Adelstein *et al.* (1967) found that the number of tagged cells was not reduced even after 5 days of continuous hibernation. This indicates that cells are progressing steadily from G_1 to S during the bout of hibernation.

After 7 days of uninterrupted hibernation, mitoses were still present, though the mitotic index was lower than it was in the active animals. The mitotic figures had an unusual appearance and contained no definite telophases, anaphases, or normal metaphases. When hibernating dormice were infused with tritiated thymidine and examined after as much as 96 hr of continuous hibernation, no labeled mitoses were ever seen.

Intracardiac injection of thymidine resulted in arousal from hibernation and 3.5 hr later, after T_b had returned to 37°C for at least 2 hr, the specific activity in the intestine reached the range found for euthermic animals. The fraction of labeled mitoses in tongue, duodenum, and ileum rose abruptly about 2 hr after T_b reached 37°C. By this time the mitotic index in duodenum and ileum had risen above the value found in euthermic dormice, with a suggestive rise 1 hr before. Normal prophases, metaphases, anaphases, and telophases could be seen as the mitotic index started to increase (Fig. 10-2).

The absence of labeled mitotic figures in the hibernating dormouse even 96 hr after thymidine infusion indicates that there is a block in the cell cycle at G_2 or late S. Since the mitotic index rose just before the appearance of labeled mitoses after arousal, it seems likely that the block is at early G_2. A second block is

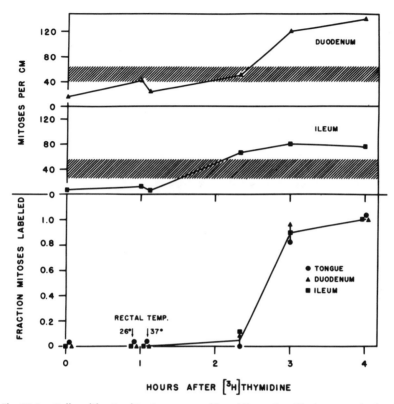

Fig. 10-2. Cell proliferation kinetics on arousal from hibernation. The lower graph plots the fraction of mitoses labeled against the time after the wakening stimulus and the administration of [³H]thymidine. Animals were killed 1 hr, 1.1 hr, 2.3 hr, 3 hr, and 4 hr after thymidine injection when the rectal temperatures had reached 26°C, 37°C, 37°C for 1 hr, 37°C for 2 hr, and 37°C for 3 hr, respectively. The upper graphs show the mitotic activity in duodenum and ileum for the same dormice. The shaded area indicates the range of values for active animals. (After Adelstein et al., 1967.)

believed to occur at mitosis. No anaphase or telophase figures have been observed in hibernating ground squirrels, hamsters, or dormice, and the metaphase figures are atypical. The mitotic index does not rise as a bout of hibernation progresses because the block of G_2 prevents the cells from reaching prophase. Recovery after arousal is rapid, for less than 2 hr later the mitotic index is above that found in control animals and all stages of the mitotic cycle are represented, as are some labeled mitoses. Figure 10-3 presents a schematic diagram of the situation. It makes an interesting contrast with the response to feeding after starvation, where the mitotic index does not rise for 7–8 hr after feeding because the cycle is blocked at G_1 (Cameron and Cleffmann, 1964). This suggests that

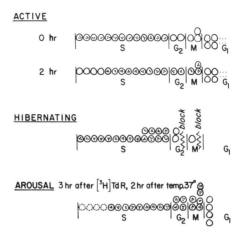

Fig. 10-3. Scheme of the proposed blocks in the divisional cycle found in hibernation and their release as a result of arousal. (After Adelstein *et al.*, 1967.)

hibernation and cold T_b, rather than the starvation that accompanies it, is the cause for the double block.

In contrast, Jaroslow *et al.* (1976) used colchicine to stop mitosis at metaphase in ileum crypt cells of hibernating and aroused thirteen-lined ground squirrels. He found that the number of mitotic figures per crypt was lowest 3–9 hr after initiation of arousal, and the count did not reach the level of active control animals until 18 hr after arousal had begun. Because of the long delay between arousal and the peak of mitosis, it was concluded that the cells were blocked at G_1 rather than at G_2 as postulated by Adelstein *et al.* (1967). The reason for this difference is not readily apparent, and repetition of the experiments with dormice can not be carried out on ground squirrels because injected thymidine is not incorporated into DNA in this species.

Using *in vitro* techniques, Manasek *et al.* (1965) found that the lymphoid cells from spleen and thymus of euthermic Syrian hamsters had a high rate of DNA synthesis when compared with similar preparations from hibernating hamsters over a temperature range of 4°–37°C. In this case, the difference was mainly due to the reduction in the hibernating animals of the number of cells that were engaged in synthesis. In contrast to the *in vivo* findings with the intestinal crypt cells of dormice (Adelstein *et al.*, 1967) there was no evidence that the rate was reduced in the individual cells that were engaged in synthesis. Furthermore, when the *in vitro* synthesis of DNA was measured in the lymphoid tissue of hamsters that were sacrificed at specific times after arousal from hibernation, it was found that the block remained 12 hr after arousal and that it had disappeared 12 hr later. Manasek *et al.* (1965) suggested that this delay in recovery after the hibernating animal aroused from hibernation was similar to the delay in recovery of cells that have been exposed *in vitro* to cold temperatures and then rewarmed. In both cases, "cold shocking" affects the divisional cycle, and "cold shocking" may also be involved in the delay in the achievement of full mitosis when the ground squirrel first arouses from hibernation as described by Jaroslow *et al.* (1976).

This rather detailed account of cellular replacement in hibernation leaves many questions unanswered, but it does provide some satisfactory conclusions. It seems clear that DNA synthesis is greatly reduced during hibernation, but it is not completely stopped. Available evidence indicates that the diminished synthesis can be due to either a reduction in the number of cells engaged in synthesis, or a reduction in the amount of synthesis in the individual cells, depending on the cell type. Blocks apparently occur at various parts of the cell replacement cycle during hibernation and the reported block at G_1 in the crypt cells of ground squirrels versus the blocks at G_2 and M in the crypt cells of dormice raises the spectre that there may be species differences as well as cell-type differences in the responses to hibernation with its accompanying low body temperature.

As far as the life of the whole animal is concerned, changes in the cell cycle during hibernation might have profound effects. The evidence to date indicates that there may be an accumulation of cells at one or more phases in the cycle. If these cells are more radioresistant than others, the animal may achieve some protection from radiation injury and this will be considered in the next chapter. However, in the natural life of a hibernator possible changes in the cell cycle must be important also. If a single block at any part of the cycle was complete, replacement cells would accumulate at that block throughout a bout of hibernation, and, if the bout were long enough, all of these cells would be in a particular phase of the replacement cycle when they were finally released by arousal from hibernation. There is no evidence that this occurs in hibernating animals, and, if it did occur, its effects on the animal are difficult to predict. In this regard, the periodic arousals from hibernation assume a new importance. If one assumes a hibernating season of 6 months and if there were a complete single block in the replacement cycle, and periodic arousals did not occur, then every cell would be at least 6 months old when the animal emerged in the spring to face the rigors of reproduction and of foraging for scarce food. Though red blood cells, at least, have been shown to live longer in the hibernator than in the active animal (Brock, 1960a), still the emerging hibernator with its population of ancient cells would presumably be at a disadvantage when compared with its nonhibernating competitors or with predators that maintained their cell populations on an even keel throughout the winter.

In point of fact, hibernators, after their emergence in the spring, are in the most active phase of their yearly cycle, and there is no reported evidence that a disrupted cell cycle is limiting this activity. Whatever disruptions may occur, they can be corrected during the periodic arousals and the return to the warm-blooded state. Since the arousals are more frequent as the hibernating season comes to a close, there is consequently more opportunity for correction at that time.

The occurrence of periodic arousals permits an adequate explanation of the continuance of cell replication during the hibernating season, including the recrudescence of the gonads in the spring. However, the concept of a complete block or blocks of cell replacement under all conditions of hibernation does not appear to be justified. Intuitively, one would assume that there would be less blockade with animals hibernating at higher T_b's. Though there appears to be no *in vivo* evidence to confirm this, Manasek *et al.* (1965) have demonstrated that thymidine uptake of hamster spleen cell suspensions is temperature-dependent. Wang's (1979) unique investigations have shown that Richardson's ground squirrel can hibernate steadily under natural conditions in the field for as long as 24 days with a T_b below 5°C. In this case, cell replacement in hibernation should be minimal if not completely blocked. But many animals hibernate under far less rigorous conditions, and acute measurements in the field show that T_b can be much warmer than 5°C. In these cases, some cell replacement may be occurring during hibernation, for there appears to be no evidence that the state of hibernation per se has an effect on cellular proliferation.

The effect of hibernation on bone has received some attention. Disuse osteoporosis is a well-known syndrome, and animals in hibernation remain virtually motionless for days at a time. The question naturally arises whether the factors that cause disuse osteoporosis function during hibernation or whether these factors are ineffective when the animal is in the hibernating state. In this case, the experiment is made more complicated by the changes in the nutritional state of animals during the season of hibernation. Whether the hibernating animal eats nothing and lives on its fat, or whether it consumes stored food during periods of arousal, it is clear that the nutritional intake during this period is curtailed and quite different from that found in the active animal in the nonhibernating season. It is reasonable to expect that species which undergo very long bouts of hibernation punctuated by short periods of arousal would be the best models to study disuse osteoporosis because the changes in the bone which could occur in a short bout of hibernation might be repaired during the euthermic period. In this regard, the microchiropteran bats appear to be the animals of choice for these experiments because their bouts of hibernation are so prolonged, but their dietary intake during the winter is extremely curtailed so that it is very difficult to separate the effects of disuse and starvation.

Information concerning blood calcium is contradictory, for it has been reported as being high, low, or unchanged during the period of hibernation. The results have been cited by Bruce and Wiebers (1969) and since that time Ferren *et al.* (1971) have reported on calcium levels in hamsters. The latter found that serum calcium was higher in the hibernating animals, as was calcium in heart and skeletal muscle, whereas calcium in bone was questionably lower. This suggested that bone may give up calcium to heart and muscle during the period of hibernation. However, the conflicting evidence on blood calcium levels in hibernating animals suggests that calcium loss from bone, if it does exist, can not be very severe.

Recently Zimmerman *et al.* (1976) have reported on the effect of hibernation on osteoporosis in the thirteen-lined ground squirrel. In order to produce a condition of disuse, they divided the sciatic nerve unilaterally in several groups of animals. The long-term effect on euthermic animals was a lowering of total calcium, percent calcium, and percent phosphorus in the bone of the operated side compared with the normal side. Histologically, the femurs on the operated side showed more enlarged lacunae. Thus, the expected changes occurred in the long bones of the paralyzed limb. It should be mentioned that dividing the sciatic nerve also cuts the autonomic nerve supply so that circulation to that leg is altered. It is possible that some of the observed changes were due to this factor rather than the immobility caused by paralysis. Normal hibernating ground squirrels showed no signs of osteoporosis at the end of either 30 or 150 days of exposure to 5°C. Remarkably, animals with one leg paralyzed also incurred no osteoporosis in either leg when exposed to cold and hibernation for 30 days. The muscle mass in the paralyzed leg of animals exposed to the same conditions for 150 days was smaller than normal, and there was some slight histological, but no chemical, evidence of osteoporosis. Thus, these experiments indicate that other degenerative changes caused by sciatic section are in some way arrested by periods of hibernation. Further, the immobility that is typical of hibernation does not result in measurable osteoporosis in normal, innervated limbs.

There is experimental evidence that osteoporosis proceeds slowly in animals with low thyroid function, and Zimmerman *et al.* (1976) suggest that this may be the reason for the lack of osteoporosis in the denervated limb of animals from the hibernating group. Alternatively, they postulate that a Q_{10} effect may retard the degenerative changes during hibernation. If this is the case, one would still expect some osteoporosis to occur in the denervated limb during the periods of arousal which took place in the 150 days of cold exposure and hibernation. The periods between arousals from hibernation varied from 1 to 13 days, with a mean of 6 days. Using the mean, an animal exposed to cold for 150 days would arouse an average of 25 times during the experimental period, which should allow enough time for some osteoporosis to occur since it was evident 30 days after the operation in the euthermic animals. Zimmerman *et al.* (1976) did not consider the effect of the lack of circulatory control in the operated limb, but this is definitely a complicating factor. It is probable that the flow of blood to the denervated limb would be greater than the circulation to the normal leg while the animal was in hibernation, but whether this would delay the onset of osteoporosis is a matter for conjecture.

Whatever the explanation for this anomalous result in the denervated limb, the data indicate that osteoporosis does not normally occur during hibernation and this is confirmed by Tempel *et al.* (1978) working with the Syrian hamster, which has shorter bouts of hibernation. In contrast, Kayser and Frank (1963) reported that osteoporosis took place during the hibernating season in the European hamster (*Cricetus cricetus*). Using radiographic and histological techniques on a small series of animals, they report that a reduction in the density of the long bones and the bones of the head occurred during the first months of hibernation. During this time, there were more osteoclasts in the bones of the hibernating animals than in the controls. After the month of February, some recovery was noted, though hibernation still occurred. The authors suggest that the observed results can be correlated with their report of seasonal activity of the parathyroids, but the effect of nutrition was not ruled out. Haller and Zimny (1977) examined interradicular alveolar bone in hibernating ground squirrels and compared it with bone from euthermic animals, using both light and electron microscopes. The experimental animals had been in hibernation for 5–8 days, but the method of measuring continuous hibernation is not stated. They concluded that the observed changes in the hibernators were due to osteocytic osteolysis, and suggested that minerals are mobilized from bone to be used elsewhere during hibernation, but again it could not be concluded from the experiments whether starvation was the important factor.

Nunez and his group made an extensive study of calcium metabolism and storage in various species of microchiropteran bats. They found that osteocytic osteolysis was present in hibernating bats and that this became increasingly evident up to the middle of the hibernating season. Radiographic studies showed that the thickness of the cortex of the femur was reduced as the osteolysis continued. During the first part of the hibernating season the parafollicular cells of the thyroid gland appeared to be inactive, with calcitonin granules accumulating as the season progressed. The parathyroid, which was active at the start of hibernation, became less active as the animals approached the period of final arousal (Whalen *et al.*, 1972). It was postulated that there was a dietary deficiency early in the hibernating season resulting in lowered blood calcium and increased bone resorption. The parafollicular and parathyroid cells were visualized as having reciprocal roles and acting synergistically in the regulation of calcium metabolism throughout the season (Haymovits *et al.*, 1976). In further work it was demonstrated that bats injected with calcitonin late in the hibernating season showed little osteolysis 21 days later when compared with untreated controls. Many of the experimental animals died before the experiment was completed, and it was presumed that hypocalcemia was the cause of death. In this experiment exogenous calcitonin was thought to have blocked recruitment of blood calcium from bone and upset the normal blood calcium balance, which occurs in hibernating bats at the expense of bone calcium (Krook *et al.*, 1977).

These observations are at odds with those of Zimmerman *et al.* (1976), but it is likely that this reflects a true species difference. The bats used in the experiments outlined above were collected in the field at various times of the year. The microchiropteran insectivorous bats of the north temperate zone are exposed to a physiologically rigorous winter. Not only do they lack food but their small body size and large surface area demand a disproportionately high metabolic rate when they are euthermic. It is reasonable to suppose that these bats would suffer from the lack of various metabolic essentials during the hibernating season even though their bouts of hibernation are longer than those of the thirteen-lined ground squirrel. The normal ground squirrel is grossly fat before entering hibernation and is generally reported to be in good condition when the season for hibernation is terminated. Dietary osteoporosis might be expected in the bats but not in the ground squirrels simply because the bats are cutting more deeply into their metabolic budget during the winter.

Up to the present, then, there appears to be no convincing evidence that disuse osteoporosis occurs during hibernation. A dietary osteocytosis has been convincingly demonstrated in some microchiropteran bats and it is reasonable to postulate that it occurs in all species of this suborder that have a long hibernating season and depend on food that is scarce or lacking in the winter. With rodents the evidence is contradictory, and this may be explained by species differences and different kinds of bone samples. However, any evidence that disuse osteoporosis occurs during hibernation must take into account that the animal in hibernation does not eat, and that the observed osteoporosis may be dietary.

As Remé and Young (1977) emphasized, the organization of vertebrate visual cells is extremely intricate, yet their complicated architecture is maintained throughout the life of the animal and factors such as age, activity, or absence of stimulation fail to produce detectable morphological changes. This is not because change is not taking place, for radioisotope studies have demonstrated that the cellular components are continually shifting and the lifetime of some of them is less than 1 day. Apparently, the processes of degradation and reconstitution are in delicate balance, so that, in spite of the constant flux, a precise, steady state is maintained.

The effect of hibernation on this steady state had not been examined until 1975 when Kuwabara (1975) reported that there were changes in the outer segments of the visual cells in hibernating frogs, bats, and thirteen-lined ground squirrels. The changes in the ground squirrels were not seen until a month of hibernation had elapsed. The ground squirrels were kept in the cold under laboratory conditions, but the number of days spent in hibernation was not monitored, so the period of hibernation that is required to produce noticeable changes is in question. Recovery was complete 1 week after hibernation was terminated.

Remé and Young (1977) have made a detailed study of the visual and pigmented epithelial cells of the active thirteen-lined ground squirrel and of the changes that occur during the hibernating state and after hibernation is terminated. The hibernating animals were kept at a T_a of 9°C, the number of days in hibernation was monitored, and the eyes were examined using both light and electron microscopes. Fundamental changes were found in the eyes of hibernating animals, including shortening of the outer segments and disruption of the mitochondria, ribosomes, and Golgi complex, which constitute the basic synthetic machinery. These observations indicate that the balanced steady state is tipped during hibernation, with the degradative mechanisms, including autophagy, in temporary dominance.

After arousal, the balance swings toward the restorative processes. The mitochondria, ribosomes, and Golgi complex are regenerated first, followed by the repair of the outer segments, and, by the seventh day, the eye is completely recovered. In spite of the sequential repair during the recovery process, there was no apparent correlation between the length of time in deep hibernation and the extent of degradation. The authors suggest that the state of hibernation creates an imbalance that proceeds until a new steady state is achieved with restoration and degradation again in balance. Were this not the case, degradation should be more apparent in periods of protracted hibernation. As Remé and Young indicate, the brief periods of arousal may also contribute to this lack of correlation. In spite of the careful monitoring of the hibernating state, the T_b of the animals throughout the experiment was unknown, and hibernation of thirteen-lined ground squirrels at a T_a of 9°C is usually not as steady as it is at 4°–5°C. To be certain that degradation is not simply a continuous Q_{10} effect, it may be necessary to monitor body temperature during hibernation, but this would involve repeating the whole time-consuming experiment.

Here, then, is clear-cut experimental evidence that an animal can experience degradation during hibernation in an organ that is essential for survival during the season of activity. It is not known whether the changes that occur during hibernation actually cause a visual deficit, but if they do, the occurrence of periodic arousals may be important in the restorative processes, and the increasing frequency of arousals as the end of the hibernating season approaches may accelerate these processes. Furthermore, there is evidence that some hibernators remain in their burrows for several days after the final spring arousal, and Rausch and Rausch (1971) report that the female northern hoary marmot (*Marmota broweri*) is pregnant before emerging in the spring. An active period of 1 week in the burrow would permit complete morphological restoration of the retina before the hibernator emerged to face the perils of daylight.

There appears to be no information on whether the thirteen-lined ground squirrel remains in its burrow after the terminal arousal from hibernation in the spring, but Wistrand (1974) presents some interesting information on their behavior under natural conditions. For the first 2–3 days after emerging in the spring the animals remained close (5–7 m) to their burrows and walked in a slinking position. If frightened, they took to their burrows and remained there for long periods. By the third to the fifth day they moved more actively and ventured as far as 20 m from the burrow, and by the tenth day this radius was increased to 35 m. There can be a multitude of explanations for this behavior, but it is tempting to ascribe it to the gradual restoration of visual acuity.

If hibernation can upset the morphological balance in the retina, it may affect other highly organized sensory or nervous tissues, but there is little experimental information on this subject. McNamara and Riedesel (1973) have tested the effect of hibernation on the retention of a learned response in the golden mantled ground squirrel and report that it was superior in animals that hibernated compared with animals that remained active in the cold. They attribute this to a reduction of the turnover of protein and other macromolecules likely to be involved in memory and a lessening of external stimuli that might alter established memory. Although these results have been criticized (Alloway, 1973; Riedesel and McNamara, 1973), there seems to be little doubt that memory is not grossly impaired during hibernation and field studies bear out this conclusion. De Vos and Gillespie (1960) observed that woodchucks emerging in March attempted to open the burrows that they had used the previous year. Apparently their memory of the exact location of the burrows was precise, for their tracks led directly to the burrows even if the latter were covered with snow. The period after the final arousal in the spring must be a challenging one for any hibernator, and loss of memory would certainly reduce the chance of survival.

The data presented in this chapter indicate that definite changes at the cellular level occur during hibernation. The question naturally arises whether these changes affect the life span of the cells themselves or the animal as a whole. The concept of suspended animation resulting in longer life has formed the basis for a multitude of fictional stories, and it seems reasonable to assume that the drastically reduced metabolism of deep hibernation would contribute to a long life in a hibernator. But low metabolism, absence of mobility, and a greatly reduced heart rate do not necessarily mean that the animal is in a state of complete and life-giving rest. Using the heart rate to illustrate this concept, it has been proposed that the number of heart beats is fixed in mammals and that the life span of all mammals is limited to the same total number of beats. Thus, the profligate shrew uses up its allotted heart beats in about a year, whereas the conservative elephant lives for almost a century before it reaches the same number of heart beats and death occurs.

Even if one accepts this unproved theory, it must be recognized that the heart beat of an animal in deep hibernation is physiologically different from a heart beat in the same animal in the euthermic state. As described in Chapter 4, the beat itself, and the time between beats, is greatly prolonged during hibernation. This would suggest that the heart was resting and hence was not aging as fast as the rapidly beating heart of the euthermic animal. On the other side of the ledger are the effects of cold on cells, tissues, and organs. At the simplest physical level, the cold heart of a hibernating animal must overcome the viscosity of its own valves, muscle, and the blood that it is pumping, in order to produce a useful stroke. The fact that electrical depolarization may occur without any change in pulse pressure (Chapter 4) indicates that the system may break down during hibernation. Is, then, the cold heart of a deeply hibernating animal actually aging at a slower rate than the heart of a euthermic animal?

There is little experimental data on this subject. Griffin (1960) has pointed out that tagging experiments with temperate zone microchiropteran bats have demonstrated that their life span may be at least 21 years, which is an extraordinarily long life for such a small mammal. There seems to be no doubt that bats may live at least this long, for Keen and Hitchcock (1980) have recently reported the survival of two male *Myotis lucifugus* for 29 and 30 years. Since these bats indulge in prolonged hibernation in the winter as well as diurnal periods of torpor in the summer, the correlation of hibernation and longevity is tempting. However, Herreid's (1964) field studies indicated that the semitropical bats, which do not hibernate, also live for surprisingly long periods, and he suggested that there was no evidence that their life span was shorter than temperate zone species.

Some experiments at the cellular level indicate that the life span of red blood cells is increased in hibernation. Brace (1953) labeled the hemoglobin of woodchucks with ^{14}C and concluded from his results that cold exposure was as effective as hibernation in prolonging the life of red blood cells. Only four animals were used in the experiment, leaving the conclusion in some question. Mann and Drips (1917) reported that the number of phagocytic endothelial cells containing red blood cells or blood pigment appeared to be reduced in the spleen of hibernating thirteen-lined ground squirrels. However, the thorough investigation by Brock (1960a) provides the most convincing evidence. She injected golden hamsters (*Mesocricetus auratus*) with ^{51}Cr via chronically implanted aortic cannulae and tagged the red blood cells. The experimental animals consisted of three groups: animals kept in the warm room, cold-exposed (T_a 8° ± 2°C) euthermic animals, and hibernating animals. Previous work from other laboratories on aging of red blood cells had shown that the disappearance of the tagged cells was dependent on three factors: elution, random destruction, and senescence. The mechanism of elution is not known, but it consists of a continual loss of ^{51}Cr after the peak in the number of tagged cells. Random destruction also accounts for the depletion of tagged cells and this is believed to be caused by physical injury to the cells. Random destruction and elution are lumped together as "random loss." As its name implies, random loss occurs at random and is independent of the age of the cell which is destroyed.

Brock found that there was no difference in the loss of tagged blood cells over time when the blood of the warm and cold room euthermic animals was compared. However, physiological random destruction and ^{51}Cr elution were virtually absent during hibernation. Furthermore, loss from senescence was greatly reduced. Similar results have been reported for tagged cells stored in cold acid-citrate-dextrose (Brock, 1960b) and it would appear that the red blood cells are virtually in "cold storage" during the hibernating period. It was also reported that there was no significant difference in the rate of loss of tagged blood cells from animals that had just aroused from long periods of hibernation, compared with the loss in euthermic animals that had not hibernated. Thus, there was no evidence that a backlog of old cells had built up during the hibernating period and then had died soon after the return to the euthermic state.

Hrůza et al. (1966) reported an experiment involving the effect of hibernation on the development of cross-linkage in collagen fibers. Tendons of rats were dissected and dipped in 2.5 M sodium perchlorate, and the relaxation and subsequent contraction was measured. After storage in vitro at either 37° or 0°C for 3 months, the tendons kept at the higher temperature showed a greater decrement in response, and this was taken as a manifestation of the effect of temperature on the aging process. Tail tendons from dormice (Glis glis) were also tested, and eight of these animals were then exposed to the cold (5°C) for 3 months, whereas control dormice were maintained at 24°C. Hibernation began after 14 days of cold exposure, with some animals hibernating for longer periods than others. At the end of the 3-month period, additional fibers were dissected from the warm and cold-exposed animals and tested for relaxation and contraction. When compared to the previous measurements, the response of the fibers from the warm room animals was greatly reduced. The reduction in the response was less in animals from the cold room, with the tendons from the animals that had hibernated the longest exhibiting the least change. It was concluded that cross-linkage of the proteins in collagen fibers was inhibited by the hibernating state. Other methods of testing for aging involve a much longer time span, and one may question whether a comparison of the changes that may occur over a maximum of 3 months truly reflects differences in the aging process in animals such as dormice that have a life span of more than 4 years.

As far as longevity in the whole mammal is concerned, we have recently reported an experiment using the Turkish hamster (Mesocricetus brandti) (Lyman et al., 1981). This species can be reared in the laboratory and hibernates for prolonged periods, and, thus, approaches the ideal for an experiment on hibernation and aging. The life span of 144 hamsters maintained at $22° \pm 3°C$ was compared with the life span of the same number of animals which were exposed to cold ($5° \pm 2°C$) for 5.5 months each year (Fig. 10-4). The amount of hibernation which occurred was monitored using the time honored method of placing material such as oats on the back of the hibernating animal (Johnson, 1931). If the oats were displaced, but the animal was hibernating at the next inspection, it was assumed that an arousal and return to hibernation had occurred. As was expected, there was great variation in the amount of hibernation that occurred among individuals of the experimental group. Therefore, this group was divided into three subgroups: "poor hibernators" or animals that hibernated 0–11% of

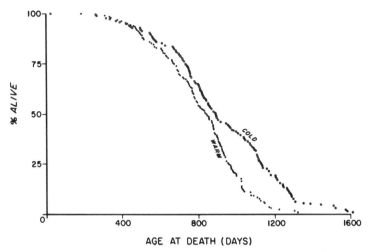

Fig. 10-4. Percentage of Turkish hamsters alive plotted against time for all animals in the study. Closed circles, warm room animals ($n = 144$). Open circles, cold-exposed animals ($n = 144$). (After Lyman *et al.*, 1981. Copyright 1981 by the American Association for the Advancement of Science.)

their lives ($n = 47$), "moderate hibernators" or animals that hibernated 12–18% of their lives ($n = 48$), and "good hibernators" or animals that hibernated 19–33% of their lives ($n = 49$).

Nine of the cold-exposed group never hibernated, and these animals died at an early age and bias the calculations. In spite of this, the average life span of 914 days for the cold-exposed group as a whole was significantly greater ($p = 0.001$) than the life span of 812 days for the warm room controls. The mean life span of 1093 days for the good hibernators was significantly greater than the life span of 916 days for the moderate hibernators ($p = 0.002$) and their life span in turn was significantly greater ($p \leq 0.001$) than the life span of 727 days for the poor hibernators. The moderate hibernators lived longer than the warm room controls, but the latter lived longer than the poor hibernators. These differences are presented graphically in Fig. 10-5. The mean life span of the good hibernators was 34% longer than that of the warm room controls and this difference has a satisfying one chance in 8×10^7 of being wrong. A graph of the linear regression of life span versus percentage of hibernation (Fig. 10-6) indicates that the more an animal hibernates the longer it lives, with the equation: life span = 653.20 + (18.002 × percent of life in hibernation).

It should be mentioned that the hamster colony consisted of two groups of laboratory-bred animals. The original stock had been trapped in two areas, one in the vicinity of Ankara and the other 125 km to the southeast. Although phenotypically identical, the former animals had a diploid number of 44 chromosomes,

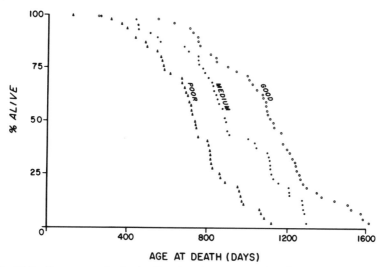

Fig. 10-5. Percentage of animals alive plotted against time for all cold-exposed hamsters. Triangles, poor hibernators (n = 47). Closed circles, moderate hibernators (n = 48). Open circles, good hibernators (n = 49). (After Lyman et al., 1981. Copyright 1981 by the American Association for the Advancement of Science.)

whereas the diploid number in the latter was 42. The percentage of life spent in hibernation was different in the two groups, for the animals from Ankara hibernated 17.1% of their lives, whereas those from the area 125 km to the southeast hibernated 12.1%. This difference is highly significant ($p < 0.001$). Further analysis revealed that the females of the group 125 km southeast of Ankara hibernated only 9.4% of their lives, which was significantly less than any other group, and the males showed the same tendency (14.1%), but the difference between this percentage and that of the males (17.6%) and the females (16.5%) from Ankara is not significant. The reason for these differences is not readily apparent, but it appears to be the only report in which there is a concurrent difference in chromosome number and tendency to hibernate within a single species (Lyman et al., 1982).

The data indicate beyond reasonable doubt that the animals that hibernated lived longer than the warm room controls that did not hibernate, and that the amount the animals hibernated correlated with the additional time the animals lived when compared with the control, warm room animals. However, the conclusion that hibernation prolongs life does not necessarily follow. Alternatively, one might suggest that the control animals in the warm room were undergoing a greater stress than the animals that were seasonally exposed to cold and hence would age faster. According to this concept, the animals that were continuously kept in the warm room were denied their natural cycle of hibernation during the

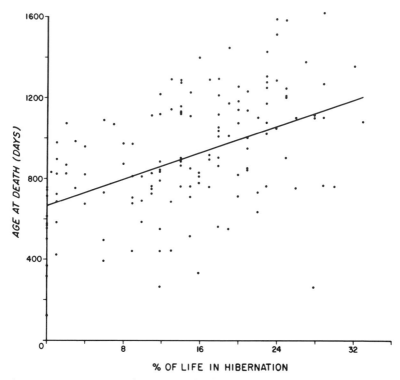

Fig. 10-6.   Regression correlating age at death as a function of percentage hibernation for all cold-exposed hamsters. Age at death = 653.20 + (18.002 × percent hibernation); $n = 144$; $\bar{y} = 914.41$; $\bar{x} = 14.510$; $r = 0.522$; $p < 0.001$. (After Lyman et al., 1981. Copyright 1981 by the American Association for the Advancement of Science.)

winter months and hence were under greater stress than the animals that were moved to the cold and thus could hibernate. Although stress is difficult to quantitate, it is generally accepted that it is correlated with a sustained, high metabolic rate. We find that the oxygen consumption doubles when Turkish hamsters are first moved from T_a 22° to 5°C (Lyman et al., unpublished observations), and, though metabolic rate declines as the animal adapts to the cold, still 5°C is well below the thermal neutral zone and metabolic rate remains higher than the rate of animals maintained at 22°C throughout the year. There is no evidence that basal metabolic rate increases as the hibernating season approaches in animals kept at a thermoneutral environment. On the contrary, Kayser (1961) and others present evidence to show that the opposite occurs. Furthermore, the hamsters that did not hibernate in the cold were the first to die, and aged animals usually died without hibernating within a week after their third seasonal exposure to cold. Also, the behavior of potential hibernators does not indicate that they seek the cold as winter approaches, for both hamsters and ground squirrels still huddle together when kept in community cages at warm temperatures, and build bigger and better constructed nests when exposed singly to the cold. Finally, in experiments that were performed in fall and winter, Gumma and South (1970) showed that Syrian hamsters preferred 26°C to a series of lower temperatures. Thus, when alternative explanations are considered, the concept that hibernation lengthens the life of Turkish hamsters is persuasive.

REFERENCES

Adelstein, S. J., and Lyman, C. P. (1968). Pyrimidine nucleoside metabolism in mammalian cells: an *in vitro* comparison of two rodent species. *Exp. Cell Res.* **50,** 104–116.

Adelstein, S. J., Lyman, C. P., and O'Brien, R. C. (1964). Variations in the incorporation of thymidine into the DNA of some rodent species. *Comp. Biochem. Physiol.* **12,** 223–231.

Adelstein, S. J., Lyman, C. P., and O'Brien, R. C. (1967). Cell proliferation kinetics in the tongue and intestinal epithelia of hibernating dormice (*Glis glis*). In "Mammalian Hibernation III" (K. C. Fisher, A. R. Dawe, C. P. Lyman, E. Schönbaum, and F. E. South, eds.), pp. 398–408. Oliver & Boyd, Edinburgh.

Alloway, T. M. (1973). Hibernation: effects on memory or performance? *Science* **181,** 86–87.

Bailey, E. D. (1965). Seasonal changes in metabolic activity of non-hibernating woodchucks. *Can. J. Zool.* **43,** 905–909.

Bailey, E. D., and Davis, D. E. (1965). The utilization of body fat during hibernation in woodchucks. *Can. J. Zool.* **43,** 701–707.

Brace, K. C. (1953). Life span of the marmot erythrocyte. *Blood* **8,** 648–650.

Brock, M. A. (1960a). Production and life span of erythrocytes during hibernation in the golden hamster. *Am. J. Physiol.* **198,** 1181–1186.

Brock, M. A. (1960b). Hibernation and cold storage effects on phosphates in hamster blood. *Am. J. Physiol.* **199,** 195–197.

Bruce, D. S., and Wiebers, J. E. (1969). Mineral dynamics during hibernation and chronic immobility: a review. *Aerosp. Med.* **40,** 855–861.

Cameron, I. L., and Cleffmann, G. (1964). Initiation of mitosis in relation to the cell cycle following feeding of starved chickens. *J. Cell. Biol.* **21,** 169–174.

Davis, D. E. (1967a). The annual rhythm of fat deposition in woodchucks (*Marmota monax*). *Physiol. Zool.* **40,** 391–402.

Davis, D. E. (1967b). The role of environmental factors in hibernation of woodchucks (*Marmota monax*). *Ecology* **48,** 683–689.

de Vos, A., and Gillespie, D. I. (1960). A study of woodchucks on an Ontario farm. *Can. Field Nat.* **74,** 130–145.

Ferren, L. G., South, F. E., and Jacobs, H. K. (1971). Calcium and magnesium levels in tissues and serum of hibernating and cold-acclimated hamsters. *Cryobiology* **8,** 506–508.

Griffin, D. R. (1960). Panel discussion. *Bull. Mus. Comp. Zool.* **124,** 530–531.

Gumma, M. R., and South, F. E. (1970). Hypothermia and behavioral thermoregulation by the hamster (*Mesocricetus auratus*). *Anim. Behav.* **18,** 504–511.

Haller, A. C., and Zimny, M. L.(1977). Effects of hibernation on interradicular alveolar bone. *J. Dent. Res.* **56,** 1552–1557.

Hansemann, D. (1898). Ueber den Einfluss des Winterschlafes auf die Zelltheilung. *Arch. Anat. Physiol., Physiol. Abt.* **103,** 262–263.

Haymovits, A., Gershon, M. D., and Nunez, E. A. (1976). Calcitonin, serotinin, and parafollicular cell granules during the hibernation activity cycle in the bat. *Proc. Soc. Exp. Biol. Med.* **153**, 383-387.

Herreid, C. F., II (1964). Bat longevity and metabolic rate. *Exp. Gerontol.* **1**, 1-9.

Hock, R. J. (1960). Seasonal variations in physiologic functions of Arctic ground squirrels and black bears. *Bull. Mus. Comp. Zool.* **124**, 155-171.

Hrůza, Z., Vrzalová, Z., Hlaváčková, V., and Hrabalova, Z. (1966). The effect of cooling on the speed of ageing of collagen *in vitro* and in hibernation of the fat dormouse (*Glis glis*). *Exp. Gerontol.* **2**, 29-35.

Huggins, C., and Blocksom, B. H., Jr. (1936). Changes in outlying bone marrow accompanying a local increase of temperature within physiological limits. *J. Exp. Med.* **64**, 253-274.

Jaroslow, B. N., Fry, R. J. M., Suhrbier, K. M., and Sallese, A. R. (1976). Radiosensitivity of ileum crypt cells in hibernating, arousing, and awake ground squirrels (*Citellus tridecemlineatus*). *Radiat. Res.* **66**, 566-575.

Johnson, G. E. (1931). Hibernation in mammals. *Q. Rev. Biol.* **6**, 439-461.

Kayser, C. (1961). "The Physiology of Natural Hibernation." Pergamon, Oxford.

Kayser, C., and Frank, R. M. (1963). Comportement des tissus calcifies du hamster d'Europe, *Cricetus cricetus,* au cours de l'hibernation. *Arch. Oral Biol.* **8**, 703-713.

Keen, R., and Hitchcock, H. B. (1980). Survival and longevity of the little brown bat (*Myotis lucifugus*) in southeastern Ontario. *J. Mammal.* **61**, 1-7.

Krook, L., Wimsatt, W. A., Whalen, J. P., MacIntyre, I., and Nunez, E. A. (1977). Calcitonin and hibernation bone loss in the bat (*Myotis lucifugus*). *Cornell Vet.* **67**, 265-271.

Kunkel, H. A., and Schubert, G. (1959). Effects of protective agents applied after irradiation. *Prog. Nucl. Energy, Ser. 6* **2**, 217-224.

Kuwabara, T. (1975). Cytological changes of the retina and pigment epithelium during hibernation. *Invest. Ophthalmol.* **14**, 457-467.

Lyman, C. P., and Dempsey, E. W. (1951). The effect of testosterone on the seminal vesicles of castrated, hibernating hamsters. *Endocrinology* **49**, 647-651.

Lyman, C. P., Weiss, L. P., O'Brien, R. C., and Barbeau, A. A. (1957). The effect of hibernation on the replacement of blood in the golden hamster. *J. Exp. Zool.* **136**, 471-486.

Lyman, C. P., O'Brien, R. C., Greene, G. C., and Papafrangos, E. D. (1981). Hibernation and longevity in the Turkish hamster, *Mesocricetus brandti. Science* **212**, 668-670.

Lyman, C. P., O'Brien, R. C., and Bossert, W. H. (1982). Differences in tendency to hibernate among groups of Turkish hamsters (*Mesocricetus brandti*). In preparation.

McNamara, M. C., and Riedesel, M. L. (1973). Memory and hibernation in Citellus lateralis. *Science* **179**, 92-94.

Manasek, F. J., Adelstein, S. J., and Lyman, C. P. (1965). The effects of hibernation on the *in vitro* synthesis of DNA by hamster lymphoid tissue. *J. Cell. Comp. Physiol.* **65**, 319-324.

Mann, F. C., and Drips, D. (1917). The spleen during hibernation. *J. Exp. Zool.* **23**, 277-286.

Mayer, W. V., and Bernick, S. (1958). Comparative histological studies of the stomach, small intestine, and colon of warm and active and hibernating arctic ground squirrels, *Spermophilus undulatus. Anat. Rec.* **130**, 747-757.

Rasmussen, A. T. (1918). Cyclic changes in the interstitial cells of the ovary and testis in the woodchuck (*Marmota monax*). *Endocrinology* **2**, 353-404.

Rausch, R. L., and Rausch, V. R. (1971). The somatic chromosomes of some North American marmots (Sciuridae), with remarks on the relationships of *Marmota broweri* Hall and Gilmore. *Mammalia* **35**, 85-101.

Remé, C. E., and Young, R. W. (1977). The effects of hibernation on cone visual cells in the ground squirrel. *Invest. Ophthalmol. Visual Sci.* **16**, 815-840.

Riedesel, M. L., and McNamara, M. C. (1973). Technical comment, Hibernation: effects on memory or performance? *Science* **181,** 87.

Sarnat, B. G., and Hook, W. E. (1942). Effects of hibernation on tooth development. *Anat. Rec.* **83,** 471–493.

Snyder, R. L., and Christian, J. J. (1960). Reproductive cycle and litter size of the woodchuck. *Ecology* **41,** 647–656.

Snyder, R. L., Davis, D. E., and Christian, J. J. (1961). Seasonal changes in the weights of woodchucks. *J. Mammal.* **42,** 297–312.

Tempel, G. E., Wolinsky, I., and Musacchia, X. J. (1978). Bone and serum calcium in normothermic, cold-acclimated and hibernating hamsters. *Comp. Biochem. Physiol. A* **61A,** 145–147.

Thomson, J. F., Straube, R. L., and Smith, D. E. (1962). The effect of hibernation on liver regeneration in the ground squirrel (*Citellus tridecemlineatus*). *Comp. Biochem. Physiol.* **5,** 297–305.

Wang, L. C. H. (1979). Time patterns and metabolic rates of natural torpor in the Richardson's ground squirrel. *Can. J. Zool.* **57,** 149–155.

Wells, L. J., and Moore, C. R. (1936). Hormonal stimulation of spermatogenesis in the testis of the ground squirrel. *Anat. Rec.* **66,** 181–200.

Whalen, J. P., Krook, L., and Nunez, E. A. (1972). A radiographic and histologic study of bone in the active and hibernating bat (*Myotis lucifugus*). *Anat. Rec.* **172,** 97–108.

Wistrand, H. (1974). Individual, social, and seasonal behavior of the thirteen-lined ground squirrel (*Spermophilus tridecemlineatus*). *J. Mammal.* **55,** 329–347.

Zalesky, M. (1934). A study of the seasonal changes in the adrenal gland of the thirteen-lined ground squirrel (*Citellus tridecemlineatus*), with particular reference to its sexual cycle. *Anat. Rec.* **60,** 291–322.

Zalesky, M. (1935). A study of the seasonal changes in the thyroid gland of the thirteen-lined ground squirrel (*Citellus tridecemlineatus*), with particular reference to its sexual cycle. *Anat. Rec.* **62,** 109–138.

Zimmerman, G. D., McKean, T. A., and Hardt, A. B. (1976). Hibernation and disuse osteoporosis. *Cryobiology* **13,** 84–94.

11

Hibernation: Responses to External Challenges

Hibernators, like other mammals, are subject to many diseases including parasitic infections and neoplastic growth. In the overall natural history of the animals concerned, the effect of hibernation with its accompanying reduction in body temperature may be a critical factor. It is important to know whether the hibernating state can mitigate or even halt an otherwise progressive disease. If hibernation cleanses the hibernator and thus increases its chances of survival, does it also reduce or eliminate useful symbionts and thus have a negative effect? Furthermore, hibernation is the only condition in mammals and birds in which a profound decrease in T_b and metabolic rate occurs naturally. As such, it provides a useful experimental model to test the effect of these variables on potentially destructive factors such as neoplasms or radiation.

At the outset, it can be stated that hibernation is not a natural cure for a deadly disease. In our experience a really sick animal never enters into the hibernating state. If exposed to cold, its T_b may drop, but the decline is irreversible unless the animal is rewarmed by exogenous heat. The delicate physiological balance that is required for the hibernating state can not be attained by an animal that is suffering from a terminal illness.

Both parasites and symbionts occur naturally in virtually all populations of wild mammals, and mammals that hibernate are no exception. The effect of hibernation, with its accompanying cold and starvation, on the intestinal flora and fauna and the endo- and ectoparasites has been reported in some detail. Chute (1964) and Schmidt (1967) review much of the work on this subject, and this and some later work are summarized here.

Barnes (1970) and Barnes and Burton (1970) examined the cecal flora of the thirteen-lined ground squirrel prior to hibernation and in the hibernating state. They found that certain of the flora were greatly reduced in the first 6 days of hibernation, particularly the anaerobic gram-positive rods, coccobacilli and the aerobic atypical lactobacilli. There was little further change in two animals examined after 42 days of hibernation, though this period must have been interrupted by periods of arousal. Whether the observed changes were due to hibernation or to the accompanying lack of food was not determined, as no comparison was made with euthermic starved animals. The gram-negative nonsporing anaerobes survived the hibernating period. As many of these are known to produce volatile fatty acids, the authors suggest that these are an available source of energy in the cecum upon arousal from hibernation.

As far as parasitic intestinal protozoa are concerned, Davis (1969) reported that *Entamoeba citelli* was no longer present in the gut of the golden mantled ground squirrel and the Belding ground squirrel (*C. beldingi*) after the animals had been exposed to hibernating conditions for as little as 15 days. In contrast, another entamoeba, which could not be identified specifically, and a flagellate (*Tritrichomonas muris*) survived for periods of intermittent hibernation lasting as long as 161 days. Davis points out that it is advantageous for the parasite to remain outside the host in an incipient state during the hibernating season, for in this condition the infection can be spread and the parasite is not dependent on the survival of the host. However, he found no evidence that either starvation or hibernation increased the amount of encystment of intestinal parasites.

Hibernation appears to have little effect on the ectoparasites of hibernating bats, according to Reisen *et al.* (1976), and they quote an unpublished doctoral thesis and a Russian reference, which come to the same conclusion. They determined the number of *Paraspinturnix globosus,* a parasitic mite, and *Trichobius major,* a parasitic fly, on the cave bat (*Myotis velifer*) between December and March and found that there was no significant change, but the population of other ectoparasites was reduced. The bats were hibernating under natural conditions with the lowest estimated T_b at 6.1°C and the highest 17°C, though the method used for obtaining these temperatures is not stated. *Trichobius major* filled with a blood meal were found on some hibernating bats, but whether the meal was obtained during a period of arousal was not determined.

From the hibernator's point of view it would be advantageous to be able to rid itself of parasites during hibernation and some early reports indicate that this may be the case. For example, Blanchard and Blatin (1907) reported that European marmots were heavily infested with parasites prior to hibernation in the autumn, but found only a few encysted nematode larvae in a series of more than 100 hibernating animals. They suggested that the parasites were rendered inert by the low temperature and expelled with the feces, but Schmidt (1967) considers this unlikely.

Simitch and Petrovitch (1954) carried out an extensive series of observations on the blood and intestinal parasites of the European ground squirrel, *Citellus citellus*. They monitored the amount of time spent in hibernation and took into consideration the possibility that a period of continuous hibernation might have more effect on the parasites than a longer period which was interrupted by frequent arousals. They found that populations of intestinal protozoa and the larval forms of cestodes showed little change with hibernation and that this also applied to blood trypanosomes. All species of adult helminthes were sensitive to the hibernating state and could be eliminated by prolonged hibernation, but the sensitivity varied from species to species.

Chute (1961) reported an experiment in which the effect of hibernation was tested on a parasite that was not normally found in the host. Using *Trichinella spiralis* as the parasite and the Syrian hamster as the host he concluded that the development of *T. spiralis* could be completely inhibited by 48–72 hr of hibernation at a T_b of 5°–7°C if the entrance into hibernation took place soon after the animals were infected.

One can reasonably conclude that some parasites are affected when the host enters the hibernating state. Experiments on the whole animal have not clarified whether the effect is due simply to the low temperature of hibernation or whether changes in the internal milieu are a causative factor.

In this regard, it is of interest that the immune response is reported to be greatly reduced during hibernation. Jaroslow and Serrell (1972) tested thirteen-lined ground squirrels and their response to injected sheep red blood cells during various periods of the hibernating cycle. The three phases of the response—latent period, proliferative phase, and declining phase—were all prolonged in different degrees by hibernation. The differences were attributed to the phase that a cell might be in when arousal occurred. Injection of the red blood cells always resulted in arousal, so that the reported results are a combination of the effect of arousal and euthermia and do not represent the effect of hibernation alone.

Sidky and Auerbach (1968) had previously tested the spleens of euthermic and hibernating Syrian hamsters for their immune response. These *in vitro* tests revealed that the spleens from euthermic animals maintained at either warm or cold temperatures showed a brisk immune response, whereas the spleens of hibernating animals reacted slightly or not at all. This indicated that there was a loss of competence during hibernation in at least one of the principal antibody-producing organs and this loss was due to some factor other than the low temperature of hibernation. Sidky and his group focused on brown fat as a possible source of a substance that might suppress the immune reaction. They found that axillary brown adipose tissue from hibernating hamsters produced this suppression when tested *in vitro* with hamster splenic tissue, whereas brown fat from euthermic animals showed only a slight suppressive action. However, brown fat from hibernating hamsters had no effect when mouse spleen was used in the *in vitro* test, indicating that the putative suppressive agent did not have a general effect among rodents as a group (Sidky *et al.,* 1969).

Using a seasonal hibernator (*Citellus richardsonii*) maintained in the warm room throughout the year, it was found that extracts of brown fat had a suppressing effect during the winter months when the animal would normally be hibernating, but not in the summer months during the period of activity. Thus, there appeared to be an annual cycle (called "circennial" by the authors) in the effectiveness of some substance in the brown fat (Sidky *et al.,* 1972a).

In further experiments, the brown fat of a whole gamut of mammals under various conditions was examined for its suppressive ability. These included *Citellus richardsonii*, the white-footed mouse (*Peromyscus maniculatus*), the little brown bat (*Myotis lucifugus*), the Syrian hamster, and the newborn domestic rabbit. The extract of brown fat from animals that were expected to have maximal brown fat thermogenesis produced a suppressive effect on the immune response when tested *in vitro* with hamster spleen. For example, the brown fat from the newborn rabbit, the cold-adapted deer mouse, the hibernating summer and euthermic winter ground squirrel, and the euthermic and hibernating bats were suppressive, whereas brown fat from deer mice kept at room temperature, active summer ground squirrels, and euthermic winter hamsters were not. Thus Sidky *et al.* (1972b) concluded that brown fat has another function, and that an immunosuppressant accumulates in this tissue during the preparation for intensive thermogenic activity. Actually, the relation between thermogenic and immunosuppressive activity is not absolute, for the spleen cells from euthermic cold-exposed hamsters showed only slight evidence of immunosuppression, whereas the extract of brown fat from deer mice and ground squirrels under the same condition had an immunosuppresive action. As Sidky *et al.* (1972b) point out, the suppression of the immune response in the fetal or newborn rabbit may prevent a reaction to a maternal antigen, but there appears to be no clear rationale for the high titre of the immunosuppressant in brown fat as reported in the other experiments. Certainly further work is necessary before the role of brown fat as an immunosuppressant is firmly established.

The effect of hibernation on the growth of neoplastic homologous tissue was examined by us in 1954 (Lyman and Fawcett, 1954). We were intrigued with the concept that this tissue might be more temperature-sensitive than its host, and thus lead to a possible treatment for cancer. In this experiment, a homologous methylcholanthrene-induced tumor was implanted in the cheek pouch of a series of Syrian hamsters which had been hibernating at $5° \pm 2°C$, and the same tumor was implanted into animals kept in a warm room. The growth of the implants was measured from time to time and the number of days the animals hibernated in the cold over the next 35 days was noted. Animals that hibernated more than 15 of the 35 days exhibited little, if any, growth in the tumor. It appeared that this inhibition of growth was not due to failure of vascularization of the tumor because of cold or hibernation. Some animals failed to hibernate for several days after implantation of the tumor. Presumably vascularization of the tumor took place during this time, but growth was as retarded as in animals that hibernated almost at once. There was no evidence that the tumor could be eradicated by long-term hibernation, for the tumors always enlarged once hibernation was terminated. Tumors in cold-exposed hamsters that did not hibernate grew at a slightly slower rate than the tumors in warm room controls. Since the cheek pouch of the cold-exposed hamster averaged about 1°C colder than that of the warm room controls, it was concluded that the slowing of growth in the cold room controls and in the hibernating animals was due to cold alone. This concept was reinforced by a study of tumor growth in frog kidney kept *in vitro* at various temperatures, for it was found that growth varied with temperature with a Q_{10} of 2.5 (Lucké *et al.*, 1953).

The experiment with hamsters showed that homologous neoplastic tissue from a hibernator was no more sensitive than normal homologous tissue when it was exposed to hibernating conditions. However, it is well established that organs and tissues from hibernators are unique in their ability to function at low temperatures (Chapter 9). The effect of hibernation on a heterologous tumor from a species that was incapable of hibernation remained to be determined. In this experiment, five human tumors were implanted in hibernating hamsters and the effect of hibernation on their growth was observed over a 4- to 6-week period (Patterson *et al.*, 1957). These tumors had all been used in oncological research and their growth rates and final fate when transplanted into the hamster cheek pouch had been previously established. The host animals hibernated 49–76% of the test period. In one case the tumor, an adenocarcinoma of the endometrium, failed to establish itself. With the other tumors, growth was greatly reduced in the hibernating animals, but it accelerated when hibernation was terminated by moving the animals to the warm room. Thus, in spite of the fact that 18°C is about the lowest tolerable *in vivo* temperature for man (Laufman, 1951), human tumors can survive in hamsters hibernating at a T_b of about 5°C.

A logical extension in the study of hibernation, cold tolerance, and tumor growth would be a comparison of the effects of chemotherapy in active and hibernating animals. It might be found that hibernation resulted in a disproportionate increase in the sensitivity of the tumor to a chemotherapeutic agent when compared with the sensitivity of the host. Alternatively, hibernation might impart some degree of protection from the chemotherapeutic agent to the host which was not imparted to the tumor.

A series of experiments with ionizing radiation suggests that hibernation may indeed alter the sensitivity of the whole animal to a potentially lethal physiological insult. In an extensive set of experiments, Musacchia and associates (see, e.g., Musacchia and Barr, 1969) have studied the protective effect of hibernation and hypothermia against lethal radiation. Both golden hamsters and thirteen-lined ground squirrels were irradiated while in hibernation and then moved to the warm room and observed for the expected lethal result. Using various doses of radiation it was found that the mean survival time of the ground squirrels irradiated in hibernation was consistently longer than the survival time of animals that were irradiated during the active state. Since the wide level of radiation dosages in these experiments spanned the hematopoietic, gastrointestinal, and central nervous system syndromes of radiation damage, it was concluded that the protective mechanisms were operative on all three of these systems. The protective effect of hibernation at various T_b's was tested by irradiating ground squirrels at T_b's of 5°, 13°, and 23°C and then moving them to a warm room. The survival time of every group of experimental animals was again longer than that of the active animals irradiated at a T_b of 37°C. In one set of experiments, the animals which were irradiated while hibernating at a T_b of 13°C proved to be more radiation-resistant than the animals hibernating at the lower temperature of 5°C. There is no obvious explanation for this unexpected finding, and similar results have not been reported for animals in hypothermia where T_b's of 0°–5°C afforded some protection, whereas there was much less at T_b's above 9°C (Musacchia and Barr, 1969).

The experiments indicate that the state of hibernation itself affords some protection against exposure to radiation, and that this protection is not simply dependent on body temperature. Earlier experiments by other workers had suggested that radiation damage was reduced if the animal remained in hibernation after receiving the radiation. The degree of protection afforded by the hibernating state differed in the various experiments and among the different species. For example, Smith and Grenan (1951) reported that woodchucks that remained in hibernation after irradiation lived longer than active animals, but suggested that some radiation damage occurred while the animals were actually in hibernation. Similar results were reported by Doull and DuBois (1953) using thirteen-lined ground squirrels. Barr and Musacchia (1969) reported longer survival times in ground squirrels maintained at 23°C after irradiation than in animals maintained at 5°C, whereas a postradiation temperature of 13°C resulted in the shortest survival times. Jaroslow *et al.* (1969) found maximum protection occurred if ground squirrels remained in hibernation 24 hr after irradiation and then were exposed to a T_a of 23°C. These animals suffered fewer deaths than ground squirrels kept at a T_a of 5°C for 2–3 weeks after irradiation. It was suggested that the spontaneous arousals from hibernation in animals kept without food or water during the 2- to 3-week period resulted in a negative energy balance that may have contributed to the higher death toll.

Working with the pallid and the Yuma bat (*Antrozous pallidus* and *Myotis yumanensis*), Osborn and Kimeldorf (1957) found that the postirradiation survival time was increased if the bats remained in hibernation. However, some of the animals died in hibernation and, as with the woodchucks, it was concluded that the hibernating state did not afford complete protection. On the other hand, Kunkel and Schubert (1959) reported that the state of hibernation afforded virtually complete protection from radiation injury in the edible dormouse (*Glis glis*). Animals that were exposed to radiation and kept in the cold for 21 days did not die until after they were moved to the warm room and hibernation had been terminated. Their survival time after being moved to the warm room was the same as the postirradiation survival time of the active controls. Furthermore, if the hibernating dormice were injected with cysteine at the time they were moved to the warm room, the protective action of the drug was the same as found in control animals injected prior to irradiation. It was concluded that damage from radiation was in abeyance during hibernation, and the effect of exposure to radiation only manifested itself when the animal returned to the active state.

Some of the discrepancies in these various experiments might be due to the differences in the reaction of the various species, but a more important factor may be the inherent difficulty of the experiments themselves. The end point in each group of experiments was the time of death of the animal, and time of death will vary greatly even in a homogeneous population. An animal may be extremely ill and recover. This animal is scored as being alive yet clearly it is not homologous with an animal that remained in good health throughout the experiment. The observed results must often involve not only radiation damage, but also repair of that damage. Presumably, this rate of repair could be influenced both by T_b and by the hibernating state. In addition, the comparisons were drawn in these experiments between hibernating and active animals, but the actual amount of hibernation that occurred in each animal was not reported. The hibernating groups must have contained animals that were euthermic part of the time, but the effort of arousal, the length of the euthermic period, and the effect of the subsequent return to the hibernating state were not included in the studies. Smith (1960) has shown that animals tend not to return to the hibernating state if they are aroused from hibernation after irradiation. In order to achieve the maximum days in hibernation after irradiation, it is essential to avoid disturbing the animals in any way and this makes the task of monitoring them for hibernation extremely difficult. To obtain unequivocal results on the effect of hibernation on radiation damage, a very large number of animals should be used and the time in hibernation should be monitored. This is a formidable task.

Musacchia and his group have suggested that the resistance to radiation damage in hibernating animals may be due to the increase of extra-adrenal catecholamines that may occur during arousal from hibernation. This might cause intense vasoconstriction resulting in local tissue hypoxia in the bone marrow and this in turn might increase the protection by means of the "oxygen effect" (Prewitt and Musacchia, 1975). Data had previously been presented to indicate that the catecholamines were high during arousal in the ground squirrel (Musacchia *et al.*, 1963) and that injection of large doses of catecholamines prior to exposure to radiation afforded the animals some protection (Prewitt and Musacchia, 1969). In order to block the effect of the catecholamines Prewitt and Musacchia (1975) injected a group of arousing ground squirrels with the α-adrenergic blocking agent phentolamine. These animals proved to have less radioresistance than saline-injected controls when exposed to lethal radiation. Hamsters rendered hypothermic with a body temperature of 7°-8°C by exposure to 80% helium and 20% oxygen at a T_a of 0°C were found to be more radiation resistant than active control animals. This method of inducing hypothermia does not produce the severe hypoxia that occurs when the Giaja technique (Giaja and Andjus, 1949) is employed, and it was argued that the degree of protection that was reached was comparable to the degree of hypoxia. Using the Giaja technique with mice, Hornsey (1957) achieved T_b's of 0°C and measured tissue P_{O_2} values of only 1-2 mmHg. Musacchia *et al.* (1963) point out that the greater radioresistance reported by Hornsey (1957) can be correlated with the greater hypoxia in his experimental animals. In spite of this emphasis on hypoxia, the explanation of the protective effect of hibernation against radiation damage is yet to be resolved. To our knowledge, there are no unequivocal data which indicate that the tissues of animals are hypoxic either in hibernation or awaking from that state (Musacchia and Barr, 1969).

Jaroslow *et al.* (1976) counted microcolonies of ileal crypt cells in ground squirrels that had been irradiated during hibernation or at various times after arousal. He found that hibernating animals and those that were irradiated 1 hr after arousal had more surviving crypts than animals that were irradiated later. He reviewed the possible reasons for this protection, including hypoxia, and suggested that the animal in hibernation may accumulate a population of cells in one phase of the mitotic cycle which are more resistant to radiation damage. As was detailed in Chapter 10, we have presented evidence which indicates that there are two blocks in the mitotic cycle in hibernating dormice, one at G_2 and the other at M (Adelstein *et al.*, 1967). The mitotic index rose markedly within 2 hr after rectal temperature reached 37°C. In contrast, the mitotic index in the ground squirrels did not increase until more than 9 hr after arousal. Jaroslow *et al.*

(1976) suggested that this long delay in recovery indicated that cells had accumulated in G_1 during hibernation, for an accumulation of cells in S or G_2 would result in an earlier peak in the mitotic index. They present evidence that cells in certain phases of the mitotic cycle are more resistant to radiation damage. If resistant cells accumulate throughout a bout of hibernation, then the resistance to radiation should be correlated with the length of time the animal remained in uninterrupted hibernation. To our knowledge, this experiment has not been attempted.

On the other hand, a variety of naturally occurring factors have been reported to increase the resistance of an animal to radiation damage. For example, Ghys (1963) has maintained that cold-adapted rats are significantly more resistant than warm room controls and that this resistance remains for a week after the animals have been removed from the cold. Ghys suggested that acclimatization to a relatively mild physiological stress afforded protection from radiation, but the specific mechanisms that were involved were not identified.

REFERENCES

Adelstein, S. J., Lyman, C. P., and O'Brien, R. C. (1967). Cell proliferation kinetics in the tongue and intestinal epithelia of hibernating dormice (*Glis glis*). *In* "Mammalian Hibernation III" (K. C. Fisher, A. R. Dawe, C. P. Lyman, E. Schönbaum, and F. E. South, eds.), pp. 398–408. Oliver & Boyd, Edinburgh.

Barnes, E. M. (1970). Effect of hibernation on the intestinal flora. *Am. J. Clin. Nutr.* **23**, 1519–1524.

Barnes, E. M., and Burton, G. C. (1970). The effect of hibernation on the caecal flora of the thirteen-lined ground squirrel (*Citellus tridecemlineatus*). *J. Appl. Bacteriol.* **33**, 505–514.

Barr, R. E., and Musacchia, X. J. (1969). The effect of body temperature and postirradiation cold exposure on the radiation response of the hibernator *Citellus tridecemlineatus. Radiat. Res.* **38**, 437–448.

Blanchard, R., and Blatin, M. (1907). Immunité de la marmotte en hibernation à l'égard des maladies parasitaires. *Arch. Parasitol.* **11**, 361–378.

Chute, R. M. (1961). Infections of *Trichinella spiralis* in hibernating hamsters. *J. Parasitol.* **47**, 25–29.

Chute, R. M. (1964). Hibernation and parasitism: recent developments and some theoretical consideration. *Ann. Acad. Sci. Fenn., Ser. A4* **71**, 113–122.

Davis, S. D. (1969). Hibernation: intestinal protozoa populations in ground squirrels. *Exp. Parasitol.* **26**, 156–165.

Doull, J., and DuBois, K. P. (1953). Influence of hibernation on survival time and weight loss of X-irradiated ground squirrels. *Proc. Soc. Exp. Biol. Med.* **84**, 367–370.

Ghys, R. (1963). Radioprotection by acclimatization to cold. *Nature (London)* **198**, 603.

Giaja, J., and Andjus, R. K. (1949). Sur l'emploi de l'anesthésie hypoxique en physiologie opératoire. *C. R. Hebd. Seances Acad. Sci.* **229**, 1170–1172.

Hornsey, S. (1957). The effect of hypothermia on the radiosensitivity of mice to whole-body X-irradiation. *Proc. R. Soc. London, Ser. B* **147**, 547–549.

Jaroslow, B. N., and Serrell, B. A. (1972). Differential sensitivity to hibernation of early and late events in development of the immune response. *J. Exp. Zool.* **181**, 111-116.

Jaroslow, B. N., Smith, D. E., Williams, M., and Tyler, S. A. (1969). Survival of hibernating ground squirrels (*Citellus tridecemlineatus*) after single and fractionated doses of cobalt-60 gamma radiation. *Radiat. Res.* **38**, 379-388.

Jaroslow, B. N., Fry, R. J. M., Suhrbier, K. M., and Sallese, A. R. (1976). Radiosensitivity of ileum crypt cells in hibernating, arousing, and awake ground squirrels (*Citellus tridecemlineatus*) *Radiat. Res.* **66**, 566-575.

Kunkel, H. A., and Schubert, G. (1959). Effects of protective agents applied after irradiation. *Prog. Nucl. Energy, Ser. 6* **2**, 217-224.

Laufman, H. (1951). Profound accidental hypothermia. *J. Am. Med. Assoc.* **147**, 1201-1212.

Lucké, B., Berwick, L., and Nowell, P. (1953). The effect of temperature on the growth of virus-induced frog carcinoma. II. The temperature coefficient of growth *in vitro. J. Exp. Med.* **97**, 505-509.

Lyman, C. P., and Fawcett, D. W. (1954). The effect of hibernation on the growth of sarcoma in the hamster. *Cancer Res.* **14**, 25-28.

Musacchia, X. J., and Barr, R. E. (1969). Comparative aspects of radio-resistance with depressed metabolism. *In* "Depressed Metabolism" (X. J. Musacchia and J. F. Saunders, eds.), pp. 569-607. Am. Elsevier, New York.

Musacchia, X. J., Jellinek, M., and Cooper, T. (1963). Effects of X-irradiation during hibernation on tissue catecholamine contents. *Experientia* **19**, 418-419.

Osborn, G. K., and Kimeldorf, D. J. (1957). Some radiation responses of two species of bats exposed to warm and cold temperatures. *J. Exp. Zool.* **134**, 159-169.

Patterson, W. B., Lyman, C. P., and Patterson, H. R. (1957). Growth of human tumors in hibernating hamsters. *Proc. Soc. Exp. Biol. Med.* **96**, 94-97.

Prewitt, R., and Musacchia, X. J. (1969). Mechanisms of radio-protection by sympathomimetics. *Physiologist* **12**, 330.

Prewitt, R. L., and Musacchia, X. J. (1975). Radio-protection of arousing ground squirrels (*Citellus tridecemlineatus*) by endogenous catecholamines. *Experientia* **31**, 230-232.

Reisen, W. K., Kennedy, M. L., and Reisen, N. T. (1976). Winter ecology of ectoparasites collected from hibernating *Myotis velifer* (Allen) in southwestern Oklahoma (Chiroptera: Vesperitilionidae). *J. Parasitol.* **62**, 628-635.

Schmidt, J. P. (1967). Response of hibernating mammals to physical, parasitic and infectious agents. *In* "Mammalian Hibernation III" (K. C. Fisher, A. R. Dawe, C. P. Lyman, E. Schönbaum, and F. E. South, eds.), pp. 421-438. Oliver & Boyd, Edinburgh.

Sidky, Y. A., and Auerbach, R. (1968). Effect of hibernation on the hamster spleen immune reaction *in vitro. Proc. Soc. Exp. Biol. Med.* **129**, 122-127.

Sidky, Y. A., Daggett, L. R., and Auerbach, R. (1969). Brown fat: its possible role in immunosuppression during hibernation. *Proc. Soc. Exp. Biol. Med.* **132**, 760-763.

Sidky, Y. A., Hayward, J. S., and Ruth, R. F. (1972a). Seasonal variations of the immune response of ground squirrels kept at 22-24° C. *Can. J. Physiol. Pharmacol.* **50**, 203-206.

Sidky, Y. A., Hayward, J. S., and Ruth, R. F. (1972b). Immunosuppression *in vitro* by brown fat. *Immunol. Commun.* **1**, 579-595.

Simitch, T., and Petrovitch, Z. (1954). Ce qu'il advient avec les helminthes du *Citellus citellus* au cours du sommeil hibernal de ce rongeur. *Riv. Parassitol.* **15**, 655-662.

Smith, D. E. (1960). The effects of ionizing radiation in hibernation. *Bull. Mus. Comp. Zool.* **124**, 493-506.

Smith, F., and Grenan, M. M. (1951). Effect of hibernation upon survival time following whole-body irradiation in the marmot (*Marmota monax*). *Science* **113**, 686-688.

12

Hibernation and the Endocrines

INTRODUCTION

The role of endocrines in hibernation has been reviewed several times in recent years (see, e.g., Lyman and Chatfield, 1955; Popovic, 1960; Kayser, 1961; Hoffman, 1964b; Wimsatt, 1969; Hudson, 1973; Raths and Kulzer, 1976; Johansson, 1978; Hudson and Wang, 1979). There appears to be general agreement that polyglandular involution is characteristic prior to the onset of hibernation in seasonal hibernators (e.g. ground squirrels, woodchuck, hedgehog): the involution of the pituitary–gonad, thyroid, and adrenal axes being the most conspicuous. Generally endocrine functions are depressed during the first half of the hibernation season. This is followed by a reactivation of all functions sometime after the mid-hibernation season, and peak activities are reached at the time of or shortly after the spring emergence. Evidence supporting such a scheme has come largely from weight changes and histological observations of particular glands, and to a lesser extent, measurements of hormone levels in glands and/or blood. Kinetic studies on turnover rates of hormones at different phases of the annual hibernation cycle have been very scarce. It is also apparent from these reviews that our understanding of a direct cause-and-effect relationship between endocrine functions and hibernation has not yet been established.

Recent advances in the understanding of neuroendocrine function in nonhibernating mammals (see, e.g., Reichlin *et al.*, 1978) has obscured the division of classically defined nervous and endocrine systems. The role of the hypothalamus in linking environmental stimuli to the manifestation of endocrine functions has been firmly established. However, the various peptide hormones synthesized and secreted by neurosecretory neurons in the hypothalamus not only closely regulate secretions of the pituitary but serve as neurotransmitters (see, e.g., Snyder, 1980) as well as having direct systemic effects. In the reverse situation, biogenic amines have been recognized as neurotransmitters involved in body temperature regulation in both hibernating and nonhibernating species (see, e.g., Feldberg and Myers, 1964; Glass and Wang, 1978a,b), and also to be directly and indirectly involved in the control of secretion of hypothalamic releasing hormones and the pituitary hormones (Moore and Bloom, 1979). Since the regulation of body temperature requires the pituitary and its target glands (Gale, 1973) and the autonomic nervous system, the cyclic changes of body temperature regulation between the euthermic and hibernating states clearly require the interaction between the nervous and the endocrine systems. Our current understanding of hypothalamic neuroendocrine regulation of hibernation has been very much limited to only a few histological (Yurisova and Polenov, 1979) and immunohistological observations (Richoux and Dubois, 1976).

The study of endocrine functions in relation to hibernation is a difficult proposition. First, it takes a considerable amount of time to follow the annual changes of endocrine functions. Second, it requires a quantitative awareness of separating the depressed endocrine functions during hibernation due to depressed body temperatures versus those due to endogenous endocrine cycles independent of body temperature change. Third, the limitation of the techniques employed in investigations must be recognized: histological evidence cannot differentiate synthetic activity from storage and release, and measurement of blood level of a hormone cannot differentiate its rate of secretion from its rate of utilization. Turnover studies, although likely to provide the best information on endocrine activity, are difficult to perform. Blood sampling techniques are sometimes critical in obtaining ''true'' physiological levels of hormones since the stress of anesthesia or handling quickly change the hormone output in many glands, resulting in misleading ''control'' values. Fourth, extirpation of glands without replacement therapy could disturb the normal feedback mechanism of endocrine regulation, resulting in activation of normally absent activities; replacement therapy, on the other hand, involves problems with dosage and absorption, stress due to handling, and other unknown effects. Fifth, and possibly the most commonly overlooked point, is the recognition that in view of the polyphyletic origin of hibernation, not all hibernators are alike; each may have adapated to its particular ecological niche and may have evolved unique physiological adaptations. Thus, endocrine changes in a seasonal hibernator (e.g., marmot) may be quite different from those of a nonseasonal hibernator (e.g., golden hamster) due to differences in pattern of energy storage, and ground dwellers (e.g., ground squirrel) may be different from tree dwellers (e.g., dormouse) because of difference in sheltering from temperature fluctuations encountered during hibernation. Therefore, generalization of endocrine changes among all hibernators may not be always possible.

The descriptions of endocrine functions presented in this chapter have taken these precautions in mind. Emphasis will be made if certain glandular activities are maintained or even increased during hibernation relative to euthermia, since this implies active regulation. Attention will also be given if cyclic changes of endocrine function persist even if the animal is prevented from hibernating since it suggests endogenous cyclicity. Integration, interpretation, deduction, and speculation have been attempted wherever the evidence warrants; however, they reflect personal prejudice more than unified consensus.

HYPOTHALAMO-HYPOPHYSEAL INTERACTIONS IN HIBERNATION

Polenov and Yurisova (1975) studied the hypothalamo-hypophyseal function of two ground squirrels, *Citellus* (*Spermophilus*) *erythrogenys* Brandt and *C. undulatus* over the hibernation period using three criteria: (1) the quantity of neurosecretory material (NSM) in the neurosecretory cells of the nucleus supraopticus (NSO), nucleus paraventricularis (NPV), and nucleus preopticus (NPO), in the hypothalamo-hypophyseal tract in the region of the median eminence, and in the posterior pituitary; (2) the percentage of different types of the neurosecretory cells; and (3) the volume of the nuclei and nucleoli in the neurosecretory and the glial cells. In *C. erythrogenys,* which according to these authors typically hibernates over prolonged periods without an arousal, the cells of hypothalamic nuclei and the posterior pituitary are filled with NSM during December–January. The volume of the nuclei and nucleoli of the hypothalamic secretory nuclei is small, indicating decreased synthetic activity and decreased release of NSM. Small amounts of NSM are found in the neurosecretory fibers of the hypothalamic nuclei as well as in the hypothalamo–hypophyseal tract in the median eminence, indicating decreased transport of NSM in the hypothalamo–hypophyseal tract. This is also borne out by the low vascular activities (lack of blood cells) in the capillaries of the hypothalamo–hypophyseal neurosecretory system (HHNS) and those of the mantel plexus. These observations substantiate those observed earlier by Yurisova (1971) in the same species. In *C. undulatus,* which arouses periodically throughout its hibernation season, a qualitatively similar pattern to that of *C. erythrogenys* was observed, the differences being slightly larger cell nuclei and nucleoli, less accumulated NSM in the hypothalamic nuclei, the hypothalamo–hypophyseal tract, and the posterior pituitary, and the higher vascular activity (presence of red blood cells) in the capillaries. These quantitative differences in the activity of the HHNS between the two species could be due to the difference in frequency of arousal from hibernation. It is well known that urine formation occurs during arousal (Moy *et al.*, 1972) and that an increase in plasma ADH concentration can be detected shortly after the onset of arousal (Burlet *et al.*, 1973). It is, therefore, not surprising to find less accumulation of NSM in the posterior pituitary in *C. undulatus.*

In a more recent study, Yurisova and Polenov (1979) extended their observations in *C. erythrogenys* to cover the whole year with monthly samples. In addition to the changes of cell types and the amount of NSM in the HHNS they also measured the body weight, testicular weight, and the adrenal weight. During the hibernation season (between October and March), the decrease of synthetic activity and the inhibition of release of NSM from the HHNS and the accumulation of NSM in the posterior pituitary were similar to those observed previously (Polenov and Yurisova, 1975). However, while the body weight showed continuous decline due to the absence of arousal and feeding, both the testicular and adrenal weights showed steady increases as the hibernation season progressed. These latter observations and that of Mikhnevich and Kostyrev (1971) on the reactivation of thyroid gland during the hibernation season in the same species, suggest a reactivation of the HHNS sometime during the second half of the hibernation season. The reactivation of neurosecretory functions in the hypothalamic nuclei became histologically evident during late hibernation season in March, and by April, when the hibernation season terminated, both the testicular and adrenal weights had peaked and the animal was in reproductive readiness (Yurisova and Polenov, 1979). The activity of the HHNS was highest in April to May during which time reproduction was complete. Beginning in June and during summer and autumn months, a progressive decrease in HHNS activity was evident as the animal prepared for another hibernation season (Yurisova and Polenov, 1979).

Using similar histological criteria, namely, the amount of Gomori-positive granules and the volume of the cellular nuclei, Jasinski (1970) studied the HHNS of the mouse-eared bat, *Myotis myotis,* in the late hibernation season (March) and during summer (June) from animals freshly collected in the field. In the hibernating bats, the amounts of NSM in the hypothalamic nuclei, the hypothalamo–hypophyseal tract, and the posterior pituitary were greater than those observed in the respective areas in active animals. However, the volume of hypothalamic neurosecretory nuclei was comparable between the two groups. These observations suggest that during hibernation synthesis and release of NSM are reduced, but not entirely inhibited. The seemingly different conclusions of this study and those of Polenov and Yurisova (1975) and Yurisova and Polenov (1979) could be explained based on (1) the different patterns of periodic arousal from hibernation between *M. myotis* and *C. erythrogenys* and (2) the relatively late sampling in the hibernating season in *M. myotis* since by March, *C. erythrogenys* already showed signs of reactivation of the HHNS even though it is still in hibernation (Yurisova and Polenov, 1979).

In the garden dormouse *Eliomys quercinus* hibernating at 4°C (Legait *et al.,* 1970), the content of ADH in the posterior pituitary was 1.4 UI in December, just before hibernation and increased to 6.35 UI in January–February, during the middle of the hibernating season. In a dynamic study during April, the ADH was 4.19 UI shortly before arousal and 2.10 UI 1 hr after arousal. These observations suggest low activity of the hypothalamo–neurohypophyseal system during hibernation, but increased activity and the release of ADH during arousal. In a subsequent study in the same animal (Burlet *et al.,* 1973) using ^{131}Cs- and ^{51}Cr-tagged red blood cells to study the capillary exchange in the adenohypophysis, neurohypophysis, and thyroid, the vascular space was observed to have increased 25%, 75%, and 80%, respectively, 17 min after the initiation of arousal from hibernation. After 35 min, the increases were 75%, 225%, and 280%, respectively. The rapid increase in vascularization of the neurohypophysis during arousal was also associated with the increase of ADH concentration in the peripheral plasma, suggesting the quick release of this hormone from storage. Although these authors did not measure other hormone titers in the plasma, it is probable that similar increases of adenohypophyseal and thyroid hormones could also occur.

Electron microscopic studies on the contact zone between nervous tissue and perivascular space of the hypothalamo–hypophyseal portal plexus (Fleischhauer and Wittkowski, 1976) indicate a functional association between the activity of the hypothalamo–hypophyseal system and the amount of nerve endings making contact with the capillaries. In the hedgehog, *Erinaceus europaeus,* the percentage of nerve endings composing the surface of the neurohaemal contact zone was 7% during December while the animal was in hibernation as compared with 14% in September when the animal was active. The synaptoid contacts between the processes of the neuroglial and nerve cells in the infundibulum have also been implicated in the secretory functions of the hypothalamo–hypophyseal system. In the hedgehogs during hibernation, the number of synaptoid contacts decreased in parallel with the decrease of nerve endings in the neurohemal contact zone in the portal plexus. These observations suggest that during hibernation, the secretion of hypothalamic releasing and inhibiting hormones (e.g., LH-RH, GH-IH or somatostatin, CRH and TRH) is decreased. However, in view of the control of release of hypothalamic releasing factors by aminergic innervations in the neurosecretory cells (Fuxe *et al.,* 1974b), it is conceivable that release of hypothalamic hormones could be accomplished via neural stimulations without substantial changes in morphology of the synaptoid and neurohemal contact zones. In the hibernating woodchuck (Young *et al.*, 1979a), TRH increased markedly in the hypothalamus, septum, and striatum over prehibernation values; this could be related to the greatly reduced thyroid activity during hibernation in this species (Krupp *et al.*, 1977).

In summary, studies on the influence of hypothalamic neurosecretory functions in hibernation have been few in number and fragmentary in coverage. Most of the available studies have been employing morphological criteria in deciphering the functional state of the hypothalamo–hypophyseal system. Within the limits of such methodology, it can be generalized that prior to the onset of hibernation season, a general reduction in the synthesis, transport, and release of NSM is evident. This is followed by a marked reduction of transport and release during hibernation resulting in accumulation of NSM in the HHNS and posterior pituitary. In those species which arouse periodically, there are temporary increases of hormonal release from storage to facilitate certain physiological processes such as urine formation during arousal. The continual increase of testicular and adrenal weights throughout the hibernation season implies a spontaneous reactivation of the hypothalamo–hypophyseal system exerting its influence through the adenohypophysis to prepare the animal for reproductive readiness prior to permanent arousal in the spring. The area of neuroendocrine control and hibernation is in dire need of studies employing more quantitatively precise methodologies, such as radioimmunoassay and radioreceptor assay techniques, to measure levels and turnover rates of hormones in the HHNS, plasma, and target glands, and the behavior of receptor sites at different seasons such that the dynamics of neuroendocrine functions can be further assessed in relation to hibernation.

ROLE OF BRAIN MONOAMINES ON CONTROL OF ENDOCRINE FUNCTIONS AND THERMOREGULATION

In recent years, the role of brain monoamines in neuroendocrine control has been gaining emphasis. Extensive catecholaminergic and serotonergic innervations of the median eminence, the hypothalamus, and the limbic system have been established (Fuxe *et al.*, 1974a). The physiological roles of these innervations involve the inhibition or facilitation of secretion of hypothalamic releasing hormones, which, consequently, affect the pattern of hormonal secretion from the pituitary. For example, generally activation of the dopaminergic mechanism in the median eminence inhibits the secretion of LH-RH, whereas activation of the norepinephrine mechanism facilitates LH-RH secretion. The serotonergic neurons of the hypophysiotrophic region of the hypothalamus are involved in the control of CRH secretion (Popova *et al.*, 1972) and the secretion of LH, FSH, and prolactin from the anterior pituitary (Fuxe *et al.*, 1974b). Studies aiming to elucidate the roles of central monoamines in controlling neuroendocrine functions in the hibernators have been lacking. The few indirect studies (Popova *et al.*, 1974; Koryakina, 1976) that employed peripheral injections of adrenomimetics and the measurement of plasma concentrations of adrenal steroids are insufficient to assess the influence of brain monoamines on hypothalamo-pituitary–adrenal axis function due the presence of the blood–brain barrier for the adrenomimetics. More direct approaches such as central injection of monoamines with concurrent measurements of pituitary and plasma hormone levels must be utilized in order to obtain further evidence for regulation of neuroendocrine function in hibernation by the monoamines.

In addition to their involvement in the control of neuroendocrine functions, brain monoamines have also been implicated in the regulation of body temperature (Feldberg and Myers, 1964). Although a detailed review of this topic is beyond the realm of this chapter, suffice it to say that norepinephrine (NE) and serotonin (5-HT) have been demonstrated to be the chief central neurotransmitters mediating thermoregulatory responses. NE, when administered centrally, increases thermogenesis and body temperature in many species of ground squirrels (Beckman and Satinoff, 1972; Glass and Wang, 1978a,b), but decreases body temperature in the marmot (Jacob *et al.*, 1971). Injection of 5-HT results in an increase in heat loss and a decrease of body temperature in the ground squirrel (Glass and Wang, 1979a) and golden hamster (Hlavicka, 1976; see also Jansky and Novotoná, 1976). The functional involvement of these amines in hibernation has been gaining steady interests in recent years (see, e.g., Beckman, 1978; Jansky, 1978; Glass and Wang, 1979a,b). Of particular significance is the role of 5-HT in hibernation. It has been observed that the turnover rate of brain 5-HT increases some 14 times with the onset of hibernation and 24 times during hibernation in the golden hamster (Novotoná *et al.*, 1975). The concentration of 5-HT in the brain was found to increase prior to or during hibernation in the hedgehog (Uuspää, 1963a), bat (Shaskan, 1969), golden hamster (Novotoná *et al.*, 1975), and in the ground squirrel, *C. major* (Popova and Voitenko, 1974). However, decreased 5-HT concentration was found in *C. lateralis* during entry into and in deep hibernation (Spafford and Pengelley, 1971) and in the European hamster during hibernation (Canguilhem *et al.*, 1977). The general consensus is that 5-HT is involved in the entry into hibernation since the 5-HT level is lower during hibernation than it is prior to entry into hibernation, and 5-hydroxyindolacetic acid, a metabolite of 5-HT, is higher during hibernation (Novotoná *et al.*, 1975; Canguilhem *et al.*, 1977; Duncan and Tricklebank, 1978). Intraperitoneal injection of pCPA (parachlorophenylalanine) or lesioning of the medial raphe nucleus, both of which serve to reduce the endogenous synthesis of brain 5-HT, result in an inability of the animal to enter hibernation (Spafford and Pengelley, 1971). On the other hand, feeding of a tryptophan-rich diet facilitates the occurrence of hibernation in the golden hamster, presumably through an increase of substrates for 5-HT synthesis in the brain (Jansky, 1978). Very recently, 5-HT has been shown to differentially increase heat loss and suppress heat production in *S. richardsonii* when injected centrally during the hibernating season of the animal, resulting in a relatively large depression in body temperature (1.6°C) (Glass and Wang, 1979a). When the same injection was performed during the noninhibernating season of the animal, 5-HT first increased heat loss and was followed by a compensatory increase in heat production, resulting in a minimum decrease of body temperature (0.4°C). Such a differential response to 5-HT appears to be dependent on the endogenous state of the animal for hibernation, suggesting a seasonal reorganization of the 5-HT mediated thermoregulatory functions. This study seems also to provide a physiological explanation for the observed drastic increase of brain 5-HT concentration and turnover rates during the hibernating season. It is possible that 5-HT may exert a steady inhibition or suppression of heat production, thus allowing the entry into hibernation as well as the continuation of hibernation. The very high turnover rates observed during hibernation (Novotoná *et al.,* 1975) are in agreement with this interpretation. However, much more evidence is needed to support this speculation.

The role of brain NE in relation to hibernation has received modest attention. In the thirteen-lined ground squirrel, *C. tridecemlineatus,* brain NE content was similar between the active and hibernating states, but the turnover rate was essentially nil during hibernation, indicating a cessation of adrenergic activity during hibernation (Draskóczy and Lyman, 1967). Since brain NE turnover was nil even at euthermia some 10 hr prior to hibernation, it is probable that a cessation of brain adrenergic activity precedes the onset of entry into hibernation. In the European hedgehog, brain NE content was the same between June in an active animal and in those sampled in December and February after being in hibernation for 4 and 12 weeks, respectively (Sauerbier and Lemmer, 1977). The turnover rate of brain NE was also found, however, to be greatly reduced in these hibernating animals, similar to that reported in the thirteen-lined ground squirrel. Measurements of brain NE content in other ground squirrels (Twente *et al.,* 1970; Feist and Galster, 1974) and the European hamster (Canguilhem *et al.,* 1977) indicated that the level was similar between the hibernating and euthermic states during winter, although a lower brain NE level was found in the hibernating European hedgehog (Uuspää, 1963a) and a lower hypothalamic NE level was found in the hibernating woodchuck (Young *et al.,* 1979a). During arousal from hibernation, there seems to be an increase (Feist and Galster, 1974) or no change (Canguilhem *et al.,* 1977) of brain NE content; however, no turnover studies have been made. Intracerebroventricular injection of NE during arousal from hibernation in *S. richardsonii* resulted in a greater rate of heat production and a faster rate of arousal while injection of 5-HT had the opposite effects (Glass and Wang, 1979b). In a series of recent studies summarized by Beckman (1978), microinjections of NE, 5-HT, and ACh into the preoptic/anterior hypothalamic area triggered arousal in hibernating *C. lateralis,* whereas only ACh injection into the midbrain reticular formation triggered arousal. These observations suggest that aminergic and cholinergic neurons in these areas may be involved with mechanisms which trigger arousal from hibernation.

In summary, the role of central monoamines in hibernation may be manifested, on the one hand, via neuroendocrine influences in which the release of hypothalamic releasing hormones and, consequently, the secretory patterns of the pituitary may be modified. On the other hand, monoamines may act as neurotransmitters involved in thermoregulation. Thus, increased 5-HT activities and decreased NE activities may facilitate heat loss and suppress heat production for the entry into and maintenance of hibernation, and increased NE activities may facilitate the increase of heat production during arousal from hibernation.

PITUITARY GLAND

The role of the adenohypophysis in hibernation has not been well studied. With the exception of ACTH, which shows cyclic changes in the pituitary (details to follow) (Krass and Khabibov, 1975) and plasma TSH (thyroid stimulating hormone), which increases from 110–250 ng/ml in euthermic *S. richardsonii* to 200–375 ng/ml during hibernation (Demeneix and Henderson, 1978c); no other pituitary hormones have been measured in relation to the annual cycle of hibernation. Most of the available studies on seasonal changes of pituitary function involve histological examinations of the various cell types, the change of glandular weight, and the changes of the target glands' histology and their hormones as an inference to secretory patterns of the pituitary.

In the thirteen-lined ground squirrel (Hoffman and Zarrow, 1958a), the pituitary gland showed a definite annual weight cycle: a maximum in April–May after emergence from hibernation, a minimum in December–January in early hibernation, and a rapid increase between January and March in the latter half of the hibernation cycle. Basophiles, the source of TSH and gonadotropin, showed maximum cytological signs of activity between February and July, and a progressive decrease in activity through the summer and fall to reach a minimum in October. The activity increased after December, and reached its peak in February. This was a particularly interesting study since, between April and August, all determinations were from animals freshly captured in the field, and between September and March from animals kept at 22°–25°C in the laboratory. Although the authors did not specify the frequency of hibernation in these animals kept at room temperature, it is significant to note that in another group of animals kept at 2°C during the winter to facilitate hibernation, a similar high activity of the basophiles was also observed in March. It is, therefore, evident that, based on the weight change and the apparent activity of the basophiles, there exists an annual cycle in pituitary function. This cycle is apparently independent of the occurrence of hibernation and the consequences of depressed body temperature. Similar patterns of changes in pituitary function have also been observed in other hibernators, for instance in the European hamster, *Cricetus cricetus* (Petrovic and Kayser, 1957), the red-cheeked suslik, *C. erythrogenys* (Krass and Khabibov, 1975), and in the bat, *Myotis myotis* (Herlant, 1956).

With respect to using the function of the target glands and their hormones as reflections of the pituitary status, the hedgehogs do not exhibit the alarm reaction by increasing their plasma ACTH when exposed to cold in the fall. It was interpreted that a blockage of the pituitary–adrenal axis is essential in preparation for hibernation (Hoo-Paris, 1971). The level of plasma ACTH during hibernation is low but increases significantly during periodical arousal from hibernation, indicating that the pituitary–adrenal axis can be reactivated during the hibernating season. In *C. erythrogenys* (Krass and Khabibov, 1975), pituitary ACTH content showed distinct seasonal fluctuations with the highest values found in the spring (April–June) and the lowest in early fall (September). During the hibernation season, an increase in ACTH was found between October and November, followed by a decrease in December. Between December and February the pituitary ACTH content was low and relatively constant. This was followed by another increase in March prior to termination of the hibernation season. The weight of the pituitary gland, on the other hand, was relatively constant throughout the year, indicating that the fluctuation of pituitary ACTH is independent of the change in weight. The blood level of 11-hydroxycorticosteroids showed a similar seasonal change to that of the pituitary ACTH, indicating the functional relationship of the pituitary–adrenal axis in this species.

The familiar pattern of thyroidal and gonadal involution during prehibernation and early hibernation followed by spontaneous recrudescence of their activities during mid- to late hibernation is another example of target gland functions reflecting pituitary secretory patterns. The details of these changes will be discussed later under ''Thyroid'' and ''Gonads.'' For the present discussion, it is sufficient to indicate that both are further examples of the cyclic nature of the pituitary functions in the hibernator.

Changes of neurohypophyseal functions in conjunction with hibernation have been discussed under ''Hypothalamo-Hypophyseal Interactions in Hibernation'' and therefore will not be elaborated here.

THYROID

There are numerous studies on the role of the thyroid gland in hibernation, but the results have been conflicting. Among the many seasonal hibernators, such as the woodchuck (Krupp *et al.,* 1977), ground squirrels (Hoffman and Zarrow,

1958b; Hudson and Wang, 1969; Bauman and Anderson, 1970), and hedgehog (Pinatel *et al.*, 1970), histological observations indicate that the gland is most active during the breeding season in spring, relatively inactive during summer, involuted during the late fall and early winter, and quiescent during the early part of hibernation; it resumes secretory activity sometime prior to the termination of the hibernation season. Measurements of plasma thyroxine (T_4) and triiodothyronine (T_3) levels in ground squirrels and woodchucks at different times of the year indicate a high level during the spring, decreasing levels during the summer and fall, and an increasing and high level during hibernation (Wenberg and Holland, 1973a; Demeneix and Henderson, 1978a; Young *et al.*, 1979b). The quantification of total versus free forms of plasma T_4 in *S. richardsonii* (Demeneix and Henderson, 1978b), and T_4 and T_3 in the woodchuck (Young *et al.*, 1979b) indicates that although both total T_4 and T_3 are high during hibernation, the amounts of free T_4 and T_3 are extremely low when measured at the hibernating body temperature of 6° or 7°C. In *S. richardsonii*, the free T_4 was 0.025–0.032% of total (Demeneix and Henderson, 1978b), and in the woodchuck, the free T_4 and T_3 were 0.005 and 0.008%, respectively of their total plasma levels (Young *et al.*, 1979b). Part of the reason for these very low free/total hormone ratios could be attributed to the effects of temperature alone, since low temperature facilitates while high temperature inhibits binding of free hormones to their plasma protein carriers. Thus, when the free/total hormone ratio was measured at 37°C, the same hibernating samples from these animals showed a 2.5–4 times increase in values resulting in greater amounts of free hormones present at this temperature (Demeneix and Henderson, 1978b; Young *et al.*, 1979b). This temperature effect on binding of free hormones could be of physiological significance. During hibernation, free hormones can be reduced to decrease their metabolic effects. During arousal from hibernation, as the body temperature gradually increases, increasing amounts of free hormones could be released from the large plasmic pools of T_4 and T_3 without involving changes in the secretory activities of the thyroid. Therefore, although the thyroid is histologically quiescent during the early part of the hibernating season, the animal could conceivably rely on the stored hormones in plasma for physiological needs prior to reactivation of thyroid function (Young *et al.*, 1979b).

In the thirteen-lined ground squirrel, the rate of release of organic iodine (thyroid hormones) from the thyroid was essentially nil between August and December even though the rate of 24 hr uptake from inorganic iodide was relatively high (27%; Hulbert and Hudson, 1976). Similarly, a high uptake rate of 20–25% was also found in the euthermic woodchuck during the hibernating season, as compared with 5–15% during the nonhibernating season (Wenberg and Holland, 1973a). These functional studies lend support to earlier histological observations in the thirteen-lined ground squirrel (Hoffman, 1964a) where the colloidal content was found to increase during the fall, whereas cell height was found to decrease and remained at a low level throughout the fall and winter (indicating decreased secretory activity). Based on the dynamics of radioiodide turnover in the thyroid, a very low thyroid releasing rate occurs prior to and during the hibernation season in the thirteen-lined ground squirrel (Hulbert and Hudson, 1976). This could be due to the lack of TSH from the pituitary rather than the nonresponsiveness of thyroid to this trophic hormone. In the round-tailed ground squirrel (Hudson and Wang, 1969), the thyroid releasing rate was nil during the hibernating season, but its basal metabolic rate could be elevated significantly by exogenous TSH or T_4. This indicates that both the thyroid and the target tissues of thyroid hormones remain responsive during the hibernating season (Hudson, 1968).

A low thyroid secretory activity during hibernation may have significance in membrane function. Recent studies (Hulbert *et al.*, 1976; Hulbert, 1978) have suggested that low thyroxine level may increase the membrane fluidity at low temperature through an increased ratio of unsaturated to saturated fatty acids of the membrane lipids. This would enhance the function of membrane-bound enzymes at low body temperature for hibernation. To date, most evidence has been derived from studies in white rats (Steffen and Platner, 1976; Hulbert *et al.*, 1976) on changes of fatty acid saturation ratio in mitochondrial and microsomal membranes and the change of phase transition temperatures. The evidence in hibernators has been only correlative (Hudson and Deavers, 1976).

Although thyroid activity is generally considered to be low in the seasonal hibernators, two notable exceptions seem to be evident. First, in the Richardson's ground squirrel (Demeneix and Henderson, 1978a,b), although the yearly profile of total serum T_4 and T_3 were similar to those reported in the woodchuck (Young *et al.*, 1979b), its free T_4 to total T_4 ratio during the hibernating season was 5–6 times greater than that found in the woodchuck. Substantial fluctuations of serum T_4 and T_3 levels were also observed during a single bout of hibernation with T_4 decreasing progressively with hibernation and T_3 decreasing sharply during arousal (Demeneix and Henderson, 1978b). The thyroid degradation rate measured during hibernation was identical to that found when the animal was euthermic in the nonhibernating season, in strong contrast to the 1000-fold reduction of this rate found in the hibernating thirteen-lined ground squirrel (Bauman and Anderson, 1970). These observations suggest an active thyroid during hibernation in *S. richardsonii*, in contrast to what has been found in the many other ground squirrels. Second, in the eastern chipmunk, *Tamias striatus* (Hudson, 1980), the rates of ^{125}I release from the thyroid gland were nearly identical between active, euthermic animals and those in hibernation, indicating that an active thyroid is characteristic even during hibernation. In view of both these late discoveries employing more functionally related methodologies, it might be worthwhile to re-examine the other hibernators for the role of the thyroid in hibernation.

With respect to the nonseasonal hibernators, such as the golden hamster, functional indices (radioiodide uptake and conversion ratio, disappearance rate of radioactive L-thyroxine, and others) indicate that the thyroid is active after exposure to 5°C for 60 days, a period typically required for this species to prepare for hibernation (Tashima, 1965). Although the functional indices decreased during hibernation, these were thought to be due to the influence of low body temperature per se. In another study on the golden hamster (Bauman *et al.*, 1969), the thyroxine release rate in the hibernating animal was essentially the same as that of the euthermic animal, indicating an active thyroid during hibernation. In the European hamster, hibernation was abolished after functional thyroidectomy with ^{131}I, but injection of thyroxine restored normal frequency of hibernation (Canguilhem, 1970), indicating an active thyroid is required for hibernation. In the garden dormouse, which is a seasonal hibernator in nature but can be induced to hibernate at any time of the year in the laboratory by food deprivation and cold exposure (Lachiver, 1964), hibernation could occur in the presence of high thyroid activity. The notable difference in having an active thyroid during hibernation in these animals in contrast to having an inactive thyroid in many ground squirrels and woodchucks (Hulbert and Hudson, 1976; Young *et al.*, 1979b) may be related to the ecology of the animals (Hudson and Wang, 1979). In the golden hamster, which hoards food before hibernation and feeds between hibernation bouts, thyroid hormones may be required for the assimilation of energy. In the garden dormouse, an arboreal species which hibernates in relatively unprotected places unlike the protected underground burrows of the ground squirrels and woodchucks, the animal may need to elevate its thermogenesis periodically to prevent the body temperature from decreasing below a critical level during hibernation (Hudson and Wang, 1979).

PARAFOLLICULAR (C) CELLS OF THYROID AND PARATHYROIDS

Calcitonin is secreted by the parafollicular or "C" cells of the thyroid in response to a high plasma calcium level. Its action is to decrease calcium release from bone and to increase calcium and phosphorus excretion, thus decreasing the plasma calcium level. The parathyroid hormone, which is secreted in response to a low plasma calcium level, has the opposite effects; it increases calcium release from bone, decreases calcium and phosphorus excretion, thus elevating the plasma calcium level. The balancing action of these two hormones regulates blood calcium level within narrow limits. The role of these hormones during hibernation can, therefore, best be assessed by the changes of blood and tissue calcium levels and bone histology.

Hypercalcemia during hibernation has been observed in the hedgehog (Johansson and Senturia, 1972), the European hamster (Raths, 1962, cited in

Ferren *et al.*, 1971), and the golden hamster (Ferren *et al.*, 1971). Increases in tissue calcium have been found in the heart and skeletal muscle of the hibernating golden hamster and in the heart (Aloia and Pengelley, 1979) and liver (Behrisch, 1978) of the hibernating ground squirrel. Osteoporosis, or the loss of bone material, has been observed during hibernation in the bat, *M. lucifugus* (Whalen *et al.*, 1971), in the dental bone of the arctic ground squirrel (Mayer and Bernick, 1963), but not in the limb bone of the thirteen-lined ground squirrel (Zimmerman *et al.*, 1976). Histological examination of the parathyroids from natural populations of the bats *M. lucifugus* and *Pipistrellus pipistrellus* (Nunez *et al.*, 1972) indicated increased synthetic and secretory activities between prehibernation and mid-hibernation, followed by a decrease of activities toward late hibernation. Similar increases in parathyroid activities during hibernation have also been observed in the hedgehog and the European hamster (Kayser, 1961). Taken together, these observations suggest increased secretion of parathyroid hormones during hibernation in comparison to the nonhibernating season. In view of the fact that dietary calcium is absent in animals that do not feed between hibernation bouts, calcium must be supplied from bone deposits to replenish excretory losses. It is perhaps not surprising, therefore, to find an active parathyroid function during hibernation.

The parafollicular cells, which are found to be more abundant among hibernating species such as bats (Nunez *et al.*, 1967), garden dormouse (Gabe and Martoja, 1969), and hedgehog (Pearse and Welsch, 1968) than in the nonhibernators, show maximum activity during prehibernation, but a progressive decrease in activity throughout hibernation. The activity may return to prehibernation level or even greater at the end of hibernation. The low activity of the follicular cells coupled with the higher activity of the parathyroids during hibernation may explain the observed hypercalcemia and osteoporosis in bats and other species. A recent study (Krook *et al.*, 1977) in which calcitonin was given to *M. lucifugus* during hibernation showed that osteoporosis could be reversed, substantiating the fact that reduced activity of the parafollicular cells is the cause for osteoporosis during hibernation. Since osteoporosis in bats is also reversed upon terminal arousal, it is possible that a burst of parafollicular activity with increased release of calcitonin could occur at a time when the parathyroids are at minimum activity (Nunez *et al.*, 1972). In the thirteen-lined ground squirrel (Kenny and Musacchia, 1977), thyroid calcitonin content was significantly lower in November and December in comparison to that found in September and February, regardless of whether the animals were hibernating or euthermic. This seems to suggest a seasonal cycle of parafollicular cell activity, which is independent of body temperature changes.

The significance of a high calcium level in blood and certain tissues during hibernation is presently conjectural. The increased calcium levels in the heart and skeletal muscle in the golden hamster (Ferren *et al.*, 1971) have been interpreted as related to the maintenance of contractility and irritability in these tissues during hibernation. The high calcium level in liver and the compensatory decreases of acid phospholipids and diphosphatidylglycerol may lead to greater membrane fluidity at low temperature for functioning during hibernation (Aloia and Pengelley, 1979). Measurements of calcium-stimulated respiration in liver mitochondria from active and hibernating *S. columbianus* (Pehowich and Wang, 1981) indicated a critical temperature for lipid phase transition near 20°C in the active state, but no phase transition in the hibernating state, suggesting the maintenance of membrane fluidity at low body temperature during hibernation. Since in the white rat calcium-stimulated respiration can be observed at lower temperature (down to −10°C) than ADP-stimulated respiration (absent between 0°-4°C; Chance *et al.*, 1979) and in *S. columbianus* the rate of calcium-stimulated respiration is greater than ADP-stimulated respiration at 5°C, it is possible that calcium metabolism could be functionally involved with energy metabolism during hibernation.

ENDOCRINE PANCREAS

The role of the endocrine pancreas in hibernation has not been well studied. Earlier reports involved primarily histological observations of the ratio of A cells, which secrete glucagon, to B cells, which secrete insulin, as indications of endocrine activity during hibernation. The results have been conflicting. The ratio of A to B cells during hibernation has been found to increase or not change in bats (Hinkley and Burton, 1970), increase in the hedgehog and European hamster (Kayser, 1961), and not change in the golden hamster and European ground squirrel (Kayser, 1961). The histological evidence of an activation of B cells during the fall has been suggested as related to increased fat and glycogen storage in preparation for hibernation (Raths and Kulzer, 1976).

The blood glucose level during hibernation can also vary between species. In animals that do not feed between hibernation bouts (e.g., ground squirrel, woodchuck, hedgehog, dormouse), there is generally a reduction of blood glucose level during hibernation; in those that do feed (e.g., golden hamster), the blood glucose level is relatively constant between euthermia and hibernation (Kayser, 1961). However, exceptions are common, for instance, in the thirteen-lined ground squirrel, blood glucose is constant regardless of hibernating status, whereas that of the garden dormouse decreases during hibernation (Agid *et al.*, 1978); in the big brown bat, blood glucose level increases 3.5 times during hibernation (Hinkley and Burton, 1970).

Measurements by radioimmunoassay of plasma insulin in the hedgehog (Laurila and Suomalainen, 1974) indicate reduced levels during fall and winter compared with spring and summer; however, during periodical arousal and euthermia, insulin levels could be even higher than those found during summer. Pancreatic insulin was highest in November, shortly after the beginning of hibernation season in the hedgehog. The low plasma level of insulin coupled with the high pancreatic insulin content seems to suggest an increase in storage and a decrease of release at this time. In another study on the hedgehog (Johansson and Senturia, 1972), immunoreactive insulin could not be detected when the animal was in hibernation even though high insulin content was found in the pancreas. Intravenous challenge with glucose during hibernation resulted in hyperglycemia with limited success in detecting plasma immunoreactive insulin. These results also suggest reduced release of insulin during hibernation. Very recently, Hoo-Paris and Sutter (1980) have found that exogenous insulin exerted no hypoglycemic effect and insulin antiserum had no hyperglycemic effect when the hedgehog was in hibernation. However, the same treatments elicited hypoglycemia and hyperglycemia, respectively, when given during arousal from hibernation as body temperature exceeded 20° and 15°C, respectively. The lack of response to insulin during hibernation could be due to low capacity for binding to membrane receptors at depressed body temperature and/or nonresponsiveness of the tissue; the lack of response to insulin antiserum during hibernation suggests the endogenous plasma insulin level could be very low to nil under this state. In the thirteen-lined ground squirrel, plasma immunoreactive insulin decreased during hibernation, whereas that of the garden dormouse remained unchanged (Agid *et al.*, 1978). The functional significance of these changes remains to be elucidated. In the ground squirrel, *C. suslicus* (Daudova and Soliternova, 1972), the stimulatory effect of glucose absorption by insulin in the adipose tissue was essentially the same between active and hibernating animals, suggesting no change in tissue sensitivity to insulin during hibernation.

The information on changes of glucagon during hibernation is scarce. The A cells are inactive during hibernation and glucagon is probably stored during hibernation (Raths and Kulzer, 1976). In view of the similar stimulatory effects on lipolysis and glycogenolysis by the catecholamines, which are amply present during arousal from hibernation, it is perhaps not surprising to find diminished glucagon secretion during hibernation.

PINEAL

The role of the pineal in hibernation is presently unknown. But since gonadal involution typically precedes hibernation and gonadal recrudescence precedes spring emergence, the pineal gland could be involved in regulation of the pituitary-gonad axis in the annual hibernation cycle. In highly photosensitive hibernators, such as the golden hamster, exposure to 2L:22D photoperiod at 8°C resulted in atrophy of the gonads and accessory sexual glands (Hoffman and

Reiter, 1965), but pinealectomy resulted in maintenance of sexual activity (Smit-Vis, 1972). Melatonin, a pineal hormone, exerts an antigonadotrophic action in the golden hamster if daily injections were made during the late hours of the light phase, but had no effect if administered at any other time of day (Tamarkin *et al.*, 1976). However, if melatonin was given via subcutaneous implants of beeswax pellet or silastic capsule, so that it was continuously available to the animal, it produced a "counter anti-gonadotropic effect" (Reiter *et al.*, 1974), i.e., the darkness-induced gonadal atrophy did not occur even though the pineal was intact. It, therefore, appears that exogenous melatonin has dual effects depending on the route of administration. Although melatonin is synthesized predominantly in the pineal, the retina and the Harderian gland also produce melatonin (Cardinali and Wurtman, 1972). Pinealectomy does not entirely eliminate plasma melatonin (Ozaki and Lynch, 1976), and it is not known if compensatory hyperactivity would result in melatonin synthesis in nonpineal sources after pinealectomy.

With this potentially complicated picture in mind, experimental manipulations of pineal functions in relation to hibernation must be viewed with caution. In the male Uinta ground squirrel, *S. armatus* (Ellis and Balph, 1976), pineal weight was relatively constant between March and August, the breeding and early estivation/hibernation season for this species. The testicular weight of adults showed continued decline during the same period. The activity of the enzyme, hydroxyindole–*O*-methyltransferase (HIOMT), which converts *N*-acetylserotonin to melatonin in the pineal, showed a marked increase during postbreeding in the adults and during prehibernation in the juveniles, at a time when testicular recrudescence is about to commence. Although the authors did not elaborate on the significance of this increase of HIOMT activity, it is possible that this could mean an increased melatonin supply for the hibernation season. If one follows the change of testicular weight as an index for gonadal development, the continuously increasing testicular weight observed in the hibernating ground squirrels (Yurisova and Polenov, 1979) indicates continued gonadal development throughout hibernation. Since under natural conditions, the constant darkness of the burrow could possibly induce gonadal atrophy via the pineal, to counter such an effect, a "functional pinealectomy" during hibernation may be essential to allow full gonadal development in preparation for spring emergence. To this end, the increased HIOMT activity during prehibernation and the possible, continuously available endogenous melatonin during hibernation could exert the counter anti-gonadotropic effect described above. The observation of an inactive pineal during hibernation in the ground squirrel, *C. suslica* (Popova *et al.*, 1975), appears to be consistent with this speculation.

The increase of frequency and duration of hibernation in *C. lateralis* after daily subcutaneous injections of melatonin during the 10-day experimental

period (Palmer and Riedesel, 1976) suggests a functional involvement of melatonin in hibernation. The effect is short-lived since replacing melatonin injection
with saline nullified the increase in the same individual. *Peromyscus leucopus,*
whose frequency of daily torpor can be increased by exposure to a short photoperiod (9L:15D), showed increased frequency of torpor under long photoperiod
(16L:8D) after receiving for 10 days subcutaneous implants of beeswax pellets
containing melatonin (Lynch *et al.,* 1978). In both studies, the mechanism of
action of melatonin-induced torpor is unknown. In view of the short duration of
the experiments and the possible dual actions of exogenous melatonin, depending
on its route of administration, it is difficult to decipher whether the observed
effects are mediated via, or independent of, the pineal–gonad axis.

GONADS

In seasonal hibernators such as ground squirrels, marmots, and the hedgehog,
reproductive activity is also seasonal and is either monoestrous or polyestrous
(Wimsatt, 1969). Involution of the gonads occurs in summer and the animals
enter hibernation with atrophied gonads (Kayser, 1961; Wimsatt, 1969). By
spring emergence, however, the gonads are fully active, ready for breeding. The
reactivation of gonadal activities, therefore, must have taken place during the
hibernation season even though the animals spend much of their time under
depressed body temperature and in total darkness in the burrow. In the woodchuck, *Marmota monax* (Wenberg and Holland, 1973b), 24-hr urinary excretion
of estrogen and 17-ketosteroids begins to increase after January, long before
spring emergence, indicating onset of gonadal recrudescence midway through
the hibernation season. Numerous reports on changes of testicular weight in the
ground squirrels (see, e.g., Yurisova and Polenov, 1979), the European hamster
(Kayser, 1961), the hedgehog (Johansson and Senturia, 1972), and the garden
dormouse (Dussart and Richoux, 1973) indicate there is continuous growth during hibernation in spite of the low body temperature. Histometric studies in the
golden hamster (Smit-Vis and Akkerman-Bellaart, 1967) and the European
hamster (Reznik-Schuller and Reznik, 1973), however, have revealed that the
increase in testicular weight was probably due only to cell proliferation since cell
differentiation from spermatocytes and spermatids to spermatozoa was blocked
in animals hibernating at $2°-11°C$. Since the differentiation to spermatozoa requires FSH from the pituitary, these studies imply a suppression of secretion of
gonadotropins at the low body temperature of hibernation. In the garden dormouse (Dussart and Richoux, 1973), it has been suggested that secretion of
gonadotropins could take place during periodical arousal. Since the frequency of
periodical arousal typically increases toward the end of the hibernation season,

the increased secretion of gonadotropins during these more frequent euthermic episodes could facilitate the differentiation and maturation of the gamates prior to spring emergence.

In bats that hibernate (mainly from the families Rhinolophidae and Vespertilionidae), breeding commences in the fall. In *Myotis, Eptesicus,* and *Pipistrellus,* copulation occurs prior to hibernation but fertilization is delayed until spring emergence. Sperm is retained in the uterus and remains viable for the entire hibernation season (Wimsatt, 1969). Delayed maturation and ovulation of oocytes is probably due to the suppression of gonadotropin secretion from the pituitary as well as the decreased tissue responsiveness to the hormones due to low body temperature. Injection of FSH in torpid bats did not cause ovulation in the cold, but bringing the injected bat to room temperature caused ovulation (Wimsatt, 1960). The most striking example among the hibernating bats is found in *Miniopterus* (Wimsatt, 1969). In this animal, copulation in fall is followed immediately by ovulation, fertilization, and embryogenesis, and the female enters hibernation in a pregnant condition. However, implantation of the blastocyst is delayed until spring emergence. Therefore, embryogenesis is essentially arrested during hibernation.

In male members of the above genera, there appears to be some dissociation of the functional morphology between the testis and the accessory glands during hibernation; the testis is involuted whereas the accessory glands are hypertrophied (Wimsatt, 1960). Radioimmunoassay of plasma testosterone levels in *M. lucifugus* (Gustafson and Shemesh, 1976) indicated a sharp surge of testosterone just prior to hibernation; this increase may be required to stimulate the accessory glands fully. The appearance of hypertrophied accessory glands during hibernation may be due to their retarded involution by low body temperature, or due to sufficient residual androgens available from nontesticular sources, such as the brown fat, to maintain hypertrophy during hibernation (Wimsatt, 1960).

Questions often asked concerning reproductive activity and hibernation are: (1) whether the involution of gonads is a prerequisite for hibernation, and (2) whether the reactivation of gonadal functions is the cause for spring emergence? In the golden hamster, kept at 8°C under 2L:22D photoperiod, hibernation occurred in animals with active testes capable of producing spermatozoa (Smit-Vis and Smith, 1970), indicating that involution of the testis is not a prerequisite for hibernation in this species. However, gonadal atrophy does facilitate hibernation in this species under the same experimental conditions (Smit-Vis, 1972). Although castration has been found to increase the duration of individual hibernation bouts in the late hibernation season, hibernation cannot be induced by castration during the nonhibernation season of the year. In fact, the circannual periodicity for hibernation is essentially the same in castrated and intact male and female *C. lateralis* (Pengelley and Asmundson, 1969). In ground squirrels and hamsters that are prevented from hibernating by high ambient temperature (20°–

27°C) involution of gonads occurs in summer and fall, recrudescence sometime after mid-winter, and full reproductive readiness by spring (Kayser, 1961). It, therefore, appears that the seasonal gonadal cycle is independent of the occurrence of hibernation and the termination of the hibernation cycle is not due to the reactivated reproductive activity.

ADRENAL GLAND

Winter involution of the adrenal gland seems to be characteristic of most hibernators (Kayser, 1961). The weight of the adrenal was lowest in winter in the red-cheeked suslik (Khabibov and Krass, 1974) and in the hedgehog (Johansson, 1978). However, Suomalainen (1960) observed an increase in relative weight of the adrenal and the thickness of the cortex in the hedgehog during hibernation. He interpreted these increases, substantiated by histological observations, as due to the adaptation syndrome of Selye, at a time when the animal is facing environmental stressors such as cold and fasting. Popovic (1960) reports that hibernation can not occur in adrenalectomized ground squirrels and European hamsters but a cortical graft in the anterior chamber of the eye or injection of cortical hormones (e.g., deoxycorticosterone) restores the ability to hibernate. On the other hand, Kayser (1957) found that a small percentage of European hamsters hibernated in spite of the complete absence of adrenal tissue.

Cortex

Two types of steroid hormones are secreted by the adrenal cortex. Aldosterone, which increases renal retention of sodium, is secreted from the zona glomerulosa in the outer cortex. Its secretion is regulated by the renin-angiotensin system. The glucocorticoids, which include cortisol, corticosterone, cortisone, and other 11-oxygenated steroids, are involved mainly in carbohydrate metabolism. They are secreted from the inner zona fasciculata-reticularis and are regulated by ACTH.

During hibernation, increased width of the zona glomerulosa and the increased width and size of cell nuclei in the zona glomerulosa have been observed in the golden hamster (Raths and Kulzer, 1976). The concentration of aldosterone in the adrenals, and the renal excretion of aldosterone also increased with the increasing duration of hibernation in the European hamster (Bloch and Canguilhem, 1966). In the marmot, *M. flaviventris* (Kastner *et al.,* 1978), plasma aldosterone concentrations increased from one-half of the euthermic level on day 1 to equal the euthermic level on day 9 in a hibernation bout. Plasma renin activity has been observed to be normal in the thirteen-lined ground squirrel (Edmonson, 1976), but increased in the marmot (Kastner *et al.,* 1978) during hibernation. In the red-cheeked suslik, *in vitro* secretion of aldosterone as stimulated by bovine angiotensin II was twice as great in adrenals from hibernating animals as compared with those from active animals (Kolpakov and Samsonenko, 1970), indicating greater tissue responsiveness during hibernation. Taken together, these studies indicate that the secretion and utilization of aldosterone are maintained or even increased during hibernation. In view of the lack of feeding and the lack of salt intake during the hibernation season in these seasonal hibernators, and the disturbance of Na/K distribution across the cell membrane after prolonged hibernation, the normal or increased aldosterone secretion during hibernation is consistent with the physiological needs of electrolyte conservation.

Measurements of glucocorticoids (cortisol and corticosterone) in the hibernating garden dormouse and hedgehog (Boulouard, 1972) indicate that the plasma concentrations of these hormones are essentially constant, regardless of whether the animals are hibernating or euthermic following periodic arousal. Adrenals sampled during hibernation did not secrete steroids when incubated at 8°C with or without ACTH present in the medium. Incubation at 37°C, however, yielded secretions which were one-fourth of those found in aroused animals; addition of ACTH eradicated this difference (Boulouard, 1972). These observations indicate that, in the garden dormouse and the hedgehog, the secretion of glucocorticoids is diminished during hibernation, and the reactivation of secretion requires both a high body temperature and ACTH. In the thirteen-lined ground squirrel (Huibregtse *et al.*, 1971), adrenals sampled during hibernation did not secrete corticosterone, the major glucocorticoid in this species, even when incubated at 37°C. ACTH did not stimulate secretion in adrenals from either hibernating or aroused animals, but increased the secretion some 30 times in adrenals from active, nonhibernating animals. Thus, in this species, tissue responsiveness to ACTH is lost during the hibernating season. In the red-cheeked suslik (Krass and Khabibov, 1975), the range of plasma concentrations of 11-hydroxycorticosteroids during hibernation was similar to that found in the nonhibernating season; the monthly fluctuations of plasma steroid concentration were in parallel with the ACTH content in the pituitary, indicating a functional correlation. In the hibernating marmot (Kastner *et al.*, 1978), plasma cortisol ranged between 6 and 14 ng/ml, much less than the 54 ng/ml found in the nonhibernating season. Since blood samples were taken via cardiac puncture, the high levels of cortisol in the nonhibernating animals could be due to the stress of handling and anesthesia. In another recent study on the woodchuck, *M. monax* (Florant and Weitzman, 1980), using a chronically implanted catheter for blood sampling without disturbing the animal, the plasma cortisol level was typically between 4 and 12 ng/ml in the nonhibernating animals in the spring and fall. These values are similar to those found during hibernation in the marmot. Since stress-induced release of ACTH and cortisol would be much less during hibernation as evidenced by the studies on adrenal secretion *in vitro* described above, it is possible that the cortisol levels reported during hibernation in the marmot are normal values. In *M. monax,* monthly measurements of 24-hr urinary output of 17-hydroxysteroids (Wenberg and Holland, 1973b), which reflect the adrenocortical function, indicated decreased outputs between August and December, minimum excretion in January and February, but sharp increases in March and April. These observations indicate a reactivation and increase of the adrenocortical function prior to spring emergence.

To summarize, it appears that plasma glucocorticoid levels are constant between the hibernating and the active states. The constant steroid level during hibernation could be due to inhibition of secretion by low body temperature and diminished tissue utilization. The increased secretion after arousal is countered by an increased utilization, most probably due to the stimulatory effect of glucocorticoids on the enzymes involved in gluconeogenesis which is observed to increase following arousal (Galster and Morrison, 1975).

Medulla

Of the catecholamine hormones, the adrenal medulla is the major source for epinephrine (E) whereas NE can also be synthesized extramedullarily by adrenergic neurons of the sympathetic nervous system. Physiologically, it is not possible to distinguish the action of NE based on its origin since both sources are released by sympathetic stimulation. Due to the diffusive sources of NE it has been difficult to elucidate the exclusive role of the adrenal medulla for this hormone in hibernation.

The adrenal medulla is apparently not necessary for hibernation in the ground squirrel since demedullated animals can enter and arouse from hibernation (Popovic, 1960). There is an increase in the adrenal E and NE content during hibernation in the golden hamster (Lew, 1972), but the ratio of E to NE remains unchanged at one. In the hedgehog, adrenal NE content was highest during summer and early autumn, and decreased to a minimum in late winter and early spring (Uuspää, 1963b), but the total adrenal catecholamines increased during hibernation, primarily due to an increase of E (Helle *et al.,* 1980). This is reflected in the activity of phenylethanolamine-N-methyltransferase, the enzyme which catalyzes the conversion of NE to E. It was highest during the fall when peak ratio of E to NE was reached (Johansson, 1978). Although NE was undetectable in the medulla of the hibernating hedgehog (Helle *et al.,* 1980), there was an increase in storage of dopamine-β-hydroxylase, the enzyme that catalyzes NE synthesis from dopamine during hibernation. This increase may be interpreted as a reserve for the rapid switching-on of NE synthesis upon terminal arousal from hibernation. In the European ground squirrel, a similar increase in total catecholamines and a higher E to NE ratio were observed during hibernation; these changes were preceded by an increase in the activity of tyrosine hydroxylase, which converts L-tyrosine to DOPA, the rate limiting step in the biosynthesis of catecholamines, prior to hibernation (Petrović *et al.*, 1978). Enzymes involved in the catabolism of catecholamines, the monoamine oxidase, and the catechol-O-methyltransferase, have highest activities in winter and spring and lowest in summer and fall (Johansson, 1978). This is probably an indication of the seasonal differences in sympathetic activities, as these are high in winter during periodic arousals and in spring after emergence when facing cold temperature.

The activation of the sympathoadrenomedullary system during arousal from hibernation is well documented (Chapter 7). Recent studies on changes of plasma catecholamines in the woodchuck (Florant, 1981) indicated that NE increased more than 30-fold, and E more than 100-fold during arousal from hibernation. The marked increase of NE that mediates nonshivering thermogenesis in the hibernators is not surprising. However, the very significant increase of E indicates that the adrenal medulla contributes substantially to the cardiovascular and calorigenic actions which these catecholamines exert during the arousal process.

PERSPECTIVES

Research on endocrine functions in relation to hibernation has been based largely on histological observations, which do not always reflect the functional status of the endocrine glands. A limited number of studies has involved measurements of plasma concentrations of hormones. However, the plasma concentration is the balance between the rates of secretion and utilization, and does not necessarily indicate the true hormonal activity. Turnover studies, which provide the most accurate indication of hormonal dynamics, have not been done, except for a very few cases. Modern techniques must be used for both sampling of blood and analysis of hormones. Comparative studies employing a hibernator and a closely related species that does not hibernate (e.g., *Ammospermophilus leucurus*) such as that by Kenagy and Bartholomew (1979) on the seasonal cycle of reproduction should also be attempted to clarify whether some of the endocrine changes observed are simply seasonal or are of special significance for hibernation.

The strong seasonal connotation in the occurrence of torpor among seasonal hibernators indicates a change of physiological state which in many species is initiated endogenously. The conspicuous fall fattening, the involution of the gonads, and the general reduction of endocrine activities are characteristic of the preparation for hibernation. However, the neuroendocrine and endocrine regulation of this process, and the changes at the molecular and biochemical levels largely remain to be elucidated. To gain more insight into this problem, one must study the neuroendocrine functions during the preparative phase for hibernation and use much more refined time frames, such as weekly or daily rather than just quarterly, or monthly samplings.

The influence of low body temperature during hibernation on hormonal activity must be separated from changes that are independent of body temperature change. *In vitro* techniques could be employed to elucidate the effects of temperature on receptor physiology, in particular, the binding and dissociation of hormone to receptors as modulated by temperature. The seasonal difference in binding characteristics of receptors to hormones in euthermic animals should also be investigated.

Another intriguing aspect is the apparent spontaneous reactivation of the neuroendocrine system sometime during the hibernation season while the animal is still in the burrow and there are no apparent temperature changes and total darkness. Even if the reactivation of neuroendocrine functions occurs only when the animal is euthermic between hibernation bouts, this endogenously driven change is still strikingly different from those of most nonhibernating, seasonal breeders that cue on photoperiod.

It is apparent from these observations that the field of endocrine functions in hibernation offers rich opportunities for future research to those who are interested in the physiology of hibernation.

ACKNOWLEDGMENTS

Literature survey was aided by a NSERC operating grant No. A6455 to L. Wang. I thank R. E. Peter for critical comments on the manuscript.

REFERENCES

Agid, R., Ambid, L., Sable-Amplis, R., and Sicart, R. (1978). Aspects of metabolic and endocrine changes in hibernation. *In* "Strategies in Cold: Natural Torpidity and Thermogenesis." (L. C. H. Wang and J. W. Hudson, eds.), pp. 499–540. Academic Press, New York.

Aloia, R. C., and Pengelley, E. T. (1979). Lipid composition of cellular membranes of hibernating mammals. *In* "Chemical Zoology" (M. Florkin and B. T. Scheer, eds.), Vol. 11, pp. 1–47. Academic Press, New York.

Bauman, T. R., and Anderson, R. R. (1970). Thyroid activity of the squirrel (*Citellus tridecemlineatus*) using a cannula technique. *Gen. Comp. Endocrinol.* **15**, 369–373.

Bauman, T. R., Anderson, R. R., and Turner, C. W. (1969). Thyroid hormone secretion rates and food consumption of the hamster (*Mesocricetus auratus*) at 25.5° and 4.5°. *Gen. Comp. Endocrinol.* **10**, 92–98.

Beckman, A. L. (1978). Hypothalamic and midbrain function during hibernation. *In* "Current Studies of Hypothalamic Function" (W. L. Veale and K. Lederis, eds.), Vol. 2, pp. 29–43. Karger, Basel.

Beckman, A. L., and Satinoff, E. (1972). Arousal from hibernation by intrahypothalamic injections of biogenic amines in ground squirrels. *Am. J. Physiol.* **222**, 875–879.

Behrisch, H. W. (1978). Metabolic economy at the biochemical level: the hibernator. *In* "Strategies in Cold: Natural Torpidity and Thermogenesis" (L. C. H. Wang and J. W. Hudson, eds.), pp. 461–497. Academic Press, New York.

Bloch, R., and Canguilhem, B. (1966). Cycle saisonnier d'élimination urinaire de l'aldostérone chez un hibernant, *Cricetus cricetus*. Influence de la température. *C. R. Seances Soc. Biol. Ses Fil.* **160,** 1500–1502.

Boulouard, R. (1972). Adrenocortical function in two hibernators: the garden dormouse and the hedgehog. *Proc. Int. Symp. Environ. Physiol.: Bioenerg. Temp. Regul.* (R. E. Smith, J. L. Shields, J. P. Hannon, and B. A. Horwitz, eds.), pp. 108–112. Fed. Am. Soc. Exp. Biol., Bethesda, Maryland.

Burlet, C., Robert, J., and Legait, E. (1973). Double tag study (Cesium 131, Cr[51]-tagged red blood cells) of capillary exchanges in the neurohypophysis of a lerot (*Eliomys quercinus* L.) during his awakening from hibernation. *In* "Neurosecretion—The Final Neuroendocrine Pathway" (F. Knowles and L. Vollrath, eds.), p. 298. Springer-Verlag, Berlin and New York.

Canguilhem, B. (1970). Effets de la radiothyroïdectomie et des injections d'hormone thyroïdienne sur l'entrée en hibernation du Hamster d'Europe (*Cricetus cricetus*). *C. R. Seances Soc. Biol. Ses Fil.* **164,** 1366–1369.

Canguilhem, B., Kempf, E., Mack, G., and Schmitt, P. (1977). Regional studies of brain serotonin and norepinephrine in the hibernating, awakening or active European hamster *Cricetus cricetus,* during winter. *Comp. Biochem. Physiol. C* **57C,** 175–179.

Cardinali, D. P., and Wurtman, R. J. (1972). Hydroxyindole-O-methyl transferases in rat pineal, retina and Harderian gland. *Endocrinology* **91,** 247–252.

Chance, B., Nakase, Y., and Itshak, F. (1979). Membrane energization at subzero temperatures: calcium uptake and oxonol-V response. *Arch. Biochem. Biophys.* **198,** 360–369.

Daudova, G. M., and Soliternova, I. B. (1972). Influence of insulin on the absorption of glucose by adipose tissue of the ground squirrel *Citellus suslicus* during hibernation and arousal. *J. Evol. Biochem. Physiol. (Engl. Transl.)* **8,** 399–401. [*Zh. Evol. Biokhim. Fiziol.* **8,** 449–541.]

Demeneix, B. A., and Henderson, N. E. (1978a). Serum T_4 and T_3 in active and torpid ground squirrels, *Spermophilus richardsoni. Gen. Comp. Endocrinol.* **35,** 77–85.

Demeneix, B. A., and Henderson, N. E. (1978b). Thyroxine metabolism in active and torpid ground squirrels. *Gen. Comp. Endocrinol.* **35,** 86–92.

Demeneix, B. A., and Henderson, N. E. (1978c). Thyroid hormone metabolism over the annual cycle of *Spermophilus richardsoni. J. Therm. Biol.* **3,** 89–90. (Abstr.)

Draskóczy, P. R., and Lyman, C. P. (1967). Turnover of catecholamines in active and hibernating ground squirrels. *J. Pharmacol. Exp. Ther.* **155,** 101–111.

Duncan, R. J. S., and Tricklebank, M. D. (1978). On the stimulation of the rate of hydroxylation of tryptophan in the brain of hamsters during hibernation. *J. Neurochem.* **31,** 553–556.

Dussart, G., and Richoux, J. P. (1973). Regulation of genital function in the garden dormouse. Action of the gonadotropic hormones on the diencephalic monoamine oxidase activities and on the genital glands during hibernation. *Ann. Endocrinol.* **34,** 45–132. (In Fr.)

Edmonson, E. J. (1976). Plasma renin activity and plasma electrolyte concentration in a hibernator. *Fed. Proc., Fed. Am. Soc. Exp. Biol.* **35,** 705. (Abstr.)

Ellis, L. C., and Balph, D. F. (1976). Age and seasonal differences in the synthesis and metabolism of testosterone by testicular tissue and pineal hydroxyindole-O-methyl transferase activity of Uinta ground squirrels, *Spermophillus armatus. Gen. Comp. Endocrinol.* **28,** 42–51.

Feist, D. D., and Galster, W. A. (1974). Changes in hypothalamic catecholamines and serotonin during hibernation and arousal in the Arctic ground squirrel. *Comp. Biochem. Physiol. A* **48A,** 653–662.

Feldberg, W., and Myers, R. D. (1964). Effects on temperature of amines injected into the cerebral ventricles. A new concept of temperature regulation. *J. Physiol. (London)* **173,** 226–326.

Ferren, L. G., South, F. E., and Jacobs, H. K. (1971). Calcium and magnesium levels in tissues and serum of hibernating and cold-acclimated hamsters. *Cryobiology* **8,** 506-508.

Fleischhauer, K., and Wittkowski, W. (1976). Morphological aspects of the formation, transport and secretion of releasing and inhibiting hormones. *Acta Endocrinol. (Copenhagen), Suppl. No. 202,* 11-12.

Florant, G. L. (1981). Plasma cortisol and catecholamine concentration in euthermic and hibernating woodchucks (*Marmota monax*). *Cryobiology* **18,** 95 (Abstr.)

Florant, G. L., and Weitzman, E. D. (1980). Diurnal and episodic pattern of plasma cortisol during fall and spring in young and old woodchucks (*Marmota monax*). *Comp. Biochem. Physiol. A* **66A,** 575-581.

Fuxe, K., Goldstein, M., Hökfelt, T., Jonsson, G., and Löfström, A. (1974a). New aspects of the catecholamine innervation of the hypothalamus and the limbic system. *In* "Neurosecretion— The Final Neuroendocrine Pathway" (F. Knowles and L. Vollrath, eds.), pp. 223-228. Springer-Verlag, Berlin and New York.

Fuxe, K., Hökfelt, T., Jonsson, G., and Löfström, A. (1974b). Aminergic mechanisms in neuroendocrine control. *In* "Neurosecretion—The Final Neuroendocrine Pathway" (F. Knowles and L. Vollrath, eds.), pp. 269-275. Springer-Verlag, Berlin and New York.

Gabe, M., and Martoja, M. (1969). Histological data on the calcitonin cells of the thyroid of the garden dormouse, *Eliomys quercinus*. *Arch. Anat. Micros. Morphol. Exp.* **58,** 107-122. (In Fr.)

Gale, C. C. (1973). Neuroendocrine aspects of thermoregulation. *Annu. Rev. Physiol.* **35,** 391-430.

Galster, W., and Morrison, P. R. (1975). Gluconeogenesis in arctic ground squirrels between periods of hibernation. *Am. J. Physiol.* **228,** 325-330.

Glass, J. D., and Wang, L. C. H. (1978a). Thermoregulatory effects of central injection of noradrenaline in the newborn Columbian ground squirrel (*Spermophilus columbianus*). *J. Therm. Biol.* **3,** 207-211.

Glass, J. D., and Wang, L. C. H. (1978b). Thermoregulatory effects of central injection of noradrenaline in the Richardson's ground squirrel (*Spermophilus richardsonii*). *Comp. Biochem. Physiol. C* **61C,** 347-351.

Glass, J. D., and Wang, L. C. H. (1979a). Thermoregulatory effects of intracerebroventricular injection of serotonin and a monoamine oxidase inhibitor in a hibernator, *Spermophilus richardsonii*. *J. Therm. Biol.* **4,** 149-156.

Glass, J. D., and Wang, L. C. H. (1979b). Effects of central injection of biogenic amines during arousal from hibernation. *Am. J. Physiol.* **236,** R162-R167.

Gustafson, A. E., and Shemesh, M. (1976). Changes in plasma testosterone levels during the annual reproductive cycle of the hibernating bat *Myotis lucifugus lucifugus* with a survey of plasma testosterone levels in adult male vertebrates. *Biol. Reprod.* **15,** 9-24.

Helle, K. B., Bolstad, G., Pihl, K. E., and Knudsen, R. (1980). Catecholamines, ATP and dopamine-β-hydroxylase in the adrenal medulla of the hedgehog in the prehibernating state and during hibernation. *Cryobiology* **17,** 74-92.

Herlant, M. (1956). Corrélations hypophyso-génitales chez la femelle de la chauvesouris, *Myotis myotis* (Borkhausen). *Arch. Biol.* **67,** 89-180.

Hinkley, R. E., and Burton, P. A. (1970). Fine structure of the pancreatic islet cells of normal and alloxan treated bats (*Eptesicus fuscus*). *Anat. Rec.* **166,** 67-86.

Hlavicka, P. (1976). Unpublished observations. Cited in Jansky and Novotoná (1976).

Hoffman, R. A. (1964a). Speculations on the regulation of hibernation. *Ann. Acad. Sci. Fenn., Ser. A4* **71,** 199-216.

Hoffman, R. A. (1964b). Terrestrial animals in cold: hibernators. *Handb. Physiol., Sect. 4: Adapt. Environ.* pp. 379-403.

Hoffman, R. A., and Reiter, R. J. (1965). Pineal gland: Influence on gonads of male hamsters. *Science* **148,** 1609-1611.

Hoffman, R. A., and Zarrow, M. X. (1958a). Seasonal changes in the basophilic cells of the pituitary gland of the ground squirrel (*Citellus tridecemlineatus*). *Anat. Rec.* **131,** 727-735.

Hoffman, R. A., and Zarrow, M. X. (1958b). A comparison of seasonal changes and the effect of cold on the thyroid gland of the male rat and ground squirrel (*Citellus tradecemlineatus*). *Acta Endocrinol. (Copenhagen)* **27,** 77-84.

Hoo-Paris, R. (1971). Hibernation and ACTH in the hedgehog, *Erinaceus europaeus*. *Ann. Endocrinol.* **32,** 743-752. (In Fr.)

Hoo-Paris, R., and Sutter, B. C. J. (1980). Blood glucose control by insulin in the lethargic and arousing hedgehog (*Erinaceus europaeus*). *Comp. Biochem. Physiol. A* **66A,** 141-143.

Hudson, J. W. (1968). Ineffectiveness of exogenous L-thyroxine in preventing torpor in *Citellus tereticaudus*. *Am. Zool.* **8** (4), Abstr. No. 159.

Hudson, J. W. (1973). Torpidity in mammals. *In* "Comparative Physiology of Thermoregulation" (G. C. Whittow, ed.), Vol. 3, pp. 97-165. Academic Press, New York.

Hudson, J. W. (1980). The thyroid gland and temperature regulation in the prairie vole, *Microtus ochrogaster* and the chipmunk, *Tamias striatus*. *Comp. Biochem. Physiol. A* **65A,** 173-179.

Hudson, J. W., and Deavers, D. R. (1976). Thyroid function and basal metabolism in the ground squirrels *Ammospermophilus leucurus* and *Spermophilus* spp. *Physiol. Zool.* **49,** 425-444.

Hudson, J. W., and Wang, L. C. H. (1969). Thyroid function in desert ground squirrels. *In* "Physiological Systems in Semiarid Environments" (C. C. Hoff and M. L. Riedesel, eds.), pp. 17-33. Univ. of New Mexico Press, Albuquerque.

Hudson, J. W., and Wang, L. C. H. (1979). Hibernation: endocrinologic aspects. *Annu. Rev. Physiol.* **41,** 287-303.

Huibregtse, W. H., Gunville, R., and Ungar, F. (1971). Secretion of corticosterone *in vitro* by normothermic and hibernating ground squirrels. *Comp. Biochem. Physiol. A* **38A,** 763-768.

Hulbert, A.J. (1978). The thyroid hormones: a thesis concerning their action. *J. Theor. Biol.* **73,** 81-100.

Hulbert, A. J., and Hudson, J. W. (1976). Thyroid function in a hibernator, *Spermophilus tridecemlineatus*. *Am. J. Physiol.* **230,** 1138-1143.

Hulbert, A. J., Augee, M. L., and Raison, J. K. (1976). The influence of thyroid hormones on the structure and function of mitochondrial membranes. *Biochim. Biophys. Acta* **455,** 597-601.

Jacob, H. K., South, F. E., Hartner, W. C., and Zatzman, M. L. (1971). Thermoregulatory effects of administration of biogenic amines into the third ventricle and preoptic area of the marmot (*M. flaviventris*). *Cryobiology* **8,** 313-314.

Jansky, L. (1978). Time sequence of physiological changes during hibernation: The significance of serotonergic pathways. *In* "Strategies in Cold: Natural Torpidity and Thermogenesis" (L. C. H. Wang and J. W. Hudson, eds.), pp. 299-326. Academic Press, New York.

Jansky, L., and Novotoná, L. (1976). The role of central aminergic transmission in thermoregulation and hibernation. *In* "Regulation of Depressed Metabolism and Thermogenesis" (L. Jansky and X. J. Musacchia, eds.), pp. 64-80. Thomas, Springfield, Illinois.

Jasinski, A. (1970). Hypothalamic neurosecretion in the bat, *Myotis myotis* Borkhausen, during the period of hibernation and activity. *In* "Aspects of Neuroendocrinology" (W. Bargmann and B. Scharrer, eds.), pp. 301-309. Springer-Verlag, Berlin and New York.

Johansson, B. W. (1978). Seasonal variations in the endocrine system of hibernators. *In* "Environmental Endocrinology" (I. Assenmacher and D. S. Farner, eds.), pp. 103-110. Springer-Verlag, Berlin and New York.

Johansson, B. W., and Senturia, J. B. (1972). Seasonal variations in the physiology and biochemistry of the European hedgehog (*Erinaceus europaeus*) including comparisons with nonhibernators, guinea pig and man. *Acta Physiol. Scand., Suppl.* No. 380.

Kastner, P. R., Zatzman, M. L., South, F. E., and Johnson, J. A. (1978). Renin-angiotensin-aldosterone system of the hibernating marmot. *Am. J. Physiol.* **234**, R178–R182.

Kayser, C. (1957). Le sommeil hivernal et les glandes surrénales. Etude faite sur le Hamster ordinaire, *Cricetus cricetus. C. R. Seances Soc. Biol. Ses Fil.* **151**, 982–985.

Kayser, C. (1961). ''The Physiology of Natural Hibernation.'' Pergamon, New York.

Kenagy, G. J., and Bartholomew, G. A. (1979). Effects of day length and endogenous control of the annual reproductive cycle of the antelope ground squirrel, *Ammospermophilus leucurus. J. Comp. Physiol.* **130**, 131–136.

Kenny, A. D., and Musacchia, X. J. (1977). Influence of season and hibernation on thyroid calcitonin content and plasma electrolytes in the ground squirrel. *Comp. Biochem. Physiol. A* **57A**, 485–489.

Khabibov, B., and Krass, P. M. (1974). Seasonal dynamics of the absolute and relative weight of the adrenals and the level of 11-hydroxy corticosteroids in the peripheral blood of the red-cheeked suslik *Citellus erythrogenys. Dokl. Biol. Sci. (Engl. Transl.)* **216**, 1433–1435.

Kolpakov, M. G., and Samsonenko, R. A. (1970). Reactivity of the adrenals of the red-cheeked suslik at various periods of vital activity. *Dokl. Akad. Nauk SSSR, Ser. Biol.* **191**, 1424–1426.

Koryakina, L. A. (1976). Seasonal changes in the response of the hypophyseal-adrenal system of the suslik (*Citellus erythrogenys*) to adrenomimetics. *J. Evol. Biochem. Physiol. (Engl. Transl.)* **12**, 444–447.

Krass, P. M., and Khabibov, B. (1975). Seasonal rhythms of functional activity of the adrenocorticotropic function of the pituitary in the red-cheeked suslik. *Dokl. Biol. Sci. (Engl. Transl.)* **225**, 474–476.

Krook, L., Wimsatt, W. A., Whalen, J. P., MacIntyre, I., and Nunez, E. A. (1977). Calcitonin and hibernation bone loss in the bat (*Myotis lucifugus*). *Cornell Vet.* **67**, 265–271.

Krupp, P. P., Young, R. A., and Frink, R. (1977). The thyroid gland of the woodchuck, *Marmota monax:* a morphological study of seasonal variations in the follicular cells. *Anat. Rec.* **187**, 495–513.

Lachiver, F. (1964). Thyroid activity in the garden dormouse (*Eliomys quercinus* L.) studied from June to November. *Ann. Acad. Sci. Fenn., Ser. A4* **71**, 283–294.

Laurila, M., and Suomalainen, P. (1974). Studies in the physiology of the hibernating hedgehog: 19. The changes in the insulin level induced by seasons and hibernation cycle. *Ann. Acad. Sci. Fenn., Ser. A4*, **201**, 1–40.

Legait, E., Burlet, C., and Marchetti, J. (1970). Contribution to the study of the hypothalamo-neurohypophyseal system during hibernation. *In* ''Aspects of Neuroendocrinology'' (W. Bargmann and B. Scharrer, eds.), pp. 310–321. Springer-Verlag, Berlin and New York. (In Fr.)

Lew, G. M. (1972). Circadian rhythms in hamster adrenal, heart, spleen and brain contents of catecholamines. *Anat. Rec.* **172**, 355. (Abstr.)

Lyman, C. P., and Chatfield, P. O. (1955). Physiology of hibernation in mammals. *Physiol. Rev.* **35**, 403–425.

Lynch, G. R., White, S. E., Grundel, R., and Berger, M. S. (1978). Effects of photoperiod, melatonin administration and thyroid block on spontaneous daily torpor and temperature regulation in the white-footed mouse, *Peromyscus leucopus. J. Comp. Physiol.* **125**, 157–163.

Mayer, W. V., and Bernick, S. (1963). Effect of hibernation on tooth structure and dental caries. *In* ''Mechanisms of Hard Tissue Destruction,'' Publ. No. 75, pp. 285–296. Am. Assoc. Adv. Sci., Washington, D.C.

Mikhnevich, O., and Kostyrev, O. (1971). Seasonal changes on morphology of thyroid gland in *Citellus erythrogenys. In* ''Hibernation and Seasonal Rhythms of Physiological Functions'' (A. D. Slonim, ed.), pp. 33–43. Nauka, Sib. Branch, Novosibirsk.

Moore, R. Y., and Bloom, F. E. (1979). Central catecholamine neuron systems: anatomy and physiology and the norepinephrine and epinephrine systems. *Annu. Rev. Neurosci.* **2,** 113–168.

Moy, R. M., Lesser, R. W., and Pfeiffer, E. W. (1972). Urine concentrating ability of arousing and normothermic ground squirrels (*Spermophilus columbianus*). *Comp. Biochem. Physiol. A* **41A,** 327–337.

Novotoná, R., Jansky, L., and Drahota, Z. (1975). Effect of hibernation on turnover of serotonin in the brain stem of golden hamster (*Mesocricetus auratus*). *Gen. Pharmacol.* **6,** 23–26.

Nunez, E. A., Gould, R. P., Hamilton, D. W., Hayward, J. S., and Holt, S. J. (1967). Seasonal changes in the fine structure of the basal granular cells of the bat thyroid. *J. Cell Sci.* **2,** 401–410.

Nunez, E. A., Whalen, J. P., and Krook, L. (1972). An ultrastructural study of the natural secretory cycle of the parathyroid gland of the bat. *Am. J. Anat.* **134,** 459–480.

Ozaki, Y., and Lynch, H. J. (1976). Presence of melatonin in plasma and urine of pinealectomized rats. *Endocrinology* **99,** 641–644.

Palmer, D. L., and Riedesel, M. L. (1976). Responses of whole-animal and isolated hearts of ground squirrels, *Citellus lateralis,* to melatonin. *Comp. Biochem. Physiol. C* **53C,** 69–72.

Pearse, A. G. E., and Welsch, U. (1968). Ultrastructural characteristics of the thyroid C cells in the summer, autumn and winter states of the hedgehog (*Erinaceaus europaeus* L.) with some references to other mammalian species. *Z. Zellforsch. Mikrosk. Anat.* **92,** 596–609.

Pehowich, D. J., and Wang, L. C. H. (1981). Temperature dependence of mitochondrial calcium transport in a hibernating and nonhibernating ground squirrel. *Acta Univ. Carolinae-Biologica* **1979,** 291–293.

Pengelley, E. T., and Asmundson, S. J. (1969). Free-running periods of endogenous circannian rhythms in the golden-mantled ground squirrel, *Citellus lateralis. Comp. Biochem. Physiol.* **30,** 177–183.

Petrovic, A., and Kayser, C. (1957). L'activité gonadotrope de la préhypophyse du hamster (*Cricetus cricetus*) au cours de l'année. *C. R. Seances Soc. Biol. Ses. Fil.* **151,** 996–998.

Petrović, V. M., Janić-Šibalić, V., Aminot, A., and Roffi, J. (1978). Adrenal tyrosine hydroxylase activity in the ground squirrel—effect of cold and arousal from hibernation. *Comp. Biochem. Physiol. C* **61C,** 99–101.

Pinatel, M. C., Durand, N., and Girod, C. (1970). Etude des variations de l'iodémie et de l'iode thyroïdien au cours du cycle annuel chez le Hérisson, *Erinaceus europaeus* L.; comparaison avec l'histologie thyroïdienne. *C. R. Seances Soc. Biol. Ses Fil.* **164,** 1719–1722.

Polenov, A. L., and Yurisova, M. N. (1975). The hypothalamo-hypophysial system in the ground squirrels, *Citellus erythrogenys* Brandt and *Citellus undulatus* Pallas. I. Microanatomy and cytomorphology of the Gomori-positive neurosecretory system with special reference to its state during hibernation. *Z. Mikrosk.-Anat. Forsch.* **89,** 991–1014.

Popova, N. K., and Voitenko, N. P. (1974). Serotonin metabolism during hibernation. *Dokl. Biol. Sci. (Engl. Transl.)* **218,** 1488–1490.

Popova, N. K., Maslova, L. N., and Naumenko, E. V. (1972). Serotonin participates directly in hypothalamic regulation of secretions of hypophyseal-adrenal system. *Brain Res.* **47,** 61–67.

Popova, N. K., Koryakina, L. A., and Naumenko, E. V. (1974). A comparison of the functional activity of the hypophyseal-adrenal system of hibernating animals. *J. Evol. Biochem. Physiol. (Engl. Transl.)* **10,** 598–602.

Popova, N. K., Kolaeva, S. G., and Dianova, I. I. (1975). State of the pineal gland during hibernation. *Bull. Exp. Biol. Med. (Engl. Transl.)* **79,** 116–117.

Popovic, V. (1960). Endocrines in hibernation. *Bull. Mus. Comp. Zool.* **124,** 104–130.

Raths, P., and Kulzer, E. (1976). Physiology of hibernation and related lethargic states in mammals and birds. *Bonn. Zool. Monogr.* **9,** 1–93.

Reichlin, S., Baldessarini, R. J., and Martin, J. B. (1978). "The Hypothalamus." Raven, New York.

Reiter, R. J., Vaughan, M. K., Blask, D. E., and Johnson, L. Y. (1974). Melatonin: Its inhibition of pineal antigonadotrophic activity in male hamsters. *Science* **185**, 1169–1171.

Reznik-Schuller, H., and Reznik, G. (1973). Comparative histometric investigations of the testicular function of European hamsters *Cricetus cricetus* with and without hibernation. *Fertil. Steril.* **24**, 698–705.

Richoux, J. P., and Dubois, M. P. (1976). Détection immunocytologique de peptides immunologiquement apparentés au LRH et au SRIF chez le Lérot dans différentes conditions. *C. R. Seances Soc. Biol. Ses Fil.* **170**, 860–864.

Sauerbier, I., and Lemmer, B. (1977). Seasonal variations in the turnover of noradrenaline of active and hibernating hedgehogs (*Erinaceus europaeus*). *Comp. Biochem. Physiol. C* **57C**, 61–63.

Shaskan, E. G. (1969). Hypothalamic and whole brain monoamine levels in bats: some aspects of central control of thermoregulation. Ph.D. Thesis, Univ. of Arizona, Tucson.

Smit-Vis, J. H. (1972). The effect of pinealectomy and of testosterone administration on the occurrence of hibernation in adult male golden hamsters. *Acta Morphol. Neerl.-Scand.* **10**, 269–281.

Smit-Vis, J. H., and Akkerman-Bellaart, M. A. (1967). Spermiogenesis in hibernating golden hamsters. *Experientia* **23**, 844–846.

Smit-Vis, J. H., and Smit, G. J. (1970). Hibernation and testis activity in the golden hamster. *Neth. J. Zool.* **20**, 502–506.

Snyder, S. H. (1980). Brain peptides as neurotransmitters. *Science* **209**, 976–983.

Spafford, D. C., and Pengelley, E. T. (1971). The influence of the neurohumor serotonin on hibernation in the golden-mantled ground squirrel, *Citellus lateralis*. *Comp. Biochem. Physiol. A* **38A**, 239–250.

Steffen, D. G., and Platner, W. S. (1976). Subcellular membrane fatty acids of rat heart after cold acclimation of thyroxine. *Am. J. Physiol.* **231**, 650–654.

Suomalainen, P. (1960). Stress and neurosecretion in the hibernating hedgehog. *Bull. Mus. Comp. Zool.* **124**, 271–283.

Tamarkin, L., Westrom, W. K., Hamill, A. I., and Goldman, B. D. (1976). Effect of melatonin on the reproductive systems of male and female Syrian hamsters: A diurnal rhythm in sensitivity to melatonin. *Endocrinology* **99**, 1534–1541.

Tashima, L. S. (1965). The effects of cold exposure and hibernation on the thyroidal activity of *Mesocricetus auratus*. *Gen. Comp. Endocrinol.* **5**, 267–277.

Twente, J. W., Cline, W. H., Jr., and Twente, J. A. (1970). Distribution of epinephrine and norepinephrine in the brain of *Citellus lateralis* during the hibernating cycle. *Comp. Gen. Pharmacol.* **1**, 47–53.

Uuspää, A. V. J. (1963a). The 5-hydroxytryptamine content of the brain and some other organs of the hedgehog (*Erinaceus europeus*) during activity and hibernation. *Experientia* **19**, 156–159.

Uuspää, A. V. J. (1963b). Effects of hibernation on the noradrenaline and adrenaline contents of the adrenal glands in the hedgehog. *Ann. Med. Exp. Biol. Fenn.* **41**, 349–354.

Wenberg, G. M., and Holland, J. C. (1973a). The circannual variations of thyroid activity in the woodchuck (*Marmota monax*). *Comp. Biochem. Physiol. A* **44A**, 775–780.

Wenberg, G. M., and Holland, J. C. (1973b). The circannual variations of some of the hormones of the woodchuck (*Marmota monax*). *Comp. Biochem. Physiol. A* **46A**, 523–535.

Whalen, J. P., Krook, L., and Nunez, E. A. (1971). Bone resorption in hibernation and its relationship to parafollicular and parathyroid cell structures. *Invest. Radiol.* **6**, 342–343.

Wimsatt, W. A.(1960). Some problems of reproduction in relation to hibernation in bats. *Bull. Mus. Comp. Zool.* **124**, 249–270.

Wimsatt, W. A. (1969). Some interrelations of reproduction and hibernation in mammals. *Symp. Soc. Exp. Biol.* **23**, 511–559.

Young, R. A., Robinson, D. S., Vagenakis, A. G., Saavedra, J. M., Lovenberg, W., Krupp, P. P., and Danforth, E. Jr. (1979a). Brain TRH, monoamines, tyrosine hydroxylase, and tryptophan hydroxylase in the woodchuck, *Marmota monax*, during the hibernation season. *Comp. Biochem. Physiol. C* **63C**, 319–323.

Young, R. A., Danforth, E., Jr., Vagenakis, A. G., Krupp, P. P., Frink, R., and Sims, E. A. H. (1979b). Seasonal variation and the influence of body temperature on plasma concentrations and binding of thyroxine and triiodothyronine in the woodchuck. *Endocrinology* **104**, 996–999.

Yurisova, M. N. (1971). Seasonal dynamics of the state of the hypothalamo-hypophysial neurosecretory system in the ground squirrel, *Citellus erythrogenys* Brandt. *In* "Hibernation and Seasonal Rhythms of Physiological Functions" (A. D. Slonim, ed.), pp. 8–27. Nauka, Sib. Branch, Novosibirsk.

Yurisova, M. N., and Polenov, A. L. (1979). The hypothalamo-hypophysial system in the ground squirrel, *Citellus erythrogenys* Brandt. II. Seasonal changes in the classical neurosecretory system of a hibernator. *Cell Tissue Res.* **198**, 539–556.

Zimmerman, G. D., McKean, T. A., and Hardt, A. B. (1976). Hibernation and disease osteoporosis. *Cryobiology* **13**, 84–94.

13

Respiration and Acid–Base State
in Hibernation

The primary role of respiration is to subserve the needs of energy metabolism by supplying oxygen to the cells and removing carbon dioxide. From the atmosphere to the cytochromes, oxygen is transported in succession by ventilatory convection, by diffusion across the alveolar-capillary membrane, by circulatory convection and by diffusion again from capillary blood to mitochondria. Steady-state conditions require that all the corresponding fluxes be kept equal to the metabolic oxygen demand. The same condition holds for carbon dioxide, which follows a reverse path. The regulation of ventilation ensures a constant matching between O_2 consumption and ventilatory uptake, and between CO_2 production and ventilatory elimination (Dejours, 1975). In this respect, the main question raised by mammalian hibernation is whether, and how, this homeostasis can be maintained throughout the hibernation cycle, in the face of wide changes of metabolic rate and body temperature.

Transition from water- to air-breathing, in the evolution of vertebrates, has freed ventilation from the constraints imposed by the scarcity of oxygen in water. From then on, steady state could be achieved for variable levels of the ratio of ventilatory flow rate to CO_2 production. Air-breathers were, thus, endowed with the ability to use ventilation to control alveolar and arterial CO_2 partial pressure, P_{CO_2}, and thereby the dissociation of the CO_2–bicarbonate buffer system. This permits an accurate and fast-acting control of extra- and intracellular acid–base states, supplementing the slow ionic regulations of kidney and cell membranes.

Recent progress concerning the interaction of body temperature with acid–base state in ectotherms has shown that pH regulation is more directed to the maintenance of protein functions than to ionic homeostasis as such. Air-breathing ectothermic vertebrates achieve this control mainly through the ventilatory adjustment of P_{CO_2} (for review, see Reeves, 1977). This has led to a new understanding of the regulation of respiration and acid–base state in hibernating mammals, which now appears as an adjustment of ventilation to a new level, giving rise to a pronounced respiratory acidosis. The functional significance of this acidosis, which does not affect all tissues equally, is not yet fully understood, but it is likely to play an important role in the control of nervous and metabolic functions.

Another intriguing feature of deep hibernation is the occurrence of intermittent breathing, with periods of apnea extending to over an hour in some species, like the hedgehog. The combination of low metabolic rate and high oxygen affinity of hemoglobin creates favorable conditions for such a breathing pattern. However, close scrutiny of the oxygen supply in an apneic hedgehog shows that these factors do not suffice, and reveals striking similarities with diving mammals.

The transient states, entrance and arousal, also raise many questions: how and when are the changes of respiratory and acid–base variables achieved? How does the gas exchange system cope with the considerable aerobic effort of thermogenesis? Does behavioral adjustment of burrow microenvironment play a role in some species?

Hibernation, thus, appears as a highly interesting field for the comparative physiology of respiration. Respiration and acid–base state, on the other hand, are likely to play an increasing role in our understanding of hibernation physiology.

STUDIES ON DEEP HIBERNATION

Pulmonary Ventilation*

Progress in this area has long been impeded by methodological problems. Ventilatory flow rate is determined by two factors: tidal volume and period. All

*Symbols used in this chapter. The symbols listed below may carry the following subscripts: A, alveolar; E, expired; I, inspired; L, pulmonary; T, tidal; a, arterial; b, blood; g, gas; i, intracellular; v̄,

of the standard methods used on euthermic mammals involve some kind of restraint such as a mask, tracheal cannula, or gas-tight collar, which are hardly compatible with the maintenance of deep hibernation. Most authors have, therefore, resorted to so-called pneumographic methods, whereby some indication of the depth of breathing is obtained by recording the movements of the chest or abdominal wall. A good example is provided by the studies of the Finnish School on the hedgehog, *Erinaceus europaeus:* Kristoffersson and Soivio (1964, 1967) and Soivio and co-workers (1968) attached a fine thread with a clip to one of the lateral spines of the animal. The thread pulled on the lever of a kymograph. Alternatively (Tähti, 1975), a disc of polystyrene was glued to the back of the animal. Some results were also obtained by recording air temperature fluctuations in front of the nose (Tähti and Soivio, 1977).

In deep hibernation at 4°C ambient temperature (Fig. 13-1), a hedgehog presents an example of ''Cheyne–Stokes respiration'': bursts of 40–50 respirations in 3–5 min are separated by periods of apnea lasting about 60 min (maximum 150 min!) (Kristoffersson and Soivio, 1964). When ambient temperature is decreased to −5°C, respiration becomes continuous (Soivio *et al.*, 1968) with a frequency of 8–9/min. Cheyne–Stokes respiration also disappears when the animal is disturbed, e.g., by handling (Kristoffersson and Soivio, 1964). This lability may explain why it has been observed by earlier authors such as Pembrey and Pitts (1899) and Chao and Yeh (1950), but not by Biörck *et al.* (1956) nor by Clausen (1966) (for reviews of earlier findings, see Kayser, 1961; Tähti, 1978).

Cheyne–Stokes respiration was studied by the glued disk method in the garden dormouse, *Eliomys quercinus,* by Pajunen (1970, 1974). Breathing periods of 1–8 min were separated by periods of apnea lasting between 1 and 130 min, the daily means ranging from 30 to 100 min. The author's illustrations show numbers of cycles per burst of 24 and 131, respectively. Cheyne–Stokes pattern was typical of deep, undisturbed hibernation.

---

mixed venous; STPD, standard temperature, pressure, dry (0°C, 760 Torr); and BTPS, body temperature, pressure, saturated.

Symbols: B, balance of electric charges of strong electrolytes (see text); C, concentration, e.g., $C_{I_{O_2}}$, concentration of O_2 in inspired air; F, fractional concentration in dry gas phase; Hb, hemoglobin; M, quantity of substance (moles); $\dot{M}$, quantity of substance per unit time, e.g., $\dot{M}_{O_2}$, oxygen consumption, $\dot{M}_{CO_2}$, carbon dioxide production; P, gas pressure or partial pressure, e.g., $P_{A_{CO_2}}$, alveolar partial pressure of CO_2, $P_{a_{O_2}}$, arterial partial pressure of O_2; P_{50}, partial pressure of oxygen at which hemoglobin is 50% saturated with O_2; R, respiratory exchange ratio, $\dot{M}_{CO_2}/\dot{M}_{O_2}$, e.g., $R_{I,E}$, respiratory exchange ratio between inspired and expired air; R_{met}, respiratory exchange ratio at tissue (metabolic) level; S, saturation, e.g., $S_{a_{O_2}}$, saturation of hemoglobin in arterial blood, (alternatively) quantity accumulated in body stores; V, volume, e.g., V_T, tidal volume; $\dot{V}$, volume per unit time, flow rate, e.g., $\dot{V}_g$: gas flow rate, ventilatory flow rate; α, dissociation ratio, e.g., α_{Im}, dissociation ratio of imidazole buffer groups: $\alpha_{Im} = [Im]/([Im] + [HIm^+])$; pK, opposite of the logarithm of the dissociation constant K of a weak electrolyte; 2,3-DPG, 2,3-diphosphoglycerate. Other symbols, see text. For a fuller description, see Dejours (1975).

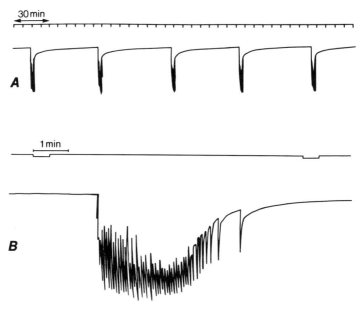

Fig. 13-1. (A) Pneumographic recording of Cheyne–Stokes breathing in a hibernating hedgehog. Inspiration is downward. Bursts of 40–50 cycles are separated by apneas of 60–70 min. (B) The time scale is expanded, showing the individual cycles within a burst. (After Kristoffersson and Soivio, 1964.)

Cheyne–Stokes breathing was also observed in golden hamsters (Lyman, 1951; Kristoffersson and Soivio, 1966), in bats (Hock, 1958; Soivio, cited in Tähti, 1978), and in a dormouse (*Glis glis* ?) by Pembrey and Pitts (1899). In examining old records, however, one sometimes wonders whether heart beats superimposed on a slow and deep breath (cf. Fig. 13-2) could not have been mistaken for Cheyne–Stokes breathing on some pneumographic records. This would explain the conflicting observations on Cheyne–Stokes breathing in species such as the marmot: this pattern was observed by Patrizi (1897) and Pembrey (1901–1902), but neither by Benedict and Lee (1938) nor Kayser (1940) nor by ourselves (Malan *et al.*, 1973) even in fully undisturbed animals.

Two methods have been employed to measure tidal volume. In the first, volumetric plethysmography, the animal lies in a closed chamber but breathes from the outside from a tube fitted to a mask covering its snout. The changes in chest volume can then be recorded with a spirometer connected to the chamber. This does impose some restraint to the animal and Endres and Taylor (1930) are the only ones who ever succeeded in recording from a hibernating animal in such conditions. At 4.8°C body temperature, inspiration lasted 3.5 sec and expiration 10 sec; this was followed by a pause of about 180 sec, so that breathing frequency was around 0.33 min⁻¹. Tidal volume was 5.6 ml. At 8.1°C frequency was up to 1 min⁻¹, while tidal volume was 13.8 ml.

Total body plethysmography, originally designed by Chapin (1954) for the euthermic hamster and by Drorbaugh and Fenn (1955) for the human newborn, was later applied to the hibernating marmot (Malan, 1973). The animal is placed in a closed chamber whose temperature is lower than body temperature. When the animal inspires, the tidal volume is warmed up and humidified, which gives rise to a small pressure change in the chamber proportional to the tidal volume and inversely proportional to the temperature difference. The chamber is aerated between the measurements, thus, permitting the measurement of oxygen consumption, $\dot{M}_{O_2}$, and carbon dioxide production, $\dot{M}_{CO_2}$, by an open-circuit method. The method looks ideal because the animal is untouched, but the theory of operation is still not fully understood (Malan, 1973; Epstein and Epstein, 1978) because of the complex adiabatic–isothermal changes occurring in the transient phases. Moreover, with a body-to-ambient temperature difference of 4°C suitable for hibernation studies, the peak-to-peak pressure variation in our set-up was 0.3 mm H_2O, i.e., 0.003% of the absolute pressure; this is equivalent to a change of 0.01°C only, which makes the operation quite difficult. However, on a lung model on which tidal volume could be measured simultaneously with a pneumotachograph, agreement between the two methods was 3%.

In most cases (Fig. 13-2), breathing was intermittent as in the records of Endres and Taylor, the breaths being interspersed with variable periods of apnea. Breathing period varied between 1 and 6 min. Periods of up to 14 min were visually observed on undisturbed animals in the cold room. Cheyne–Stokes breathing was never seen, but longer periods of apnea were sometimes followed by two consecutive breaths. Superimposed on the ventilatory trace of Fig. 13-2 one sees small upward deflections with a frequency of 4 min⁻¹. These are cardiogenic oscillations, synchronous with heart beats as could be checked by recording arterial pressure. They probably result from variations in the distribution of blood between the heart and large vessels in the chest, and between the chest and the rest of the body (Malan, 1973). Breathing period (T), 1–6 min⁻¹, was much more variable than tidal volume (V_τ), which was 32.5 ml BTPS (body temperature and pressure, saturated) on the average (Fig. 13-3). On such a V_τ versus T diagram, straight lines passing through the origin correspond to a constant ventilatory flow rate $\dot{V} = V_\tau/T$. $\dot{V}$ varied from 11.9 to 48.7 ml BTPS·min⁻¹ in deep hibernation. By the end of the hibernation period (March) another breathing pattern occurred, in which the same ventilatory flow rate was achieved by a shallow and more rapid breathing, with a small dispersion of V_τ and T (the corresponding points are located in the lower left area of Fig. 13-3). The arousal data will be discussed later (Malan, 1973; Malan *et al.*, 1973).

There is a sizable discrepancy between these two sets of data on tidal volume. Although the values of Endres and Taylor (1930) may be too low because of

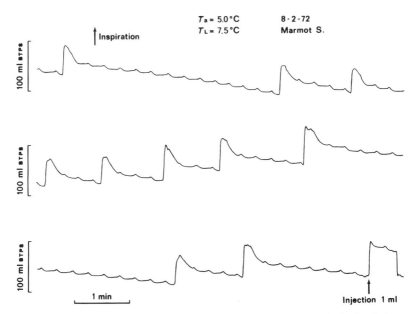

Fig. 13-2. Ventilation of a hibernating marmot monitored by whole body plethysmography. The three lines are consecutive. Inspiration is upward. Body (lung) temperature (T_L) was higher here than ambient temperature (T_a). The small deflections are cardiogenic oscillations. Calibration was achieved at the end of this tracing by injecting a known volume of air. (After Malan, 1973.)

probably unavoidable leakage around the mask, our own data might well be too high. From the ventilatory flow rate ($\dot{V}$), the oxygen consumption ($\dot{M}_{O_2}$), and the inspired O_2 concentration ($C_{I_{O_2}}$), one can calculate the extraction coefficient for oxygen (E_{O_2}) (Dejours, 1975), representing the proportion of the oxygen of the inspired air that is actually delivered to the blood: $E_{O_2} = \dot{M}_{O_2}/\dot{V}_g \cdot C_{I_{O_2}} = 0.110$. With a single exception, the figures listed by Dejours (1975) for euthermic mammals range between 0.175 and 0.218. Along the same line, the equations of pulmonary exchange (Dejours, 1975) permit calculation of the alveolar CO_2 partial pressure ($P_{A_{CO_2}}$) which results from a steady state between metabolic CO_2 production ($\dot{M}_{CO_2}$) and convective elimination of CO_2 by the alveolar ventilation ($\dot{V}_A$):

$$P_{A_{CO_2}} = RT \frac{\dot{M}_{CO_2}}{\dot{V}_A} \tag{1}$$

where R = constant of perfect gases, and T = absolute temperature. On a single occasion, arterial P_{CO_2} ($P_{a_{CO_2}}$) could be measured while the animal was recorded

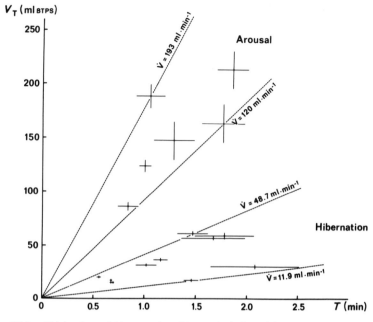

Fig. 13-3. Tidal volume (V_T) versus breathing period (T) in hibernating marmots. Dotted lines with zero intercept each correspond to a constant value of ventilatory flow rate, $\dot{V} = V_T/T$. Each point represents the mean ±S.E. of the 8–27 consecutive breaths recorded during a single 20–30 min period. Two patterns of ventilation may be seen in hibernation: a slow breathing with a highly variable period such as in Fig. 13-2, and a rapid and shallow breathing with fairly constant period (lower left corner). Early arousal is characterized by a highly increased tidal volume with no change in period at first. (After Malan et al., 1973.)

from in the body plethysmograph. Calculated $P_{A_{CO_2}}$ was 15.5 Torr, whereas $P_{a_{CO_2}}$ was 36.8 Torr. Steady-state conditions could not be ascertained, however, which restricts the validity of this check. The validity of our technique has been discussed elsewhere (Malan, 1973). The excellent agreement with pneumotachograph data on a lung model should leave little room for major errors. Effects of undue correction of baseline drift and of overestimation of water vapor pressure could reach in hibernation no more than a fraction of the 15% maximum at 37°C calculated by Epstein and Epstein (1978). Errors on body temperature measurement (i.e., deep buccal lower than core temperature) would have resulted in a too low figure for tidal volume. The answer lies either with the special features of ventilation in hibernation (intermittent breathing, etc.), which might invalidate some of the above assumptions: steady-state conditions as implied by Eq. (1), near equality of $P_{a_{CO_2}}$ and $P_{A_{CO_2}}$, etc., or with some still hidden inadequacy of the theory of the plethysmograph.

Ventilatory responses to changes in inspired gas composition will be studied later. Nothing is known about respiratory mechanics, the energy requirements of respiratory muscles, nor ventilation–perfusion inhomogeneities in the lung in hibernation.

In spite of these uncertainties, we are left with a clear-cut conclusion: deep hibernation is generally characterized by intermittent breathing, with long apneas separating either single or multiple breaths depending on the species.

Acid–Base State

Blood acid–base studies on hibernating mammals can be traced back to the last century (Dubois, 1896), but until recent years the results were hard to understand because of the interaction of temperature with acid–base variables. Beginning with Robin (1962), and mostly through the efforts of Rahn (1967) and co-workers (see review in Reeves, 1977), considerable progress has been achieved in studies on acid–base state at a variable body temperature in ectotherms, and on the underlying blood physical chemistry. Their main conclusions can now be applied to hibernating mammals; they will first be briefly summarized.

Dissociation and Protein Alpha Imidazole

Proteins play an essential role in cellular mechanisms, especially as catalysts of biochemical reactions (enzymes). Changes in pH affect the osmotic properties of all proteins, but a much higher pH dependency is found for the activity of some enzymes; the extreme case is that of phosphofructokinase (PFK), a major regulating enzyme of glycolysis, in which a pH drop of 0.1 unit can reduce the activity by 80% (Trivedi and Danforth, 1966). A more common type of activity versus pH curve is found for Na^+,K^+-ATPase (Fig. 13-4) (Park and Hong, 1976). The pH effects result from the titration of amino acid residues in the protein molecule, the most abundant of these buffer groups being histidine imidazole. Its dissociation ratio, often termed alpha imidazole (α_{Im}) is a function of the (pH–pK_{Im}) difference, where pK_{Im} is the dissociation constant of imidazole (Reeves, 1972, 1977). Since pK_{Im} increases when temperature is lowered (Fig. 13-5), the titration curve relating α_{Im} to pH is shifted to the right when going from 37° to 8°C (Fig. 13-4). The pH optimum of the enzyme, which depends on the titration of imidazole groups, is shifted by the same amount so that it corresponds to a constant value of α_{Im}. Similar results have been obtained on various enzymes from ectotherms (Hazel *et al.*, 1978) and from the cold-adapted peripheral tissues of a mammal (Behrisch *et al.*, 1977). Whatever the temperature, α_{Im} provides an excellent index of titration-independent properties of proteins.

When their body temperature varies, most ectotherms studied adjust their blood acid–base regulation so as to keep a constant (pH–pK_{Im}) difference, i.e., a constant alpha imidazole. This has been termed alphastat regulation by Reeves

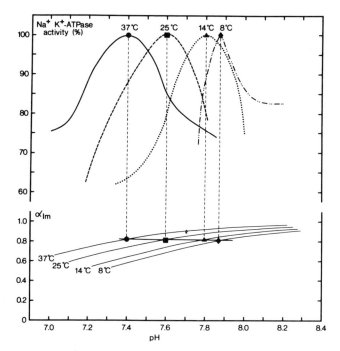

Fig. 13-4. Upper panel: Activity of toad skin Na^+,K^+-ATPase as a function of pH at various temperatures. The pH optimum is shifted to the right as temperature decreases. (Modified from Park and Hong, 1976.) Lower panel: Titration curves of protein imidazole buffer groups at the same temperatures. α_{Im} = dissociation ratio of imidazole: α_{Im} = $[Im]/([Im] + [HIm^+])$. The pH optimum of the enzyme corresponds to a fairly constant value of α_{Im}. (After Malan, 1980.)

(1972, 1977) (Fig. 13-5). Since the curve relating pK_{Im} to temperature is parallel to that of the pH of neutral water, this also amounts to keeping a constant relative alkalinity (Rahn, 1967). In the species in which it has been determined (frogs and turtles), intracellular pH is also adjusted so as to keep a constant α_{Im} (Malan *et al.*, 1976a). The regulation of acid–base state can, thus, be described as subserving the constancy of pH-dependent properties of proteins, of which most important are their enzymatic functions.

Exchanges of Matter and Acid–Base State

How is a given acid–base state achieved? At a constant temperature, it is fully determined by the molar concentrations of all the chemical species involved: weak electrolytes or buffers, strong electrolytes, and water. For the buffers the total concentrations of dissociated plus undissociated forms have to be considered, e.g., C_{CO_2} = ([CO_2 dissolved] + [HCO_3^-] + [H_2CO_3]). These concentrations result from the metabolism and from the net exchanges of matter with the

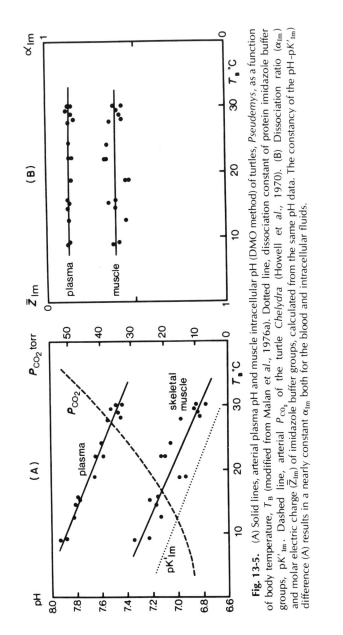

Fig. 13-5. (A) Solid lines, arterial plasma pH and muscle intracellular pH (DMO method) of turtles, *Pseudemys*, as a function of body temperature, T_B (modified from Malan *et al.*, 1976a). Dotted line, dissociation constant of protein imidazole buffer groups, pK'_{Im}. Dashed line, arterial P_{CO_2} of the turtle *Chelydra* (Howell *et al.*, 1970). (B) Dissociation ratio (α_{Im}) and molar electric charge ($\bar{Z}_{Im}$) of imidazole buffer groups, calculated from the same pH data. The constancy of the $pH-pK'_{Im}$ difference (A) results in a nearly constant α_{Im} both for the blood and intracellular fluids.

environment, and represent the true control variables of acid–base state (Rodeau and Malan, 1979). The main factors are (1) C_{CO_2}, which can be replaced by P_{CO_2} at a constant temperature; (2) the balance (B) of the electric charges of strong electrolytes (= "strong ion difference;" Stewart, 1978; often unduly called "buffer base"); (3) the concentration of protein (imidazole) buffers.

At a constant temperature, changes in these quantities that have resulted in a given change of pH can be derived from the bicarbonate–pH diagram (Figs. 13-6 to 13-10): changes in protein (e.g., hemoglobin) buffer concentration are seen as changes in the slope of the "buffer line" (i.e., the titration curve of protein buffers), whereas changes in B ("metabolic titration") generally resulting from ionic exchanges are measured by the vertical displacement of this buffer line (downward for an acidification). P_{CO_2} changes ("respiratory titration") are read from the P_{CO_2} isopleths.

Determining the Changes in the Control Variables of Acid–Base State

When the body temperature changes, as in hibernators or ectotherms, this standard analysis is no longer valid because of the interactions of temperature with acid–base equilibria. For instance, P_{CO_2} is considerably affected by the effect of temperature on CO_2 solubility which from 37° to 6°C increases from 0.03 to 0.07 mmoles $\cdot$ liter$^{-1} \cdot$ Torr^{-1}. On theoretical grounds, the best way to describe acid–base state would be to use only total molar quantities of the substances involved, such as C_{CO_2}. Some of these temperature-independent quantities, such as B, can be determined only indirectly. In most instances, a simpler approach can be sufficient: let us consider what occurs in blood that is cooled or warmed in a closed syringe ("anaerobic" conditions): pH and P_{CO_2} will change with temperature, but there is no exchange of matter with the environment. Therefore, the total numbers of moles of all the species involved in acid–base equilibria stay constant; the pH and P_{CO_2} changes observed result exclusively from temperature effects. Conversely, any deviation from the acid–base versus temperature behavior of the closed system will be due to changes in the chemical factors of acid–base state. Let us then choose a standard temperature, say 25°C, at which all acid–base quantities will be expressed. By bringing any blood sample to 25°C in a closed syringe before reading pH and P_{CO_2}, one eliminates the effect of temperature and obtains temperature-corrected values directly. The corrected pH, P_{CO_2}, and bicarbonate concentration, respectively, pH*, P_{CO_2}*, and [HCO$_3^-$]*, can be plotted on a bicarbonate–pH diagram from which the changes of the chemical factors of acid–base state can be analyzed as above (Malan, 1977, 1978a). Instead of being measured directly, pH* and P_{CO_2}* can also be calculated, using nomograms or a mathematical model of acid–base versus temperature relationships (Malan, 1977; Rodeau and Malan, 1979). This is generally the only procedure applicable to intracellular fluids; but simple approximations can be made (Malan, 1978a).

When this is applied to the turtle blood data of Fig. 13-5, one sees that the open-system acid–base regulation mimics closed-system conditions (Fig. 13-6): the constancy of pH*, P_{CO_2}*, and [HCO_3^-]* means that no net changes of B or of C_{CO_2} take place over wide excursions of body temperature. This requires adjusting the balance of elimination and production of CO_2, as expressed by P_{CO_2}, at a different level for each body temperature, so that the blood pH versus temperature curve is parallel to that of pK_{Im}. Due to the high diffusivity of CO_2, adjusting P_{CO_2} results in a rapid adjustment of intracellular pH to the requirements of a constant α_{Im} (Reeves and Malan, 1976) (see Fig. 13-5). Notice that a constant pH* is nearly equivalent to a constant α_{Im}; this results from the predominant role of protein imidazole in the buffering properties of blood and intracellular fluids (Reeves, 1976, 1977).

The reasons for selecting 25°C rather than 37°C as a reference temperature are as follows: (1) 25°C is closer to midrange both for ectotherms and hibernating

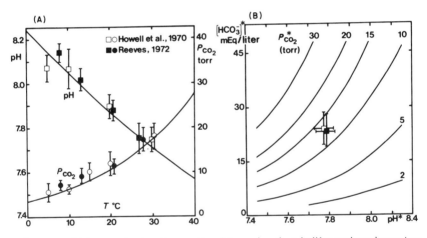

Fig. 13-6.   (A) Average *in vivo* plasma pH and P_{CO_2} values from bullfrogs adapted to various body temperatures (symbols) and *in vitro* plasma pH and P_{CO_2} in blood whose temperature is changed under closed-system conditions (solid lines: calculated from a model). The variations with temperature of the dissociation constants of blood buffers and of CO_2 solubility result in an increase of pH and a decrease of P_{CO_2} when temperature is lowered, in the absence of any exchange of matter (e.g., C_{CO_2}, ions), with the environment. (After Reeves, 1972.) (B) The same data on a temperature-corrected bicarbonate-pH diagram. The *in vivo* temperature-corrected blood pH* and P_{CO_2}* remain constant over the 5°–30°C body temperature range. Consequently, the total CO_2 concentration and the balance B of charges of strong electrolytes are constant as in blood *in vitro*. The constancy of C_{CO_2} requires increasing the relative ventilation (alveolar ventilatory flow rate/CO_2 production) when body temperature decreases, to reduce P_{CO_2} as in blood *in vitro*. (After Malan, 1977.)

mammals; (2) errors, if any, on the correction procedure will be shared between the high and low temperature data; (3) the higher blood temperature is raised, the higher the risk of measurement errors due to CO_2 diffusion; (4) 25°C is already standard for tables of dissociation constants (Malan, 1977). A pH of 7.40 and a P_{CO_2} of 40.0 Torr at 37°C are equivalent to a pH of 7.57 and a P_{CO_2} of 22.7 Torr at 25°C.

Blood Acid–Base State of Hibernating Mammals

Only data obtained with chronic indwelling catheters have been listed in Table 13-1, as the reliability of values obtained by heart puncture is highly dependent on experimental conditions, and the venous or arterial nature of the sample is often undetermined. Some remarks may be made also about certain data of Table 13-1: the pH values of Kreienbühl et al. (1976) for the euthermic dormouse look fairly acidotic; the value for B of the hibernating marmots of Goodrich (1973) is so high that it could hardly be reached in hibernation in the absence of an efficient urine acidification—but pH and P_{CO_2} were measured in an extracorporeal circulation, a method in which the electrode signals may be biased by ground leaks through the animal; the hibernation data of Kreienbühl et al. (1976) have been determined at 37°C and back-corrected by the authors to the animal temperature with temperature coefficients which are now known to be unsuitable for such a wide temperature range (Reeves, 1976).

Most of the values recorded in euthermic animals fall within the normal range for nonhibernators. In hibernation, the values read at body temperature may be confusing: should one attempt to make a direct comparison with euthermic data read at 37°C, one would conclude that on going from euthermia to hibernation pH is kept constant or slightly increased, while P_{CO_2} somewhat decreases, so that hibernation would correspond to a mild respiratory alkalosis. This completely ignores the temperature effects on pH and P_{CO_2}. Model studies (Rodeau and Malan, 1979) as well as comparison with blood closed-system behavior clearly show that when temperature decreases, keeping P_{CO_2} and pH constant requires adding to the blood large quantities of CO_2. This is obvious when the hibernators' data are plotted on a temperature-corrected $[HCO_3^-]^*$–pH* diagram (Fig. 13-7). Compared with euthermic data, all hibernation data show a large increase of $P_{CO_2}^*$, corresponding to the addition of CO_2 to the blood (notice that $[HCO_3^-]^*$ is increased). In most cases, the transition occurs along the titration curve (buffer line) of extracellular water, with a slope of 11.4 mmoles·liter^{-1}·pH unit^{-1} corresponding to blood equilibrated with interstitial fluid (Woodbury, 1965). In a few instances, the data points lie above this line, indicating the existence of some back-titration (alkalinization) of the blood by ionic exchanges at a constant P_{CO_2} ("metabolic compensation").

On the whole, then, hibernation acid–base state corresponds to a marked respiratory acidosis, in which the arterial pH* drop ranges from 0.24 to 0.48,

TABLE 13-1

Arterial Acid-Base Data of Hibernating and Euthermic Mammals[a]

Species	Body temperature (× = assumed) (°C)	Variables measured at body temperature		Temperature-independent variables		Variables corrected to 25°C		Ref.[b]
		pH	P_{CO_2} (Torr)	C_{CO_2} (mmoles/liter)	B (mEq/liter)	pH*	P_{CO_2}* (Torr)	
A. Hibernating								
Hedgehog								
Erinaceus europaeus								
Beginning of apnea	5	7.45	30.1	36.4	20.8	7.14	86.6	1
	6×	7.46	26.5	29.1	14.7	7.15	65.7	2
		(0.04)	(4.4)					
End of apnea	6×	7.42	34.6	34.8	19.2	7.12	84.1	2
		(0.05)						
Ground squirrel								
Citellus tridecemlineatus	5	7.39	35	36.9	20.0	7.08	98.7	3
		(0.03)						
	6	7.44	32.9	38.1	22.6	7.15	89.0	4
		(0.03)	(2.3)					
Marmot								
Marmota monax and	11.1	7.63	51	83.6	72.0	7.44	101.8	5
M. flaviventris	(0.5)	(0.02)	(5)					
Marmota marmota	8	7.57	33.3	50.1	37.6	7.32	79.6	6
		(0.03)	(3.4)					
European hamster								
Cricetus cricetus	9	7.57	36.1	53.3	41.0	7.33	82.9	6
		(0.02)	(9.0)					

Golden hamster *Mesocricetus auratus*	6	7.46 (0.02)	40.5 (4.6)	49.1	33.4	7.18	107.2	7
Common dormouse *Glis glis*	6	7.44 (0.03)	27.4 (2.6)	31.7	16.7	7.14	75.5	7
B. Euthermic								
Hedgehog	32	7.38 (7.29–7.51)	43.6 (31.9–55.0)	28.5	20.3	7.48	31.1	1
	36[x]	7.33 (0.03)	53.5 (3.3)	30.1	21.8	7.48	33.4	2
Ground squirrel	37	7.39 (0.06)	42	26.1	19.1	7.56	23.9	3
	37	7.40 (0.03)	47.7 (5.0)	30.7	23.7	7.57	27.4	4
Marmot	35.8	7.49 (0.02)	30	22.7	16.9	7.65	16.7	5
European hamster	37	7.40 (0.02)	45.3 (6.6)	29.2	22.2	7.57	25.8	6
Golden hamster	37[x]	7.30 (0.02)	59.7 (1.7)	30.6	22.1	7.47	34.6	7
Common dormouse	37[x]	7.24 (0.03)	38.5 (3.2)	17.3	8.7	7.42	21.8	7

[a] Mean ± S.D. (or range). C_{CO_2}, B, and temperature-corrected data have been calculated using the pK' of Rispens et al. (1968) and the model described by Malan (1977). C_{CO_2}, total CO_2 concentration in plasma; B, balance of the charges of strong electrolytes.

[b] Source of data: 1, Clausen (1966); 2, Tähti and Soivio (1975); 3, Kent and Peirce (1967); 4, Musacchia and Volkert (1971); 5, Goodrich (1973); 6, Malan et al. (1973); 7, Kreienbühl et al. (1976).

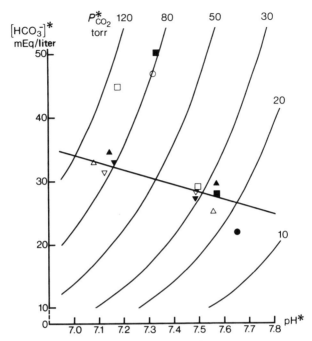

Fig. 13-7. Acid–base state of arterial blood of hibernators on a temperature-corrected bicarbonate–pH diagram. Standard temperature is 25°C. Straight line, buffer value of extracellular space, 11.4 mmoles·liter^{-1}·pH unit^{-1} (Woodbury, 1965). Compared with euthermia (lower right; P_{CO_2}* 20–35 Torr), hibernation is characterized by a pronounced respiratory acidosis (left: P_{CO_2}* 80–110 Torr), with some metabolic compensation in certain species. Symbols: ▼, hedgehog (Clausen, 1966); ▽, id. (Tähti and Soivio, 1975); △, ground squirrel (Kent and Peirce, 1967); ▲, id. (Musacchia and Volkert, 1971); ●, marmot (Goodrich, 1973); ○, id. Malan et al., 1973; ■, European hamster (Malan et al., 1973; □, golden hamster (Kreienbühl et al., 1976).

whereas P_{CO_2} is increased by a factor of 2.5–4.1. As will be discussed later, this necessarily results from a change in the relative ventilation, i.e., the ratio of alveolar ventilatory flow rate to CO_2 production, leading to a build up of body CO_2 stores.

The existence of a respiratory acidosis can also be inferred from the values of C_{CO_2} and B, both temperature-independent quantities (Table 13-1). An increase in C_{CO_2} with little or no change in B corresponds to a pure respiratory acidosis in which an addition of CO_2 molecules to the blood results from a change in the relative ventilation, $\dot{V}_A/\dot{M}_{CO_2}$ (ground squirrels, Clausen's data on hedgehogs). In the other cases this is accompanied by ionic exchanges, evidenced by changes of B, and the interpretation is not so straightforward as with the [HCO$_3^-$]*–pH* diagram. This is especially true when a C_{CO_2} increase due to respiratory acidosis is offset by a reduction in B (metabolic acidosis), such as in the hedgehogs at the beginning of apnea (Tähti and Soivio, 1975). Therefore, no final conclusion can be reached from C_{CO_2} data alone. C_{CO_2} increases in hibernation had been observed by earlier authors before the development of catheterization techniques: on marmots (Dubois, 1896; Rasmussen, 1916), ground squirrels (Stormont *et al.*, 1939; Person, 1950; Svihla and Bowman, 1952), and European hamsters (Endres, 1924); but not by Lyman and Hastings (1951) on the golden hamster and thirteen-lined ground squirrel.

The long periods of apnea undergone by hibernating mammals can bring about acid–base fluctuations. In the hedgehog, from the beginning to the end of an apnea, arterial pH* varies from 7.15 to 7.12, whereas P_{CO_2}* increases from 65.7 to 84.1 Torr (Tähti and Soivio, 1975). This small pH excursion, in spite of the 1-hr duration of the apnea, may result from a metabolic compensation: the charge balance B increases 4.5 mEq/liter. This probably does not occur in the marmot, in which pH excursions of the same amplitude (0.02–0.04) are observed for apneas of only 6 min (Malan *et al.*, 1973).

Judging from the only existing data, there would be no change of arterial pH over the time course of a bout of hibernation. No correlation was found in the marmot between the time spent since last arousal (1–20 days) and either pH ($r = 0.004$) or P_{CO_2} ($r = 0.18$) (Malan *et al.*, 1973).

Tissue Acid–Base State

Most enzymes are intracellular. What then are the repercussions on the intracellular milieus of the blood respiratory acidosis of the hibernating animal? The intracellular pH values of various tissues of the European hamster (*Cricetus cricetus*) have been measured with the DMO tracer equilibration method in four conditions: deep hibernation (2 days since last arousal), euthermia in winter (the animals were withdrawn from the cold room and kept at 20°C for 24 hr), respiratory acidosis in spring (90-min exposure to 10% CO_2), and the corresponding spring controls (Malan *et al.*, 1976b, 1982). On a temperature-corrected bicarbonate–pH diagram (Fig. 13-8) blood plasma pH presents a typical respiratory acidosis both in hibernation and in acute hypercapnia; only notice that breathing 10% CO_2 is a less severe acid load than hibernating! It is generally considered that tissue P_{CO_2} is approximately equal to venous P_{CO_2}; the data shown here thus correspond to venous blood. Intracellular fluids have a higher buffer value than mean extracellular fluid. This results in the decrease of intracellular pH (pH_i) being lower than that of extracellular pH, even for tissues such as skeletal muscle and diaphragm which undergo a simple CO_2 titration along the buffer line (Figs. 13-8 and 13-9). In brain (Fig. 13-10), the pH* drop is still higher; point H lies below the buffer line; this would indicate the existence of a superimposed metabolic acidosis, due either to build up of acidic metabolites or,

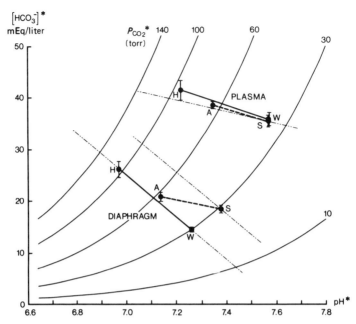

Fig. 13-8. Temperature-corrected acid –base data of venous blood (plasma) and diaphragm intracellular fluid (DMO method) in European hamsters. Standard temperature is 25°C. Solid lines: winter experiments; W, winter controls; H, hibernating animals. Dashed lines, spring experiments; S, spring controls; A, respiratory acidosis due to breathing 10% CO_2 over 90 min. Thin interrupted lines are the corresponding buffer lines with slopes of 11.4 and 40 mmoles·liter^{-1}·pH unit^{-1} for extracellular and intracellular spaces, respectively. (After Malan et al., 1982.)

more likely, to the uptake of acidic equivalents from the blood by (passive?) ionic exchanges. On the contrary, in the heart (Fig. 13-9) and more conspicuously in the liver (Fig. 13-10) the respiratory acidosis is partially compensated by a metabolic back-titration. The net extrusion of acidic equivalents is represented by the vertical distance of point H above the buffer line; it most probably corresponds to the well-known mechanism of regulation of intracellular pH by membrane "pumps" exchanging HCO_3^- against Cl^- or Na^+ against H^+ ions (Roos and Boron, 1981). For intracellular fluids whose composition is not fully known, the temperature-correction procedure is only approximate, and the mean values of pH* and [HCO_3^-]* may be somewhat biased by the nonlinearities of the equations. These conclusions have, therefore, been checked by statistical calculations on a temperature-independent quantity, the charge balance B (Malan et al., 1982); only the metabolic component of brain acidosis fails to reach statistical significance ($p = 0.1$). Decreases of pH$_i$* comparable to those occurring in muscle have also been observed in smooth muscle (esophagus) and brown fat

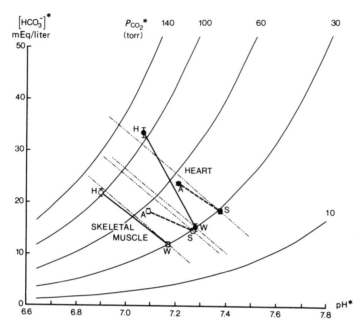

Fig. 13-9. Same as Fig. 13-8 for heart and skeletal muscle.

(Malan *et al.*, 1976b), but the DMO method is open to some difficulties in these tissues.

Most tissues are acidotic, then, in hibernation. Only liver and heart are partially protected and tend to follow alpha imidazole regulation. Acid extrusion by pumps is an energy-requiring process, since ions have to be moved against electrochemical potential gradients. The occurrence of such energy expenditures in an energy-sparing animal probably reflects the importance of reducing the acidosis in these two organs. Moreover, in the absence of an efficient urine acidification in hibernation, the acidic equivalents rejected by the cells should lead to a metabolic acidification of the blood. Since this is not the case, they must be taken up by another body compartment. This is likely to be bones: osteoporosis has been observed in jaw alveolar bone of hibernating European hamsters and thirteen-lined ground squirrels (Kayser and Franck, 1963; Haller and Zimny, 1977), in dental bone of ground squirrels (Zimny and Haller, 1978) and in skull, femur, and tibia of hamsters (Kayser and Franck, 1963). There may be differences between bone types and between species, however, since no osteoporosis was seen in the hind limb bones of ground squirrels (Zimmerman *et al.*, 1976) and golden hamsters (Tempel *et al.*, 1978) (Chapter 10).

Comparison of hibernation with hypercapnic acidosis in spring raises a question of time scale. Assuming that ionic pumps, like overall metabolic rate, are

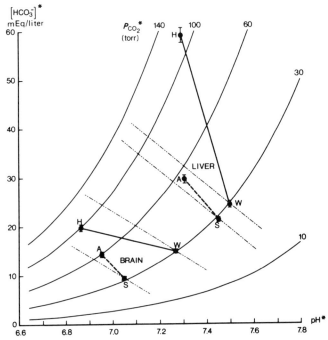

Fig. 13-10. Same as Fig. 13-8 for liver and brain. For the latter, a buffer value of 30 mmoles·liter^{-1}·pH unit^{-1} has been determined from a figure of Messeter and Siesjö (1971). Notice the extent of the metabolic compensation of the intracellular acidosis of liver cells in hibernating hamsters.

slowed down by a factor of 30 in hibernation, 90 min of euthermic hypercapnia would be equivalent to 45 hr of hibernation. After 90 min of euthermic hypercapnia, diaphragm and skeletal muscle show some metabolic acidosis superimposed on the respiratory acidosis, which is not seen in hibernation. Brain, which is the most efficiently protected tissue against pH$_i$ changes in euthermic hypercapnia (Messeter and Siesjö, 1971), shows some metabolic compensation of the acidosis; this is also observed in the liver. Liver and heart, however, show much less compensation than in hibernation. Full restoration of pH$_i$ would be observed in 24 hr in the brain (Messeter and Siesjö, 1971), but only after a week in heart and muscle (Wood and Schaefer, 1978). These differences between hibernation and euthermic hypercapnia cannot be explained only by the existence of urine acidification in the latter case, since they are in the opposite direction for brain and liver. Regulation of pH$_i$ is controlled by norepinephrine and glucagon in the heart (Fenton *et al.*, 1978). Endocrine control might thus explain the differences between hibernation and euthermic hypercapnia.

The pH_i differences between spring and winter euthermic animals might reflect the existence of a seasonal variation of intracellular pH, but they must be considered with some caution, since the experiment had not been planned to test seasonal differences.

A complementary indication on tissue acid–base state is provided by determinations of P_{CO_2} in subcutaneous gas pockets, using the method of Van Liew (1962). Tissue P_{CO_2} was 68 ± 2 Torr in euthermia and 51 ± 1 Torr in hibernation (8°C) in the European hamster, versus 66 ± 2 and 39 ± 1 (8°C), respectively, in the marmot (Piquard and Malan, 1982). Except for the hibernating marmot, these values are much higher than the arterial blood values of Table 13-1 (uncorrected), but gas pocket data may be exaggerated by low blood perfusion rate (Van Liew, 1962). Similar results had been obtained on four ground squirrels by Stormont *et al.* (1939): ranges were 66–88 Torr in euthermia and 41–70 Torr in hibernation.

Oxygen Transport and Tissue Oxygenation

Alveolar to Capillary Diffusion

The equations of alveolar ventilation (see Dejours, 1975) allow the prediction of changes in alveolar oxygen partial pressure ($P_{A_{O_2}}$) associated with the changes in acid–base state from hibernation to euthermia. The equation for oxygen exchange is as follows:

$$P_{I_{O_2}} - P_{A_{O_2}} = RT \cdot \dot{M}_{O_2}/\dot{V}_A \qquad (2)$$

where $P_{I_{O_2}}$ = inspired oxygen partial pressure. A similar equation holds for CO_2 [Eq. (1)], but there inspired pressure is assumed to be zero. By combining the two equations, and assuming further that $P_{A_{CO_2}} \simeq P_{a_{CO_2}}$ (arterial P_{CO_2}), one gets the following:

$$P_{A_{O_2}} = P_{I_{O_2}} - P_{a_{CO_2}} \cdot \dot{M}_{O_2}/\dot{M}_{CO_2} \qquad (3)$$

(The occurrence of apneic oxygenation in intermittent breathing would necessitate a corrective term.)

In the ground squirrels studied by Musacchia and Volkert (1971), $P_{a_{CO_2}}$ dropped from 47.7 Torr at 37°C to 32.9 Torr in hibernation at 6°C (Table 13-1). Assuming a respiratory quotient ($\dot{M}_{CO_2}/\dot{M}_{O_2}$) of 0.75, the ($P_{I_{O_2}} - P_{A_{O_2}}$) difference thus decreased from 63.6 to 43.9 Torr. In normoxic conditions, at a barometric pressure of 760 Torr, $P_{I_{O_2}}$ increases from 149.4 to 157.8 Torr from 37° to 6°C due to the reduced water vapor pressure. Therefore, $P_{A_{O_2}}$ should have been, respectively, 85.8 Torr in euthermia and 113.9 Torr in hibernation. Hibernating animals must then have an elevated $P_{A_{O_2}}$.

This conclusion may look surprising in connection with the respiratory acidosis described above. At a constant body temperature, the acidosis would correspond to an increased $P_{A_{CO_2}}$ associated with a lowered relative ventilation, $\dot{V}_A/\dot{M}_{CO_2}$ (Eq. 1). When body temperature is simultaneously reduced, however, the effect of the increased CO_2 solubility predominates. In order to keep a constant CO_2 content, when going from 37° to 6°C, the animal would have to reduce $P_{A_{CO_2}}$ by a factor of about 4.6 (Figs. 13-5 and 13-6) and, therefore, to increase $\dot{V}_A/\dot{M}_{CO_2}$ by the same amount, like a turtle (Jackson, 1971). By increasing $\dot{V}_A/\dot{M}_{CO_2}$ by a much smaller amount, the hibernator increases its CO_2 stores while $P_{A_{CO_2}}$ still decreases, but to a lesser degree.

The increase of $P_{A_{O_2}}$ in the hibernating animal leads to a corresponding increase of arterial partial pressure ($P_{a_{O_2}}$) from 64.9 Torr (range 58.0–78.0) to 87.7 Torr (35.0–120.0) in ground squirrels (Musacchia and Volkert, 1971). From euthermia to hibernation, a 30-fold reduction in O_2 uptake, combined with the temperature effect on diffusion constant (see below), would result in a 24-fold reduction of the alveolar-to-arterial P_{O_2} difference, other conditions being unchanged. This probably explains why values of $P_{a_{O_2}}$ up to 120 Torr can be reached in hibernation, 50% higher than the euthermic maximum.

The wide fluctuations of $P_{a_{O_2}}$ observed in hibernation are clearly linked with the existence of long apneic pauses. This is evidenced by the studies on hedgehogs: after a series of breaths and just before an apneic pause, $P_{a_{O_2}}$ reaches 120 ± 27 Torr (compare to above data); by the end of the apnea, $P_{a_{O_2}}$ is down to 10.5 ± 3.7 Torr (Tähti and Soivio, 1975).

Blood Oxygen Transport

Between the diffusion areas of lungs and tissues, oxygen is transported in the blood by convection. Under steady-state conditions diffusive and convective oxygen fluxes have to be equal. Diffusive fluxes depend on partial pressure gradients. The convective flux ($\dot{M}_{O_2}$) is related to blood flow ($\dot{V}_b$) (cardiac output) and to the arteriovenous concentration difference ($C_{a_{O_2}} - C_{\bar{v}_{O_2}}$) by the Fick equation:

$$\dot{M}_{O_2} = \dot{V}_b \, (C_{a_{O_2}} - C_{\bar{v}_{O_2}}) \tag{4}$$

By introducing the capacitance coefficient, $\beta_{O_2} = (C_{a_{O_2}} - C_{\bar{v}_{O_2}})/(P_{a_{O_2}} - P_{\bar{v}_{O_2}})$, this can also be written as follows:

$$\dot{M}_{O_2} = \beta_{O_2} \cdot \dot{V}_b \, (P_{a_{O_2}} - P_{\bar{v}_{O_2}}) \tag{5}$$

to show the interfacing of diffusive and convective processes (see Dejours, 1975). The capacitance coefficient β_{O_2} depends on hemoglobin concentration and on the shape of the hemoglobin dissociation curve, currently expressed by the pressure of half saturation (P_{50}). A low P_{50} corresponds to a high oxygen affinity and vice versa (Fig. 13-11).

Studies on hemoglobin concentration [Hb] have long given conflicting results.

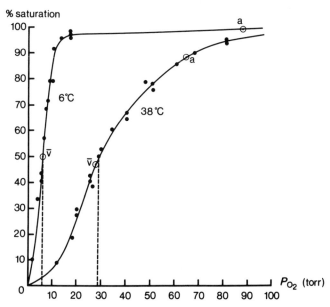

Fig. 13-11. Oxyhemoglobin dissociation curve in blood from euthermic and hibernating ground squirrels, determined at the corresponding temperatures: 38° and 6°C, respectively: (a) *in vivo* arterial blood; ($\bar{v}$) *in vivo* mixed venous blood. Vertical dashed lines show the P_{O_2} of half-saturation (P_{50}). (Modified from Musacchia and Volkert, 1971.)

Restricting ourselves to recent studies, an increase from euthermia to hibernation was found in golden hamsters by Tempel and Musacchia (1975) and in hedgehogs by Suomalainen and Rosoviki (1973), whereas significant changes were observed neither in ground squirrels (Burlington and Whitten, 1971; Larkin, 1973) nor in hedgehogs by Kramm *et al.* (1975). Harkness *et al.* (1974) recorded in the woodchuck first an increase after 2 months of hibernation, and then a decrease below the value of euthermia. A similar, but not significant time-pattern can be seen on the graph of Wenberg *et al.* (1973) for the same species. Such time-related changes might well explain many of the discrepancies among the various published data.

Temperature affects hemoglobin P_{50} mostly by a direct action on the binding of oxygen, but also by its effect on the pK values of the amino acids responsible for the binding of ligands, such as the H^+ ions giving rise to the Bohr effect (Antonini and Brunori, 1971). The effect of pK variations can be excluded by performing the temperature change at a constant (pH–pK_{Im}), or in closed-system conditions. This was done over a sufficient temperature range only in a recent study on human blood (Reeves, 1978, 1980).

At a pH of 7.4 which is close both to euthermia and hibernation values, lowering the temperature from 37° to 6°C reduces P_{50} from 29 to 6 Torr in blood from active winter ground squirrels (Musacchia and Volkert, 1971) (Fig. 13-11). In human blood, the P_{50} reduction would be still greater, since a P_{50} of 5.8 Torr has been already measured at 13°C by Reeves (1980); the 37°C value was the same as in ground squirrels. A similar conclusion was derived by Clausen and Ersland (1968) from the comparison between blood from hibernating hedgehogs and corresponding data from other mammals.

When comparisons are carried out at 37°C and pH 7.4, a lower P_{50} is found for the hibernating than for the euthermic animal. P_{50} is reduced in hibernation from 23.4 ± 0.6 to 22.3 ± 0.8 Torr in the woodchuck, *Marmota monax* (Harkness *et al.*, 1974), and from 40.9 ± 0.9 to 33.3 ± 0.5 in the hedgehog (Kramm *et al.*, 1975). Since both the Bohr effect (modification of P_{50} by proton binding) and the effect of temperature were ruled out in these experiments, another factor intervenes. This is the concentration of 2,3-diphosphoglycerate, [2,3-DPG], a side product of glycolysis in the erythrocyte which lowers O_2 affinity by binding to hemoglobin (Duhm, 1976).

[2,3-DPG] is markedly reduced in hibernation (Table 13-2). The P_{50} value of hibernation can be fairly well predicted from the change in [2,3-DPG](Harkness *et al.*, 1974; Kramm *et al.*, 1975). As to [2,3-DPG] itself, its decrease probably results from the respiratory acidosis. In human blood, *in vitro* (Rørth, 1970) as well as *in vivo* (Astrup, 1970), [2,3-DPG] is reduced when pH decreases. From these authors' data, a reduction by about 50% of [2,3-DPG] can be predicted from the observed changes in pH* from euthermia to hibernation (Table 13-1); this is in fair agreement with the [2,3-DPG] values actually measured (Table 13-2). The pH effect on [2,3-DPG] mainly results from the inhibition of phosphofructokinase, as observed in hypercapnic guinea pigs (Messier and Schaefer, 1971; Jacey and Schaefer, 1972). In these studies, as well as *in vitro* (Rørth, 1970), the [2,3-DPG] changes followed the pH changes with a time lag of several hours. This explains why P_{50} may still be found to be reduced when measured at 37°C, pH 7.4, on the blood from a hibernating animal.

The Bohr effect reduces the affinity of hemoglobin when pH is lowered; it facilitates oxygen delivery to active tissues where the concentrations of CO_2 and acidic metabolites increase. Data on the Bohr effect factor, $\Delta \log P_{50}/\Delta pH$, in hibernators, are scanty and inconclusive. For Bartels *et al.* (1969), it has a normal value, −0.49, and is unaffected by hibernation. For Clausen and Ersland (1968), it is higher, −0.62 at 38°C, but is reduced at low temperature: −0.36 at 5°C; whereas for Kramm *et al.* (1975), also on the hedgehog, it would still be −0.67 at 20°C. The low [2,3-DPG] should in fact increase the Bohr effect factor (Duhm, 1976).

We can now combine these various effects in the case of the ground squirrel data of Fig. 13-11. Assuming that the reduction of [2,3-DPG] in hibernation and

TABLE 13-2

2,3-Diphosphoglycerate Concentrations in the Blood of Hibernators

Species	Euthermia	Hibernation	Units	Ref.[a]
Ground squirrels				
Citellus tridecemlineatus	11.2 ± 0.7	6.05 ± 0.35	μmoles/g Hb	1
C. lateralis	10.2 ± 1.1	5.3 ± 0.8	μmoles/g Hb	1
C. mexicanus	11.2 ± 1.3	6.5 ± 1.6	μmoles/g Hb	2
Woodchuck				
Marmota monax	5.2 ± 0.8	3.4 ± 0.4	μmoles/ml RBC	3
Golden hamster				
Mesocricetus auratus	19.2 ± 1.2	11.7 ± 1.2	μmoles/g Hb	4
Hedgehog				
Erinaceus europaeus	1.82 ± 0.06	1.46 ± 0.05	μmoles/mole Hb$_4$	5

[a] Sources: 1, Burlington and Whitten (1971); 2, Larkin (1973); 3, Harkness *et al.* (1974); 4, Tempel and Musacchia (1975); 5, Kramm *et al.* (1975). A small decrease in [2,3-DPG] in hibernation was also reported by Brock (1967).

its effect on P_{50} are the same as in hedgehogs (Kramm *et al.*, 1975), hibernating ground squirrels would have a P_{50} of $6 \times 40.9/33.3 = 7.4$ Torr in the absence of [2,3-DPG] change. The pH* change from euthermia to hibernation is -0.42; assuming a Bohr effect factor of -0.53, P_{50} would be reduced from 7.4 to 4.4 Torr in the absence of pH* change. Temperature is by far the major factor since it would by itself account for the reduction from 29 to 4.4 Torr. The net effect of acidosis would be to increase P_{50} from 4.4 to 6.0 Torr, favoring tissue oxygenation: from 4.4 to 7.4 Torr via the Bohr effect and from 7.4 to 6.0 Torr due to [2,3-DPG] decrease.

Hibernation is thus characterized by a low P_{50} (high affinity) of hemoglobin, resulting from the combined effect of temperature and respiratory acidosis; the latter increases P_{50} via the Bohr effect more than it decreases it via its influence on red cell metabolism and [2,3-DPG]. With such a low P_{50}, arterial blood oxygen saturation remains close to maximal down to a low P_{O_2} (Fig. 13-11), endowing the animal with a high tolerance toward ambient hypoxia.

Finally, it has been claimed (Kay, 1977) that euthermic hibernators would be characterized by a low P_{50} and a high [2,3-DPG], a trait also found in endurance runners, burrowers, and high-altitude mammals. However, all the hibernators listed are burrowers, and fossorial habits with the corresponding exposure to hypoxia may be more important than hibernation as an ecological factor favoring a low P_{50} (Ar *et al.*, 1977). In fact, the hedgehog, a nonfossorial hibernator, has a higher than average P_{50}: 41 Torr at 37°C (Kramm *et al.*, 1975) against 25–30 Torr in most mammals (Kay, 1977).

Tissue Oxygenation

On the tissue side now, having a very low P_{50} in hibernation could create a risk of tissue hypoxia, due to the low pressure head for the diffusion from capillaries to mitochondria (capillary P_{O_2} is generally assumed to be close to venous P_{O_2}). In fact, owing to the decrease of Krogh's diffusion constant (Q_{10} = 1.1; Dejours, 1975), the partial pressure gradient should be increased 25% in hibernation. However, a model study on a similar case shows that the temperature effects on P_{50} and diffusion constant are almost exactly offset by the reduction of metabolic rate (Reeves, 1978). Regarding hibernators, absence of tissue hypoxia is confirmed by the following evidence: (1) venous blood saturation is somewhat higher, not lower, in hibernation than in euthermia (Fig. 13-11); (2) tissue P_{O_2} values determined by the gas pocket method in hibernating hamsters (8°C) and marmots (7°C) were 13.3 ± 0.8 and 14.8 ± 0.7 Torr, respectively, in both cases probably much higher than P_{50}; venous blood saturation would thus have been more than 50% (Piquard and Malan, 1982); a range of 8–18 Torr was observed in four ground squirrels by Stormont et al. (1939); and (3) tissue lactate concentrations are quite low in hibernation (Twente and Twente, 1968).

An Apnea of a Hibernating Animal

We can now try to understand how a hibernating hibernator can survive the long periods of apnea characteristic of intermittent breathing, by bringing together the data on the various steps of gas exchange in the best-studied case, that of the hedgehog. The average duration of the apnea is 60 min (Kristoffersson and Soivio, 1964) (Fig. 13-1). The mean O_2 consumption $\dot{M}_{O_2}$ is 0.47 mmoles·kg^{-1}· hr^{-1}. For a 650-g animal, assuming that $\dot{M}_{O_2}$ stays constant during a breathing cycle, this represents a total consumption M_{O_2} of 308 μmoles over this time period. How can the animal make up for this oxygen requirement (Table 13-3)?

Blood Stores. During the apnea, the animal draws on the O_2 stored in arterial blood, as $P_{a_{O_2}}$ drops from 120 to 10.5 Torr (Tähti and Soivio, 1975). According to standard allometric relationships (Dejours, 1975), blood volume (V_b) is given by the following:

$$V_b = 51 \text{ ml} \times B^{0.99} \simeq 33 \text{ ml}$$

(B = body mass, kg). Of this about 25% is arterial blood, i.e., 8.3 ml (Farhi and Rahn, 1955). The corresponding oxygen carrying capacity is 6.83 mmoles/liter (Clausen and Ersland, 1968), i.e., 56.6 μmoles. Assuming the same hemoglobin dissociation curve as for the ground squirrel (Fig. 13-11) after correction for the higher P_{50}, one finds an O_2 saturation of 70% at 10.5 Torr. Assuming 100% saturation at 120 Torr, the decrease of arterial HbO$_2$ during apnea is 56.6 × 0.30 = 17.0 μmoles. To this one has to add the decrease in dissolved O_2 concentra-

TABLE 13-3

Estimated Sources of Oxygen during a 1 hr Apnea of a 650 g Hedgehog[a]

Source	Quantity of oxygen supplied (μmoles)	Oxygen consumption (%)
Arterial stores	19	6.2
Lung stores	44	14.3
O_2 diffusion	60	19.5
Bulk diffusion of air	65	20.9
Total	188	60.9

[a] Oxygen consumption: 308 μmole/hr. Glottis open.

tion, 2.0 μmoles. Altogether, the depletion of arterial O_2 stores supplies 19 μmoles of O_2, i.e., only 6.2% of the total consumption during the apnea.

Lung Stores. Here the animal has a choice between two tactics:

(1) The first is to fully inflate the lungs before the apnea, and close the glottis so that the lung can remain inflated without a sustained contraction of the respiratory muscles. This would maximize lung O_2 stores. The total lung capacity (TLC) is given by the allometric relationship (Dejours, 1975):

$$\text{TLC} = 54 \text{ ml} \times B = 35.1 \text{ ml}$$

For a given change in oxygen partial pressure (ΔP_{O_2}) the corresponding concentration change is given by:

$$\Delta C_{O_2} = \Delta P_{O_2}/RT$$

(R = constant of perfect gases, T = absolute temperature; Dejours, 1975). Assuming that alveolar and arterial P_{O_2} are equal, this gives $\Delta C_{A_{O_2}} = 6.29$ mmoles/liter and for 35.1 ml the decrease of lung O_2 stores would be 221 μmoles O_2, i.e., 72% of total M_{O_2} of apnea. With the blood stores this would add up to about 78% (but not 100%) of the oxgen consumption of 1 hr of apnea.

This possibility must probably be ruled out, however, since the hedgehog apparently begins the apnea at the end of an expiration (Fig. 13-1). In the case of the marmot (Fig. 13-2) cardiogenic oscillations are recorded as changes in chamber pressure throughout the apnea and, therefore, the glottis is certainly open.

(2) The second possibility, an open glottis and an apnea starting at the end of expiration, reduces lung stores but allows the diffusion of oxygen from the atmosphere to the lungs during the apnea, a phenomenon called apneic oxygenation (Clausen and Ersland, 1968). Assuming a functional residual capacity of

20% TLC, the lung O_2 stores are now 44 μmoles only. Oxygen diffusion into the lungs occurs by two processes simultaneously.

The first is the diffusion of O_2 down its partial pressure gradient. The diffusion coefficient of O_2 in the air at $0°C$ is 0.178 $cm^2 \cdot sec^{-1}$ (Weast, 1979). Morphometric data for the airways of a hedgehog are lacking; rough estimates of the lengths and cross-sectional areas were, therefore, derived from the model of human lung of Thurlbeck and Wang (1974) assuming that length scales as $TLC^{1/3}$ and cross-sectional area as $TLC^{2/3}$. In terms of diffusional resistance the airways of a hedgehog breathing through the mouth would be equivalent to a cylinder of 5.4 cm length and 0.09 cm^2 cross-sectional area. Assuming a mean P_{O_2} difference of 100 Torr (158–58) between ambient and alveolar gas (averaged over the apneic period) the net quantity of O_2 diffusing into the lungs during a 1-hr apnea would be about 60 μmoles (Table 13-3).

The second source of oxygen is bulk diffusion of air into the lungs. The quantity of CO_2 (M_{CO_2}) generated for a M_{O_2} of 308 μmoles is 231 μmoles (with a respiratory quotient of 0.75) and its volume is 23/31 of O_2 volume. Most of this CO_2, however, is not released into the lungs during the apnea, but remains within blood and tissues because of their high capacitance for CO_2 (Dejours, 1975). From the beginning to the end of an apnea, arterial C_{CO_2} increases 10.7 mmoles/liter (Table 13-1) of which about 6.2 mmoles/liter (10.7–4.5) result from respiratory changes. Should this value apply to the whole blood volume (33 ml), a total of 205 μmoles of CO_2 would be stored in the blood only (89% of the above figure for M_{CO_2}). Conversely, the diffusional flux of CO_2 to the atmosphere is much smaller than the O_2 flux: the alveolar-to-ambient partial pressure gradient is only 30.3 Torr on the average (alveolar: 26.5–34.6 Torr: Table 13-1; ambient: 0.25 Torr), whereas the diffusion coefficient is 0.139 $cm^2 \cdot sec^{-1}$ against 0.178 for oxygen (Weast, 1979). The diffusional flux of CO_2 is thus about 14 mmoles $\cdot hr^{-1}$, i.e., only 6.1% of $\dot{M}_{CO_2}$ and less than 5% of the oxygen uptake. Neglecting then the volume of CO_2 released into the lung during the apnea, the volume of O_2 uptake from the lungs will be replaced by an equal volume of ambient air, containing 21% O_2. This bulk diffusion thus provides up to 21% of the oxygen consumption.

Notice that these two diffusive sources of oxygen differ from the depletion of O_2 stores in that they represent an unlimited supply, the diffusive O_2 flux even increasing as the apnea is prolonged so that nearly 50% of the average $\dot{M}_{O_2}$ could be covered by the sum of these two fluxes. This is most important for an animal which can remain in apnea for up to 150 min.

Clausen and Ersland (1968) have stressed that due to the very low P_{50} of hemoglobin resulting from the low body temperature, most of the changes of arterial P_{O_2} take place over a nearly flat part of the hemoglobin dissociation curve (Fig. 13-11), whereas the venous point lies on the steep part. This results in a considerable "buffering" of $P\bar{v}_{O_2}$ changes while giving room for the wide fluctuations of $P_{a_{O_2}}$ associated with the apneic cycle.

Tissue Stores.   Heart and muscles contain myoglobin, an oxygen-binding protein. This can supply O_2 during a short muscular exercise, but whether it can intervene here is doubtful. Myoglobin P_{50} is about 0.5 Torr at 10°C (Antonini and Brunori, 1971) and less at 5°C. Only toward the end of an apnea, when $P_{v_{O_2}}$ starts declining significantly, could tissue P_{O_2} approach this figure, so that myoglobin-linked O_2 could be released to subserve local needs.

Is Oxygen Consumption Cyclic? The Analogy with Diving Mammals.
Altogether the various supplies of oxygen available during an apnea add up to 188 μmoles·hr^{-1}, far less than the 308 μmoles·hr^{-1} measured on the average over several breathing cycles (Table 13-3). The figure corresponding to bulk diffusion of air can only be overestimated. O_2 Diffusion was grossly estimated but the resistance due to nostrils was not taken into account. In the absence of experimental data, the functional residual capacity (FRC) of the hedgehog has been estimated to be 20% of total lung capacity (17% in man). This could be much too high since small mammals have a very low FRC due to their high chest wall compliance (Leith, 1976). The figure for lung O_2 stores is thus probably an overestimate. Even if the first and third quantities were underestimated by 50%, which is unlikely, the total would fail by 26% to account for the oxygen uptake of the animal. The balance would be even far more negative for the 150 min apnea recorded by Kristoffersson and Soivio (1964) or for the 99.8 min once recorded as a 24-hr average breathing period in a garden dormouse, *Eliomys quercinus*, by Pajunen (1974).

We are, thus, left with the unescapable conclusion that the hibernating animal behaves somewhat like a diving mammal (see Dejours, 1975) and reduces its oxygen consumption during apnea; the value measured would, thus, be an average over the breathing and apneic phases. The analogy can probably be pushed further, since the hedgehog presents another typical diving reflex, a bradycardia at the beginning of apnea: heart frequency drops from 6.8 min^{-1} during the rapid breathing to 4.5 min^{-1} at the onset of apnea (Kristoffersson and Soivio, 1964). The fluctuations of blood pressure (Tähti and Soivio, 1975) do not parallel the changes in heart rate. Assuming that stroke volume does not change considerably (which has been verified for diving mammals), there must be significant changes in peripheral circulatory resistance: this should increase sharply at the beginning of apnea since a drop in heart rate from 6.8 to 4.5 min^{-1} is accompanied by an increase of arterial pressure from 30/6 to 50/20 Torr; conversely, the resistance should reach a minimum during the breathing period when the peak heart rate coincides with the lowest arterial pressure. In diving mammals, these changes of peripheral resistance correspond to a circulatory redistribution: the reduced perfusion of some areas (muscles, viscera, etc.) is accompanied by a reduction of O_2 consumption which is not repaid as an O_2 debt during the next breathing period (this corresponds to a reduced energy utilization). The short dives of everyday life are not accompanied by a rise in blood lactate (Kooyman et al., 1980). A similar phenomenon occurring in the hibernating mammal would result in a cyclical O_2 consumption, and explain why the sum of O_2 fluxes in apnea (Table 13-3) can be lower than the average O_2 consumption. Of course, further evidence is needed to substantiate these conclusions.

Finally, a typical feature of apneic oxygenation is the progressive accumulation of CO_2 in the blood and tissues. The excess stored CO_2 needs to be washed out periodically to avoid sizeable acid-base fluctuations. This shows up as a periodical burst of consecutive breaths (Cheyne-Stokes) in the hedgehog and other species with very long apneas (garden dormouse, hamster), and as one or two consecutive breaths in species with shorter apneas such as the marmot. The washout of CO_2 would be facilitated by an increased blood perfusion (Farhi and Rahn, 1955).

Control of Ventilation

The low body temperature of hibernating mammals does not correspond to the loss of temperature regulation, but to its adjustment to a new controlled level (Heller, 1979; see also Chapters 3 and 4). The same seems to be true of the regulation of ventilation. Two lines of evidence support this contention: first, hibernating mammals react to increased CO_2 fractional concentration in inspired air by increasing their breathing rate. The CO_2 threshold was 2.5% in golden hamsters and ground squirrels, Citellus tridecemlineatus (Lyman, 1951), and about 2% in the marmot (Endres and Taylor, 1930). In the hedgehog, shortening of the apneic period occurred already at 0.7–1.7%. Periodic breathing was replaced by continuous breathing when CO_2 fraction reached 5.4–9.1% depending on individuals. Reducing oxygen fraction in a normocapnic mixture also elicited shortening of apneic periods, with a threshold of about 16% O_2 (Tähti, 1975). These figures might express a somewhat reduced sensitivity of ventilatory regulation compared to euthermia, but they represent only a frequency response and tidal volume might be a more sensitive index.

The existence of a regulation of ventilation is also evidenced by the fact that the statistical dispersion of arterial pH, and for many species that of Pa_{CO_2} are not significantly increased, in spite of the occurrence of periodic breathing (Table 13-1). Since the animal is an open system, in which Pa_{CO_2} and pH depend on an instantaneous ratio of alveolar ventilatory flow rate to CO_2 production, much wider excursions of these variables would be observed in the absence of ventilatory control.

The mechanisms underlying Cheyne–Stokes or intermittent breathing in hibernation are still unknown. Various explanations have been proposed for clinical occurrences, but nothing seems to justify their application to hibernation (Longobardo *et al.*, 1966; Dowell *et al.*, 1971). In view of the wide excursions of P_{O_2} tolerated by the hibernating hedgehog, Tähti (1978) has suggested that P_{CO_2} and pH rather than P_{O_2} should be the control variables of Cheyne-Stokes breathing.

Carotid chemoreceptor function of euthermic marmots shows no unambiguous differences when compared with other mammals (Leitner and Malan, 1973). No data are available for hibernation.

As concerns temperature regulation, hibernation is similar to an exaggerated slow wave sleep, both corresponding to the adjustment of the regulated variables to a new, stable level (Heller, 1979). The analogy can now be extended to ventilatory control. In both hibernation and slow wave sleep (Phillipson, 1978), arterial pH and P_{CO_2} are adjusted to a new level by the regulation of alveolar ventilation with an increased P_{CO_2} and decreased pH, i.e., a respiratory acidosis.

TRANSIENT STATES: ENTRANCE AND AROUSAL

In terms of respiration and acid–base state, three major features characterize the transition from euthermia to hibernation and back. The first is the change in body CO_2 stores: building up a respiratory acidosis during the entrance into hibernation requires the accumulation within the body of metabolically produced CO_2, by reducing ventilatory CO_2 output below metabolic production. The reverse occurs in arousal.

Second, these changes in CO_2 stores involve a readjustment of the control of ventilation to a new level. The mechanism and causation of this resetting are entirely unknown, however.

Third, the thermogenetic effort of arousal imposes a considerable load on the gas exchange system. In a chipmunk (*Tamias striatus*) arousing from torpor, maximum metabolic rate of about seven times basal metabolic rate (BMR) is achieved at a deep body temperature of 27°C (this is likely to be close to mean body temperature since thermal gradients are small in this animal) (Wang and Hudson, 1971). Assuming a Q_{10} of 2.5 this is equivalent to 17.5 times BMR (Malan, 1978b). The gas exchange system must then be working near maximal aerobic capacity, which raises a whole series of important questions, from lung diffusion capacity to mitochondrial surface area. These have been studied recently on a wide range of mammalian species (see the special issue of *Respiration Physiology,* Vol. 44, No. 1, 1981, on the design of mammalian respiratory system), but data are completely lacking for hibernating mammals. The following will therefore be restricted to the reversible changes in CO_2 stores and the associated phenomena.

Amplitude of the Changes in CO_2 Stores

In the intracellular pH experiments reported above (Malan *et al.*, 1982) tissue water contents and extracellular spaces were determined as well as their respective contributions to total body mass. From these data, the net changes in C_{CO_2} and charge balance B could be integrated over the whole body water, with some assumptions for the tissues for which no data were available (viscera, bone, skin and white fat). The net increase in CO_2 stores due to ventilatory changes (ΔS_{CO_2}) would be approximately 6.0 mmoles·kg^{-1} (calculated for a 340 g European hamster). This is a rough estimate in view of all the underlying assumptions, but may be used as a starting point.

How does this figure compare with the total metabolic production of CO_2 (M_{CO_2}) over the course of entrance or arousal? From Wang's (1978) data on the energy budget of a 400 g Richardson's ground squirrel, the total oxygen consumption (M_{O_2}) of the entrance into hibernation is about 72.5 mmoles. Assuming a metabolic respiratory quotient (R_{met}) at tissue level of 0.72, the corresponding M_{CO_2} is 52 mmoles. The change in CO_2 stores (ΔS_{CO_2}) is 2.4 mmoles and represents the accumulation of only 4.6% of the metabolically produced CO_2. The proportion is even smaller in arousal: due to the intense thermogenetic effort the total M_{O_2} is now 145 mmoles. Assuming an average R_{met} of 0.85, M_{CO_2} would be about 123 mmoles. The ventilatory output of CO_2 ($M_{CO_2} + \Delta S_{CO_2}$) needs to be increased only 1.9% (from 123 to 125.4 mmoles) to eliminate all the stored CO_2.

Entrance

Can some information about changes in CO_2 stores be derived from the recordings of $\dot{M}_{O_2}$ and $\dot{M}_{CO_2}$ that represent up to now the only available data on this subject? When CO_2 stores are changing, the ventilatory gas exchange ratio (or "respiratory quotient," $R_{I,E}$) calculated from the difference between inspired and expired air (Dejours, 1975) is no longer equal to the metabolic respiratory quotient $R_{met} = M_{CO_2}/M_{O_2}$:

$$R_{I,E} = \frac{M_{co_2} + \Delta S_{co_2}}{M_{o_2} + \Delta S_{o_2}} \tag{6}$$

The changes in O_2 stores (ΔS_{O_2}) brought about by hyper- or hypoventilation are small compared with ΔS_{CO_2} (Farhi and Rahn, 1955) and will be neglected here. Therefore:

$$\frac{R_{I,E}}{R_{met}} \simeq \frac{M_{co_2} + \Delta S_{co_2}}{M_{o_2}} \tag{7}$$

(ΔS_{CO_2} can be positive or negative). From the above figures, $R_{I,E}$ should thus be lower than R_{met} by 4.6% on the average during entrance into hibernation.

Snapp and Heller (1981) have tried to detect such changes in ground squirrels (*Citellus lateralis*). Only during the initial phase of entrance could a significant drop in $R_{I,E}$ be detected: from 0.81 to 0.74 to 0.87. This was more than 4.6%, but did not last long. Considering the uncertainty on the nature of oxidized substrates and, therefore, on R_{met}, plus the analytical error, this result can be considered as a first confirmation of the expected phenomenon.

Notice that it is easier to accumulate CO_2 at the beginning of entrance, when metabolic rate is still high. Referring again to Wang's data (Wang, 1978), at least 80% of the total M_{O_2} of entrance takes place while the body temperature is still over 27°C. It would take an average decrease of $R_{I,E}$ of 4.6/0.8 = 5.8% below R_{met} to complete the respiratory acidosis within this first phase of arousal.

Arousal

The first overt physiological process in the arousal of a marmot is a considerable hyperventilation. Without any change in breathing period at first, tidal volume measured by whole body plethysmography (Fig. 13-2) increases by a factor

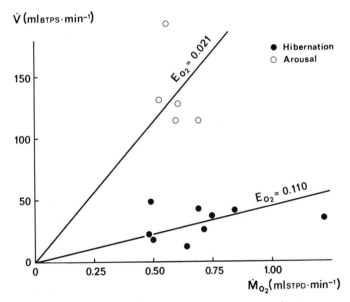

Fig. 13-12. Ventilatory flow rate ($\dot{V}$) versus oxygen consumption ($\dot{M}_{O_2}$) in marmots. In early arousal (open circles), ventilatory flow rate increases considerably prior to any increase in oxygen consumption above the hibernation rate (solid circles). (After Malan *et al.*, 1973.)

of at least 2.5. This occurs prior to any increase in metabolic rate or body temperature (Fig. 13-12) and represents, therefore, a most typical hyperventilation decreasing CO_2 stores. Since pneumographic methods usually indicate only breathing frequency, this hyperventilation had remained unnoticed to most earlier authors. It was visually observed by Kayser et al. (1954).

Preliminary data on the time course of blood acid–base variables (measured at 25°C) show a concomitant reduction of $P_{CO_2}*$ and rise in pH*. In one marmot, when a body temperature of 9.9°C (radiotelemetered from below the diaphragm) was reached, $P_{CO_2}*$ had already dropped 47% from 98.8 to 52.0 Torr, whereas pH* had increased from 7.16 to 7.26. Plasma C_{CO_2} had already decreased 35% from 43.9 to 28.5 mmoles/liter.

Measurements of P_{O_2} with a polarographic electrode in the abdominal cavity of the hedgehog (Barr and Silver, 1972) fairly agree with this. Over the first 30 min of arousal, as body temperature rose from 6° to 11°C, abdominal P_{O_2} increased from 11 to 30 Torr. This results from a respiratory change and not from the temperature effect on hemoglobin P_{50}, which according to the hedgehog data of Clausen and Ersland (1968) would rise only from 8.4 to 10.2 Torr over the same temperature range (and even less if pH* increases).

High values of gas exchange ratio $R_{I,E}$ have often been observed in early arousal by earlier authors. In the first phase of arousal (body temperature 5°–9.9°C), Kayser et al. (1954) recorded an average $R_{I,E}$ of 1.08 in the European hamster (*Cricetus cricetus*), 0.90 in the golden hamster (*Mesocricetus auratus*), and 0.91 in the ground squirrel (*Citellus citellus*). Compared with the overall $R_{I,E}$ of the rest of arousal for the same species, the relative increases are 28.5, 8.3, and 18.5%, respectively. A similar $R_{I,E}$ of 1.10 has been measured recently by Snapp and Heller (1981) on *Citellus lateralis*. With the exception of Dontcheff and Kayser (1934) and Kayser et al. (1954) these high "respiratory quotients" had been generally attributed to the switching from lipid to glucose utilization. It is now thought that during the first phase of arousal, thermogenesis is of nonshivering origin and results mainly from lipid oxidation, especially in brown adipose tissue; but see Chapter 7.

By combining the data of Kayser et al. (1954) on respiratory quotients and those of Wang (1978) on oxygen consumption in a ground squirrel, the amount of change of CO_2 stores during this first phase of arousal can be calculated. From the beginning of arousal until a body temperature of 10°C is reached, total M_{O_2} is about 8.5 mmoles, whereas $R_{I,E}$ is 0.908 compared with R_{met} of 0.766. Over this period of 70 min, the animal would eliminate 8.5 (0.908–0.766) = 1.2 mmoles of stored CO_2 [Eq. (7)], i.e., one-half the amount accumulated during the entrance into hibernation. This probably conservative estimate underlies the importance of the hyperventilation of early arousal in the acid–base modifications from hibernation to euthermia.

In the second phase of arousal, the high lactate production in shivering muscles (Ambid and Agid, 1975) can be expected to give rise to a metabolic acidosis. This is obvious in preliminary data published earlier (Malan, 1977). As for the entrance into hibernation, a systematic study of the time course of acid–base variables is still lacking.

ENVIRONMENTAL RESPIRATORY CONDITIONS AND HIBERNATION

Under field conditions, some species live and hibernate in environments far different from those used in most laboratory studies.

The first example is provided by species that live at high altitudes such as the marmots (*Marmota marmota* and *M. flaviventris*) and the ground squirrel (*Citellus lateralis*) (Bullard and Kollias, 1966). Paul Bert (1868) described a hibernation-like state induced by hypoxia in the garden dormouse at 12°C in winter; but was it true hibernation? Since then, unfortunately, no study has been done on the possible interaction of normocapnic hypoxia with hibernation.

Many hibernating species have fossorial habits. Hypoxic and hypercapnic conditions have been described in the burrows of several species of hibernators in summer: ground squirrels, *Citellus beecheyi* (Baudinette, 1974), *C. tridecemlineatus* (Studier and Baca, 1968; Studier and Procter, 1971), *C. parryi* (Williams and Rausch, 1973); chipmunks, *Eutamias quadrivittatus* (Studier and Baca, 1968); and marmots, *Marmota broweri* (Williams and Rausch, 1973). The deviations of the fractional concentrations F_{O_2} and F_{CO_2} from the normoxic and normocapnic values are rather small, however, and never exceed 2.6% on the average. This is well within the range observed in nonhibernating species such as the rabbit (Hayward, 1966); the pocket gopher, *Thomomys bottae* (Darden, 1972), or the mole rat, *Spalax ehrenbergi* (Arieli *et al.*, 1977). The last species lives in quite extreme conditions and presents a remarkable resistance to hypoxia and hypercapnia. Similar features have been described in hibernators, sometimes as "adaptations to hibernation," and since they are probably more related to fossorial habits than to hibernation they will not be reviewed here.

In the marmot, however, the preparation for hibernation is associated with a deliberate change in the burrow atmosphere (Williams and Rausch, 1973). The mouth of the burrow is tightly plugged with a mixture of soil, nesting material and feces. The same is used to seal all the cracks in the walls. This has been studied by the authors in semiartificial burrows, but a similar observation was reported to me by a trapper on the alpine marmot in natural conditions. The burrow atmosphere then becomes highly hypercapnic and hypoxic (Fig. 13-13) with F_{CO_2} reaching as high as 13.5% and F_{O_2} as low as 4% in extreme cases. High CO_2 of course facilitates the buildup of CO_2 stores characteristic of the

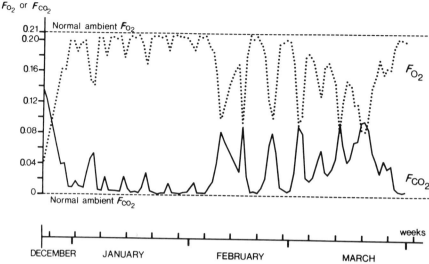

Fig. 13-13.   Fractional concentrations of oxygen (F_{O_2}) and carbon dioxide (F_{CO_2}) in the den of marmots, *Marmota broweri*. The weekly cycle obviously corresponds to the periodic arousals of one or several of the eight marmots hibernating in the den. In spring, the progressive increase in F_{CO_2} and decrease in F_{O_2} probably result from the increasing time spent in euthermia. (Redrawn from Williams and Rausch, 1973, reprinted with permission from *Comp. Biochem. Physiol.* © 1973, Pergamon Press Ltd.)

entrance into hibernation; the incidence of hypoxia is not known. Once the animals have entered hibernation, their metabolic rates become low enough to permit a slow return toward normoxia and normocapnia by diffusion of O_2 and CO_2 through the ground, or through the nest walls in the present case. The arousal will then take place in a nearly normoxic and normocapnic atmosphere, i.e., in the most favorable conditions for the intense aerobic effort of rewarming and for the release of CO_2 stores. Then the high metabolic rate of arousal will again create hypoxic and hypercapnic conditions for the next entrance, and the cycling continues throughout the winter. This is a unique example of behavioral regulation of respiratory variables in an air-breathing animal.

DISCUSSION: RESPIRATION AND THE PHYSIOLOGY OF HIBERNATION

We have seen that respiration in hibernation is characterized by two main features: intermittent breathing and respiratory acidosis. Let us now briefly summarize our conclusions and examine how they can be related to other traits of hibernation physiology, and especially to energetics.

Oxygen Transport

Oxygen transport is framed within two conditions imposed by hibernation physiology. One is the ventilatory adjustment of Pa_{CO_2} which results in an increased Pa_{O_2}, at least in normoxic conditions—thus, probably endowing the animal with an enhanced tolerance to hypoxia. The other is the high oxygen affinity of hemoglobin resulting mainly from the low body temperature. Arterial P_{O_2} can thus fluctuate over a wide range. Together with the low metabolic rate, this creates highly favorable conditions for the occurrence of intermittent breathing in which diffusive oxygenation probably plays an important role. This not only reduces evaporative water loss and energy utilization by respiratory muscles, but may also permit a periodical depression of metabolic rate during the apneas, like that found in diving mammals (with no buildup of lactate concentration). The possible mechanisms of such metabolic controls are not well understood, even for the diving mammals; they may involve the turning off of mitochondrial oxidations by a low oxygen partial pressure in tissues in which glycolytic activity is either small or inhibited.

Respiratory Acidosis

The occurrence of respiratory acidosis in hibernation is now well documented. As mentioned previously, it is a controlled phenomenon corresponding to a new adjustment of the regulation of ventilation, a phenomenon already known for temperature regulation and occurring also in slow wave sleep for both of these regulations. In some species, this autonomous regulation is accompanied by a behavioral regulation of burrow environment.

Can acidosis protect the tissues against cold injury? This has been proposed (for review, see Popovic and Popovic, 1974), but is unlikely. Exposed to similar fluctuations of body temperature, over which they have to remain active, most ectothermic vertebrates do not become acidotic but keep a constant pH* and P_{CO_2}* and a constant alpha imidazole (Figs. 13-5 and 13-6). The same is true of the peripheral tissues (leg muscles, etc.) of an endotherm, which are also exposed to a blood with a constant alpha imidazole whatever the fluctuations of local temperature (Reeves, 1976, 1977; Behrisch et al., 1977). In the hibernators themselves, the same alpha imidazole regulation is nearly achieved at the intracellular level in liver and heart, two important organs for survival, at the expense of energy-dependent ionic pumping (Figs. 13-9 and 13-10). Why should such a compensation occur if there was a benefit to the acidosis?

The acidosis may rather exert an inhibitory role (Malan et al., 1973; Malan, 1978b, 1980). As a potential inhibitory factor, CO_2 has several advantages. CO_2 concentration can be rapidly increased or decreased by ventilatory adjustments. Due to the high diffusivity of CO_2 these changes will spread over all intracellular body compartments (Reeves and Malan, 1976). The timing of the observed changes of CO_2 stores is in good agreement with such a role: an inhibitory factor has to be released prior to the major effort of rewarming—and arousal begins with a hyperventilation; the inhibition should start at the beginning of entrance and this is when a low RQ has been observed.

What is inhibited? Intracellular pH* is reduced in hibernation in most tissues including brain, in which the efficient metabolic compensation of acidosis seen in euthermia is absent. This opens the possibilities of both a general effect on metabolism and a more specific effect on nervous function.

The threshold temperature for shivering is lowered in guinea pigs breathing 15% CO_2, which brings about a respiratory acidosis similar to that of hibernation (Schaefer and Wünnenberg, 1976). This can be related to recent observations of Wünnenberg and Werner (1981): in an euthermic golden hamster, the firing rates of two types of neurons of the preoptic area of the hypothalamus (an essential area for temperature control) are depressed when the animal breathes 10% CO_2. The buildup and disappearance of respiratory acidosis might thus contribute to the shifts of thermogenic response observed in the entrance and arousal, respectively.

Acidosis affects a variety of cellular processes (for reviews, see Cohen and Iles, 1975; Roos and Boron, 1981). A low intracellular pH (pH_i) is a potent inhibitor of glycolysis, mainly by the inhibition of phosphofructokinase (Trivedi and Danforth, 1966). In the red blood cell, a 0.35 unit decrease in pH_i results in a 60% reduction of glycolytic rate (Tomoda et al., 1977). Conversely a 0.1 unit increase in pH_i suffices to explain the 40% increase in glycolytic rate elicited by insulin in frog muscle (Moore, 1979; Moore et al., 1979). In sea urchin eggs the onset of protein synthesis at fertilization is due to a 0.3 unit pH_i increase and can be reversibly suppressed by intracellular acidification (Grainger et al., 1979). In mammalian brown adipose tissue, extracellular acidosis reduces resting oxygen consumption and decreases by about 50% the thermogenic effect of norepinephrine (Friedli et al., 1978). In the whole dog in vivo, lowering blood pH by 0.31 unit inhibits the lipolytic effect of norepinephrine by 68% (Nahas and Poyart, 1967). Other effects of intracellular acidosis concern membrane ionic conductances, cell-to-cell coupling, muscle contractility, etc. (for review, see Roos and Boron, 1981). A wide range of possible metabolic effects of hibernation acidosis is thus open; inhibition of glycolysis, for instance, may explain the prevailing of gluconeogenesis (Chapter 8).

In nonmammalian vertebrates, respiratory acidosis is associated with an overall reduction of metabolic rate, beyond the normal Q_{10} effect of temperature, in several kinds of torpid states: the estivation of lungfishes, the wintering of turtles, and probably the daily torpor of birds (for review, see Malan, 1980). Similar effects have been obtained in experiments on euthermic mammals: breathing 15% CO_2 reduces oxygen consumption 31% at 25°C (Schaefer et al.,

1975), in spite of the associated release of norepinephrine. Nonshivering thermogenesis is also inhibited by hypercapnia *in vivo* (Varnai *et al.,* 1971; Pepelko and Dixon, 1974; Farkas and Donhoffer, 1975).

Can acidosis be responsible for a similar reduction of metabolic rate in hibernation? By collecting the literature data on metabolic rates in hibernators, Kayser (1964) derived two allometric relationships between metabolic rate (MR) and body mass, for euthermia and hibernation, respectively. The mass exponents being similar, an average ratio of euthermia to hibernation MR of 31:1 can be calculated for 13 species ranging from bear to bat. With a Q_{10} of 2.5, temperature alone could account for a reduction of only 12:1. It was then proposed that acidosis could contribute to the remaining 60% reduction (Malan, 1978b, 1980). This hypothesis has been criticized recently by Snapp and Heller (1981) on two main counts. First, the data of earlier authors should be reexamined because of insufficient information on body and ambient temperatures, seasonal effects on MR, etc. Second, from their own measurements on *Citellus lateralis* with EEG measurement of sleep states, they find that from slow wave sleep to deep hibernation the animal reduces its MR with body temperature with an average Q_{10} of 2.5 ± 0.3.

To ascertain whether or not there is a metabolic reduction beyond the effect of temperature, one needs a sound reference concerning the change of enzymatic reaction rates with temperature in hibernating animals. The metabolic enzymes listed by Charnock (1978)—not including ionic transport ATPases—have a mean energy of activation of 14.25 kcal/mole, which for a 36°–6°C temperature drop gives a MR reduction of 11.9:1 and a Q_{10} of 2.28. The difference between Snapp and Heller's figure of 2.5 and this one is less than the standard deviation, but suffices to make room for a 32% reduction in MR not explained by enzyme Q_{10}. Statistical noise is thus too high to let us settle the debate. Two observations should be added however: first, by taking the MR of sleep as a starting point, Snapp and Heller utilize a value which is already lowered compared to wakefulness; this reduction may already be related to the acidosis which is known to occur in diurnal sleep in man (Phillipson, 1978). Second, as was already mentioned by Snapp and Heller, large size-related interspecific differences may exist. The calculation procedure used by Kayser (1964) minimized the euthermia-to-hibernation MR reduction for smaller species. Morrison (1960) predicted that smaller hibernators could not survive long enough on their fat reserves without a MR reduction beyond Q_{10} effect. And are all reports of metabolic reduction ratios higher than 12:1 or are high Q_{10} values just experimental artifacts?

In conclusion, even though our knowledge is far from being complete, it is very likely that either by its effect on nervous control of temperature or by the metabolic consequences of intermittent breathing and acidosis, respiration plays a significant role in an essential aspect of hibernation physiology: energy conservation.

REFERENCES

Ambid, L., and Agid, R. (1975). Lactates tissulaires, pH et gaz du sang chez le lérot hibernant: évolution au cours des réveils périodiques. *C. R. Seances Soc. Biol. Ses Fil.* **169,** 1380–1385.

Antonini, E., and Brunori, M. (1971). "Hemoglobin and Myoglobin in Their Reactions with Ligands." North-Holland Publ., Amsterdam.

Ar, A., Arieli, R., and Shkolnik, A. (1977). Blood gas properties and function in fossorial mole rat under normal and hypoxic-hypercapnic atmospheric conditions. *Respir. Physiol.* **30,** 201–219.

Arieli, R., Ar, A., and Shkolnik, A. (1977). Metabolic responses of a fossorial rodent (*Spalax ehrenbergi*) to simulated burrow conditions. *Physiol. Zool.* **50,** 61–75.

Astrup, P. (1970). Dependence of oxyhemoglobin dissociation and intraerythrocytic 2,3-DPG on acid-base status of blood. II. Clinical and experimental studies. *In* "Red Cell Metabolism and Function" (G. J. Brewer, ed.), pp. 67–79. Plenum, New York.

Barr, R. E., and Silver, I. A. (1972). Oxygen tension in the abdominal cavity of the hedgehogs during arousal from hibernation. *Respir. Physiol.* **16,** 16–21.

Bartels, H., Schmelzle, R., and Ulrich, S. (1969). Comparative studies of the respiratory function of mammalian blood. V. Insectivora: shrew, mole and nonhibernating and hibernating hedgehog. *Respir. Physiol.* **7,** 278–286.

Baudinette, R. V. (1974). Physiological correlates of burrow gas conditions in the California ground squirrel. *Comp. Biochem. Physiol. A* **48,** 733–743.

Behrisch, H., Ortner, I., and Wieser, W. (1977). Temperature and the regulation of enzyme activity in heterothermic tissues of an alpine mammal. Lactate dehydrogenase from skeletal muscle of the chamois. *J. Therm. Biol.* **2,** 185–189.

Benedict, F. G., and Lee, R. C. (1938). Hibernation and marmot physiology. *Carnegie Inst. Washington, Publ.* 497.

Bert, P. (1868). Hibernation artificielle des lérots dans une atmosphère lentement appauvrie en oxygène. *C. R. Seances Soc. Biol. Ses Fil.* **5,** 13.

Biörck, G., Johansson, B. W., and Schmid, H. (1956). Reactions of hedgehogs, hibernating and non-hibernating, to the inhalation of oxygen, carbon dioxide and nitrogen. *Acta Physiol. Scand.* **37,** 71–83.

Brock, M. A. (1967). Erythrocyte glycolytic intermediates of control, cold-exposed and hibernating hamsters. *In* "Mammalian Hibernation III" (K. C. Fisher, A. R. Dawe, C. P. Lyman, E. Schönbaum, and F. E. South, eds.), pp. 409–420. Oliver & Boyd, Edinburgh.

Bullard, R. W., and Kollias, J. (1966). Functional characteristics of two high altitude mammals. *Fed. Proc., Fed. Am. Soc. Exp. Biol.* **25,** 1288–1292.

Burlington, R. F., and Whitten, B. K. (1971). Red cell 2,3-diphosphoglycerate in hibernating ground squirrels. *Comp. Biochem. Physiol. A* **38,** 469–471.

Chao, I., and Yeh, C. J. (1950). Hibernation of the hedgehog. II. Respiratory patterns. *Chin. J. Physiol.* **17,** 379–390.

Chapin, J. L. (1954). Ventilatory response of the unrestrained and unanesthetized hamster to CO_2. *Am. J. Physiol.* **179,** 146–148.

Charnock, J. S. (1978). Membrane lipid-phase transitions: a possible biological response to hibernation? *In* "Strategies in Cold: Natural Torpidity and Thermogenesis" (L. C. H. Wang and J. W. Hudson, eds.), pp. 417–460. Academic Press, New York.

Clausen, G. (1966). Acid-base balance in the hedgehog *Erinaceus europaeus* (L.) during hibernation hypothermia, cooling and rewarming. *Arbok Univ. Bergen, Mat.-Naturvitensk. Ser.* No. 6, 1–11.

Clausen, G., and Ersland, A. (1968). The respiratory properties of the blood of the hibernating hedgehog *Erinaceus europaeus* L. *Respir. Physiol.* **5,** 221–233.

Cohen, R. D., and Iles, R. A. (1975). Intracellular pH: measurement, control and metabolic interrelationships. *CRC Crit. Rev. Clin. Lab. Sci.* **6**, 101–143.

Darden, T. R. (1972). Respiratory adaptations of a fossorial mammal, the pocket gopher (*Thomomys bottae*). *J. Comp. Physiol.* **78**, 121–137.

Dejours, P. (1975). "Principles of Comparative Respiratory Physiology." North-Holland Publ., Amsterdam.

Dontcheff, L., and Kayser, C. (1934). Explication de certains quotients respiratoires aberrants chez les hibernants. *C. R. Seances Soc. Biol. Ses Fil.* **118**, 81–83.

Dowell, A. R., Buckley, C. E., Cohen, R., Whalen, R. E., and Sieker, H. O. (1971). Cheyne-Stokes respiration. A review of clinical manifestations and critique of physiological mechanisms. *Arch. Intern. Med.* **127**, 712–726.

Drorbaugh, J. E., and Fenn, W. O. (1955). A barometric method for measuring ventilation in newborn infants. *U.S. Air Force Syst. Command, Res. Technol. Div., Air Force Mater. Lab., Tech. Rep.* **WADC-TR-55-357**, 370–378.

Dubois, R. (1896). Etude sur le mécanisme de la thermogenèse et du sommeil chez les mammifères. Physiologie comparée de la marmotte. *Ann. Univ. Lyon* **25**.

Duhm, J. (1976). Dual effect of 2,3-diphosphoglycerate on the Bohr effects of human blood. *Pfluegers Arch.* **363**, 55–60.

Endres, G. (1924). Die physikalisch-chemische Atmungsregulation bei winterschlafenden Warmblütern. *Z. Gesamte Exp. Med.* **43**, 311–331.

Endres, G., and Taylor, H. (1930). Observations on certain physiological processes of marmot. II. The respiration. *Proc. R. Soc. (London), Ser. B* **107**, 231–240.

Epstein, M. A. F., and Epstein, R. A. (1978). Theoretical analysis of barometric method for measurement of tidal volume. *Respir. Physiol.* **32**, 105–120.

Farhi, L. E., and Rahn, H. (1955). Gas stores of the body and the unsteady state. *J. Appl. Physiol.* **7**, 472–484.

Farkas, M., and Donhoffer, S. (1975). The effect of hypercapnia on heat production and body temperature in the newborn guinea pig. *Acta Physiol. Acad. Sci. Hung.* **46**, 201–217.

Fenton, R. A., Gonzalez, N. C., and Clancy, R. L. (1978). The effect of dibutyryl cyclic AMP and glucagon on the myocardial cell pH. *Respir. Physiol.* **32**, 213–223.

Friedli, C., Chinet, A., and Girardier, L. (1978). Comparative measurements of *in vitro* thermogenesis of brown adipose tissue from control and cold-adapted rats. *In* "Effectors of Thermogenesis" (L. Girardier and J. Seydoux, eds.), pp. 259–266. Birkhaeuser, Basel.

Goodrich, C. A. (1973). Acid-base balance in euthermic and hibernating marmots. *Am. J. Physiol.* **224**, 1185–1189.

Grainger, J. L., Winkler, M. M., Shen, S. S., and Steinhardt, R. A. (1979). Intracellular pH controls protein synthesis rate in the sea urchin egg and early embryo. *Dev. Biol.* **68**, 396–406.

Haller, A. C., and Zimny, M. L. (1977). Effects of hibernation on interradicular alveolar bone. *J. Dent. Res.* **56**, 1552–1557.

Harkness, D. R., Roth, S., and Goldman, P. (1974). Studies on the red blood cell oxygen affinity and 2,3-diphosphoglyceric acid in the hibernating woodchuck (*Marmota monax*). *Comp. Biochem. Physiol. A* **48**, 591–599.

Hayward, J. S. (1966). Abnormal concentrations of respiratory gases in rabbit burrows. *J. Mammal.* **47**, 723–724.

Hazel, J. R., Garlick, W. S., and Sellner, P. A. (1978). The effects of assay temperature upon the pH optima of enzymes from poikilotherms: a test of the imidazole alphastat hypothesis. *J. Comp. Physiol.* **123**, 97–104.

Heller, H. C. (1979). Hibernation: neural aspects. *Annu. Rev. Physiol.* **41**, 305–321.

Hock, R. J. (1958). Hibernation. *Josiah Macy Found. Conf. Cold Injury, 5th* pp. 61–133.

Howell, B. J., Baumgardner, F. W., Bondi, K., and Rahn, H. (1970). Acid-base balance in cold-blooded vertebrates as a function of body temperature. *Am. J. Physiol.* **218**, 600–606.

Jacey, M. J., and Schaefer, K. E. (1972). The effects of chronic hypercapnia on blood phosphofructokinase activity and the adenine nucleotide system. *Respir. Physiol.* **16**, 267–272.

Jackson, D. C. (1971). The effect of temperature on ventilation in the turtle, *Pseudemys scripta elegans. Respir. Physiol.* **12**, 131–140.

Kay, F. R. (1977). 2,3-diphosphoglycerate, blood oxygen dissociation and the biology of mammals. *Comp. Biochem. Physiol. A* **57**, 309–316.

Kayser, C. (1940). Les échanges respiratoires des hibernants à l'état de sommeil hibernal. *Ann. Physiol. Physicochim. Biol.* **16**, 128–221.

Kayser, C. (1961). "The Physiology of Natural Hibernation." Pergamon, Oxford.

Kayser, C. (1964). La dépense d'énergie des mammifères en hibernation. *Arch. Sci. Physiol.* **18**, 137–150.

Kayser, C., and Franck, R. M. (1963). Comportement des tissus calcifiés du hamster d'Europe *Cricetus cricetus* au cours de l'hibernation. *Arch. Oral Biol.* **8**, 703–718.

Kayser, C., Rietsch, M. L., and Lucot, M. A. (1954). Les échanges respiratoires et la fréquence cardiaque des hibernants au cours du réveil de leur sommeil hivernal. *Arch. Sci. Physiol.* **8**, 155–194.

Kent, K. M., and Peirce, E. C., II (1967). Acid-base characteristics of hibernating animals. *J. Appl. Physiol.* **23**, 336–340.

Kooyman, G. L., Wahrenbrock, E. A., Castellini, M. A., Davis, R. W., and Sinnett, E. E. (1980). Aerobic and anaerobic metabolism during voluntary diving in Weddell seals: evidence of preferred pathways from blood chemistry and behavior. *J. Comp. Physiol.* **138**, 335–346.

Kramm, C., Sattrup, G., Baumann, R., and Bartels, H. (1975). Respiratory function of blood in hibernating and non-hibernating hedgehogs. *Respir. Physiol.* **25**, 311–318.

Kreienbühl, G., Strittmatter, J., and Ayim, E. (1976). Blood gas analyses of hibernating hamsters and dormice. *Pfluegers Arch.* **366**, 167–172.

Kristoffersson, R., and Soivio, A. (1964). Hibernation in the hedgehog (*Erinaceus europaeus* L.). Changes of respiratory pattern, heart rate and body temperature in response to gradually decreasing or increasing ambient temperature. *Ann. Acad. Sci. Fenn., Ser. A4* **82**, 3–17.

Kristoffersson, R., and Soivio, A. (1966). Duration of hypothermia periods and type of respiration in hibernating golden hamster, *Mesocricetus auratas* Waterh. *Ann. Zool. Fenn.* **3**, 66–67.

Kristoffersson, R., and Soivio, A. (1967). Studies on the periodicity of hibernation in the hedgehog (*Erinaceus europaeus* L.). II. Changes of respiratory rhythm, heart rate and body temperature at the onset of spontaneous and induced arousals. *Ann. Zool. Fenn.* **4**, 595–597.

Larkin, E. C. (1973). The response of erythrocyte organic phosphate levels of active and hibernating ground squirrels (*Spermophilus mexicanus*) to isobaric hyperoxia. *Comp. Biochem. Physiol. A* **45**, 1–6.

Leith, D. (1976). Comparative mammalian respiratory mechanics. *Physiologist* **19**, 485–510.

Leitner, L. M., and Malan, A. (1973). Possible role of the arterial chemoreceptors in the ventilatory response of the anesthetized marmot to changes in inspired P_{O_2} and P_{CO_2}. *Comp. Biochem. Physiol. A* **45**, 953–959.

Longobardo, G. S., Cherniack, N. S., and Fishman, A. P. (1966). Cheyne–Stokes breathing produced by a model of the human respiratory system. *J. Appl. Physiol.* **21**, 1839–1846.

Lyman, C. P. (1951). Effect of increased CO_2 on respiration and heart rate of hibernating hamsters and ground squirrels. *Am. J. Physiol.* **167**, 638–643.

Lyman, C. P., and Hastings, A. B. (1951). Total CO_2, plasma pH and P_{CO_2} of hamsters and ground squirrels during hibernation. *Am. J. Physiol.* **167**, 633–637.

Malan, A. (1973). Ventilation measured by body plethysmography in hibernating mammals and poikilotherms. *Respir. Physiol.* **17**, 32–44.

Malan, A. (1977). Blood acid-base state at a variable temperature. A graphical representation. *Respir. Physiol.* **31**, 259-275.

Malan, A. (1978a). Intracellular acid-base state at a variable temperature in air-breathing vertebrates and its representation. *Respir. Physiol.* **33**, 115-119.

Malan, A. (1978b). Hibernation as a model for studies on thermogenesis and its control. *In* "Effectors of Thermogenesis" (L. Girardier and J. Seydoux, eds.), pp. 303-314. Birkhaeuser, Basel.

Malan, A. (1980). Enzyme regulation, metabolic rate and acid-base state in hibernation. *In* "Animals and Environmental Fitness" (R. Gilles, ed.), pp. 487-501. Pergamon, Oxford.

Malan, A., Arens, H., and Waechter, A. (1973). Pulmonary respiration and acid-base state in hibernating marmots and hamsters. *Respir. Physiol.* **17**, 45-61.

Malan, A., Wilson, T. L., and Reeves, R. B. (1976a). Intracellular pH in cold-blooded vertebrates as a function of body temperature. *Respir. Physiol.* **28**, 29-47.

Malan, A., Daull, F., and Rodeau, J. L. (1976b). Relations entre les états acide-base extra- et intracellulaire dans l'acidose de l'hibernation. *J. Physiol. (Paris)* **72**, 105A.

Malan, A., Rodeau, J. L., and Daull, F. (1982). Intracellular pH in hibernation and respiratory acidosis in the European hamster. Submitted for publication.

Messeter, K., and Siesjö, B. K. (1971). The intracellular pH' in the brain in acute and sustained hypercapnia. *Acta Physiol. Scand.* **83**, 210-219.

Messier, A. A., and Schaefer, K. E. (1971). The effect of chronic hypercapnia on oxygen affinity and 2,3-diphosphoglycerate. *Respir. Physiol.* **12**, 291-296.

Moore, R. D. (1979). Elevation of intracellular pH by insulin in frog skeletal muscle. *Biochem. Biophys. Res. Commun.* **91**, 900-904.

Moore, R. D., Fidelman, M. L., and Seeholzer, S. H. (1979). Correlation between insulin action upon glycolysis and change in intracellular pH. *Biochem. Biophys. Res. Commun.* **91**, 905-910.

Morrison, P. R. (1960). Some interrelations between weight and hibernation function. *Bull. Mus. Comp. Zool.* **124**, 75-91.

Musacchia, X. J., and Volkert, W. A. (1971). Blood gases in hibernating and active ground squirrels: HbO_2 affinity at 6 and 38°C. *Am. J. Physiol.* **221**, 128-130.

Nahas, G. C., and Poyart, C. (1967). Effect of arterial pH alterations on metabolic activity of norepinephrine. *Am. J. Physiol.* **212**, 765-772.

Pajunen, I. (1970). Body temperature, heart rate, breathing pattern, weight loss and periodicity of hibernation in the Finnish garden dormouse, *Eliomys quercinus* L. *Ann. Zool. Fenn.* **7**, 251-266.

Pajunen, I. (1974). Body temperature, heart rate, breathing pattern, weight loss and periodicity of hibernation in the French garden dormouse, *Eliomys quercinus* L., at 4.2 ± 0.5°C. *Ann. Zool. Fenn.* **11**, 107-119.

Park, Y. S., and Hong, S. K. (1976). Properties of toad skin Na,K-ATPase with special reference to the effect of temperature. *Am. J. Physiol.* **231**, 1356-1363.

Patrizi, M. L. (1897). Contributo allo studio dei movimenti respiratori negli ibernanti. Nota critico-sperimentale. *Accad. Sci. Med. Nat. Ferrara* Apr. 22. Cited in Kayser (1961).

Pembrey, M. S. (1901-1902). Observations upon the respiration and temperature of the marmot. *J. Physiol. (London)* **27**, 66-84.

Pembrey, M. S., and Pitts, A. G. (1899). The relation between internal temperature and respiratory movements in hibernating animals. *J. Physiol. (London)* **24**, 305-316.

Pepelko, W. E., and Dixon, G. A. (1974). Elimination of cold-induced nonshivering thermogenesis by hypercapnia. *Am. J. Physiol.* **227**, 264-267.

Person, R. S. (1950). Effect of temperature on the CO_2 retention by the blood of some mammals. *Biokhimija (Moscow)* **15**, 346-353. (In Russ.)

Phillipson, E. A. (1978). Respiratory adaptations in sleep. *Annu. Rev. Physiol.* **40**, 133-156.

Piquard, F., and Malan, A. (1982). Gas pocket estimation of tissue P_{O_2} and P_{CO_2} in hibernating marmots and hamsters. Submitted for publication.

Popovic, V., and Popovic, P. (1974). "Hypothermia in Biology and in Medicine." Grune & Stratton, New York.

Rahn, H. (1967). Gas transport from the environment to the cell. In "Development of the Lung" (A. V. S. de Reuck and R. Porter, eds.), pp. 3–23. Churchill, London.

Rasmussen, A. T. (1916). A further study of the blood gases during hibernation in the woodchuck (*Marmota monax*). The respiratory capacity of the blood. *Am. J. Physiol.* **41**, 162–172.

Reeves, R. B. (1972). An imidazole alphastat hypothesis for vertebrate acid–base regulation: tissue carbon dioxide content and body temperature in bullfrogs. *Respir. Physiol.* **14**, 219–236.

Reeves, R. B. (1976). Temperature induced changes in blood acid–base status: pH and P_{CO_2} in a binary buffer. *J. Appl. Physiol.* **40**, 752–761.

Reeves, R. B. (1977). The interaction of body temperature and acid–base balance in ectothermic vertebrates. *Annu. Rev. Physiol.* **39**, 559–586.

Reeves, R. B. (1978). Temperature and acid–base balance effects on oxygen transport by human blood. *Respir. Physiol.* **33**, 99–102.

Reeves, R. B. (1980). The effect of temperature on the oxygen equilibrium curve of human blood. *Respir. Physiol.* **42**, 317–328.

Reeves, R. B., and Malan, A. (1976). Model studies of intracellular acid–base temperature responses in ectotherms. *Respir. Physiol.* **28**, 49–63.

Rispens, P., Dellebarre, C. W., Eleveld, D., Helder, W., and Zijlstra, W. G. (1968). The apparent first dissociation constant of carbonic acid in plasma between 16 and 42.5°. *Clin. Chim. Acta* **22**, 627–637.

Robin, E. D. (1962). Relationships between temperature and plasma pH and carbon dioxide tension in the turtle. *Nature (London)* **195**, 249–251.

Rodeau, J. L., and Malan, A. (1979). A two compartment model of blood acid–base state at constant or variable temperature. *Respir. Physiol.* **37**, 5–30.

Rørth, M. (1970). Dependence of oxyhemoglobin dissociation and intraerythrocytic 2,3-DPG on acid–base status of blood. I. *In vitro* studies on reduced and oxygenated blood. In "Red Cell Metabolism and Function" (G. J. Brewer, ed.), pp. 57–65. Plenum, New York.

Roos, A., and Boron, W. F. (1981). Intracellular pH. *Physiol. Rev.* **61**, 296–434.

Schaefer, K. E., and Wünnenberg, W. (1976). Threshold temperatures for shivering in acute and chronic hypercapnia. *J. Appl. Physiol.* **41**, 67–70.

Schaefer, K. E., Messier, A. A., Morgan, C., and Baker, G. T., III (1975). Effect of chronic hypercapnia on body temperature regulation. *J. Appl. Physiol.* **38**, 900–906.

Snapp, B. D., and Heller, H. C. (1981). Suppression of metabolism during hibernation in ground squirrels (*Citellus lateralis*). *Physiol. Zool.* **54**, 297–307.

Soivio, A., Tähti, H., and Kristoffersson, R. (1968). Studies on the periodicity of hibernation in the hedgehog (*Erinaceus europaeus* L.). III. Hibernation in a constant ambient temperature of −5°C. *Ann. Zool. Fenn.* **5**, 224–226.

Stewart, P. A. (1978). Independent and dependent variables of acid–base control. *Respir. Physiol.* **33**, 9–26.

Stormont, R. T., Foster, M. A., and Pfeiffer, C. (1939). Plasma pH, CO_2 content of the blood and "tissue gas" tensions during hibernation. *Proc. Soc. Exp. Biol. Med.* **42**, 56–59.

Studier, E. H., and Baca, T. P. (1968). Atmospheric conditions in artificial rodent burrows. *Southwest. Nat.* **13**, 401–410.

Studier, E. H., and Procter, J. W. (1971). Respiratory gases in burrows of *Spermophilus tridecimlineatus*. *J. Mammal.* **52**, 631–633.

Suomalainen, P., and Rosoviki, V. (1973). Studies on the physiology of the hibernating hedgehog.

17. The blood cell count at different times of the year and in different phases of the hibernation cycle. *Ann. Acad. Sci. Fenn. Ser. A4* **198,** 1–8.

Svihla, A., and Bowman, H. C. (1952). Oxygen carrying capacity of the blood of dormant ground squirrels. *Am. J. Physiol.* **171,** 479–481.

Tähti, H. (1975). Effects of changes in CO_2 and O_2 concentrations in the inspired gas on respiration in the hibernating hedgehog (*Erinaceus europaeus* L.). *Ann. Zool. Fenn.* **12,** 183–187.

Tähti, H. (1978). Periodicity of hibernation in the hedgehog (*Erinaceus europaeus* L.)—Seasonal respiratory variations with special reference to the regulations of Cheyne–Stokes respiration. Thesis, Univ. of Helsinki, Helsinki.

Tähti, H., and Soivio, A. (1975). Blood gas concentrations, acid-base balance and blood pressure in hedgehogs in the active state and in hibernation with periodic respiration. *Ann. Zool. Fenn.* **12,** 188–192.

Tähti, H., and Soivio, A. (1977). Respiratory and circulatory differences between induced and spontaneous arousals in hibernating hedgehogs (*Erinaceus europaeus* L.). *Ann. Zool. Fenn.* **14,** 197–202.

Tempel, G. E., and Musacchia, X. J. (1975). Erythrocyte 2,3-diphosphoglycerate concentration in hibernating, hypothermic and rewarming hamsters. *Proc. Soc. Exp. Biol. Med.* **148,** 588–592.

Tempel, G. E., Wolinsky, I., and Musacchia, X. J. (1978). Bone and serum calcium in normothermic, cold-acclimated and hibernating hamsters. *Comp. Biochem. Physiol. A* **61,** 145–147.

Thurlbeck, W. M., and Wang, N. S. (1974). The structure of the lungs. *MTP Int. Rev. Sci.: Physiol.* **2,** 1–30.

Tomoda, A., Tsuda-Hirota, S., and Minakami, S. (1977). Glycolysis of red cells suspended in solutions of impermeable solutes. Intracellular pH and glycolysis. *J. Biochem. (Tokyo)* **81,** 697–701.

Trivedi, B., and Danforth, W. H. (1966). Effect of pH on the kinetics of frog muscle phosphofructokinase. *J. Biol. Chem.* **241,** 4110–4112.

Twente, J. W., and Twente, J. A. (1968). Concentrations of L-lactate in the tissues of *Citellus lateralis* after known intervals of hibernating periods. *J. Mammal.* **49,** 541–544.

Van Liew, H. D. (1962). Tissue P_{O_2} and P_{CO_2} estimation with rat subcutaneous gas pockets. *J. Appl. Physiol.* **17,** 851–855.

Varnai, J., Farkas, M., and Donhoffer, S. (1971). Thermoregulatory responses to hypercapnia in the new-born rabbit. *Acta Physiol. Acad. Sci. Hung.* **40,** 145–172.

Wang, L. C. H. (1978). Energetic and field aspects of mammalian torpor: the Richardson's ground squirrel. *In* "Strategies in Cold: Natural Torpidity and Thermogenesis" (L. C. H. Wang and J. W. Hudson, eds.), pp. 109–145. Academic Press, New York.

Wang, L. C. H., and Hudson, J. W. (1971). Thermoregulation in normothermic and hibernating eastern chipmunk, *Tamias striatus*. *Comp. Biochem. Physiol.* **38,** 59–90.

Weast, R. C. (1979). "CRC Handbook of Chemistry and Physics." CRC Press, Boca Raton, Florida.

Wenberg, G. M., Holland, J. C., and Sewell, J. (1973). Some aspects of the hematology and immunology of the hibernating and non-hibernating woodchuck (*Marmota monax*). *Comp. Biochem. Physiol. A* **46A,** 513–521.

Williams, D. D., and Rausch, R. L. (1973). Seasonal carbon dioxide and oxygen concentrations in the dens of hibernating mammals (Sciuridae). *Comp. Biochem. Physiol. A* **44A,** 1227–1235.

Wood, S. C., and Schaefer, K. E. (1978). Regulation of intracellular pH in lungs and tissues during hypercapnia. *J. Appl. Physiol.* **45,** 115–118.

Woodbury, J. W. (1965). Regulation of pH. *In* "Physiology and Biophysics" (T. C. Ruch and H. D. Patton, eds.), pp. 899–938. Saunders, Philadelphia, Pennsylvania.

Wünnenberg, W., and Werner, R. (1981). Responses of single units of thermosensitive preoptic area (POA) to hypercapnia. *In* "Advances in Physiology." (Z. Szelényi and M. Székely, eds.), Vol. 32, pp. 93–95. Pergamon, New York.

Zimmerman, G. D., McKean, T. A., and Hardt, A. B. (1976). Hibernation and disuse osteoporosis. *Cryobiology* **13,** 84–94.

Zimny, M. L., and Haller, A. C. (1978). Effects of hibernation on dental tissues: a SEM analytical study. *Comp. Biochem. Physiol. A* **60,** 257–262.

14

Recent Theories of Hibernation

Twente and Twente (1978) have recently taken the stand that the maintenance of hibernation and the initiation of arousal are independent of the regulation of thermogenesis. Rather, they consider that all phases of hibernation are regulated and controlled by parasympathetic and sympathetic activity. They use the heart rate as an index of autonomic activity and undertake to demonstrate that the uneven heart rate of animals in hibernation and a preponderance of long R-R intervals indicate a high level of parasympathetic influence. Intraperitoneal injection of atropine in the hibernating bat (*Eptesicus fuscus*), Columbian ground squirrel (*Citellus columbianus*), or golden mantled ground squirrel (*Citellus lateralis*) resulted in a tachycardia and an abolishment of the uneven heart rate after a latent period of a minimum of 2 hr. If a hibernating animal started to arouse and then returned to deep hibernation, the sequence of events was referred to as an "inhibited arousal." This is characterized by tachycardia followed by bradycardia and an increase in the appearance of prolonged R-R intervals. Squirting water in the face of a hibernating animal (the "diving test") usually resulted in a brief bradycardia followed by a tachycardia, but this did not occur when the animal was atropinized.

The concept of "progressive irritability" was further tested by injecting hibernating *C. lateralis* intraperitoneally with various doses of epinephrine, norepinephrine, isoproterenol, and phenylephrine. Isoproterenol is a beta agonist, and phenylephrine is an alpha agonist. In general, as a bout of hibernation progressed, the animals became increasingly sensitive to the drugs and responded either by an "inhibited" or by a full arousal. Unlike their previous report (Twente and Twente, 1968b) norepinephrine was about as effective as epinephrine in eliciting these responses.

It was concluded that hibernation is controlled by the "autonomic hibernation maintenance response" (AHMR). The parasympathetic system was visualized as controlling the entrance into hibernation, the hibernating state, and the latter part of inhibited arousals. The increased number of prolonged R–R intervals during these times was viewed as evidence for this domination. Apparently because these changes could take place without a change in T_b, the authors concluded that the observed results were divorced from temperature regulating mechanisms, though no measurements of metabolic rate were reported.

Many of the observations reported by the Twentes are similar to those observed by us some years ago (Lyman and O'Brien, 1963), but the conclusions are more far-reaching and, since our contrary findings are not discussed in their paper, some clarifications are in order. In our attempts to unravel the functions of the autonomic nervous system during the hibernating cycle, we were constantly reminded that a multitude of substances, infused intra-arterially, could result in tachycardia in the hibernating animal, with abolishment of asystoles if they were present. By measuring heart rate and blood pressure simultaneously, it was possible to show that some vasodilatory drugs produced a drop in blood pressure followed by cardioacceleration that was interpreted as being compensatory. For these reasons, conclusions concerning the activity of the autonomic nervous system drawn from the heart rate alone were accepted with great caution.

Both the Twentes' and our work indicate that parasympathetic influence is active as the animal enters hibernation, for atropine abolishes the periodic asystoles that are so typical of this period. However, we would emphasize that the atropinized animal continues to enter hibernation and we presented evidence to indicate that the parasympathetic influence was a fine control rather than the basic mechanism behind the process of entering hibernation.

Our conclusions concerning the influence of the parasympathetic system differ from that of the Twentes. As detailed in Chapter 4, all our evidence indicated that parasympathetic influence, as gauged by its effect on the heart, fades as the animal enters the hibernating state. Electrical stimulation of the vagus or choking to elicit the vagal reflex failed to produce bradycardia. Intra-arterial infusion of parasympathomimetics did not slow the heart rate, and it was only after the heart had been beating at a faster rate for some time that the physical or pharmacological effects could be demonstrated. Furthermore, atropine in a wide variety of doses often produced no chronotropic effect. Similar results have been reported by other investigators as detailed in Chapter 4. It was concluded that the parasympathetic influence on the heart fades as the animal enters the deeply hibernating state, though it was not established whether this diminution was due only to the increased threshold to parasympathetic influence of the hibernating heart, or whether lessening of vagal tone was also involved.

Twente and Twente published evidence for the influence of the parasympathetic system on the heart rate during deep hibernation which is similar to ours. In their illustrated experiments using the "diving test" the heart rate is accelerated, varying between 9 and 14 beats a minute, and with an accelerated heart rate we found that bradycardia could be produced by choking or by stimulation of the vagus.

In the atropinized animal, the diving test produced no bradycardia, but here the figure illustrates a heart rate of six or seven beats a minute. According to our observations, vagal slowing of the heart would not take place with this slow heart rate even if the animal were not atropinized.

If the parasympathetic system maintains the animal in hibernation via AHMR, then atropinization should block AHMR, premature arousal should occur, and sensitivity to stimuli should increase, but the Twentes did not report these changes. They suggest that atropinization with the dosage used blocks the parasympathetic effect on the heart, and the heart rate increases as a result, but that atropine does not block other unidentified parasympathetic influences that maintain the animal in hibernation.

We favor the alternate proposal that the whole parasympathetic system had been indeed blocked by atropinization, but that the animal remained in deep hibernation, with an accelerated heart rate, because this system had little influence during deep hibernation. If the chronotropic effect of the parasympathetic system is more sensitive to atropine than the part of the system which controls the maintenance of hibernation, then the latter should be blocked by larger doses of atropine, but this experiment has not been reported.

The Twentes espouse the theory that "autonomic activity appears to regulate the maintenance of the state of hibernation," and reject the widely accepted concept that thermoregulation is involved in this maintenance and in the initiation of the arousal process. Using ECG patterns as the indicator of autonomic activity, they note that these patterns changed in response to pharmacological injections 30–45 min before any measurable change in body temperature. Thermogenic activity is here equated with body temperature, but even with relatively crude methods of measuring oxygen consumption it has been reported that metabolic rate and heart rate may change without a change in body temperature during hibernation (see, e.g., Lyman, 1948, 1958). With more sophisticated equipment, changes in oxygen consumption and heart rate during hibernation can be shown to be virtually simultaneous, with changes in T_b occurring later (Lyman, unpublished observations). Throwing coal into a furnace is a thermogenic act, but it takes some time for the heat to reach the parlor.

The Twentes emphasize the role of the parasympathetic system in maintaining hibernation, and there is no doubt that manifestations of parasympathetic activity can be very dramatic during arousal. When the arousal process is well under way, touching or manipulating a ground squirrel can result in prolonged apnea and bradycardia before tachycardia and polypnea are resumed. The apnea and bradycardia are akin to, if not identical with, the "startle reflex," which is so well developed in wild animals, particularly prey mammals, which "freeze" at the approach of danger. Because apnea and bradycardia may occur during arousal, it does not necessarily follow that the parasympathetic system is attempting to force the arousing animal back into the hibernating state. We favor the view that the autonomic system is an important part of the complex process of temperature regulation both in the active and the hibernating animal, but that this system must act in concert with the other temperature-regulating mechanisms in maintaining the balanced state which is typical of hibernation.

Another point of view is offered by Beckman and co-workers (Beckman and Satinoff, 1972). They implanted bilateral cannuli in the midbrain reticular formation (MRF) or in the preoptic–hypothalamic area (POAH) of golden mantled ground squirrels. Microinjections of ACh, norepinephrine, or 5-hydroxy-tryptamine could be infused into one or the other of these areas and the results noted. They found that infusion of any one of the three substances into the POAH area resulted in arousal from hibernation, but only ACh could produce the arousal response when infused into the MRF. This suggested to them that there was a greater diversity of synaptic input to the POAH than to the MRF area (Beckman *et al.*, 1976a; Beckman, 1978).

During the course of their experiments, they studied the responses that took place when the animals were stimulated by handling or when microinjections of ACh were made, or when spontaneous changes occurred in the thermogenic pattern. They distinguished three types of response to these stimuli. In type I response, midbrain temperature increased slowly to a variable level, then decreased again as the animal returned to deep hibernation. With type II response, the increase in midbrain temperature was rapid and the rate of increase varied little from experiment to experiment. At a rather precise brain temperature, the animal started to cool and then returned to hibernation. In the third type of response (type III), the increase in temperature was similar to the type II response, but reversal did not occur and the animals went on to full arousal. Since the initial increase in T_b was the same in the type II and type III response, they suggest that these were produced by a common neuronal event, and that this arousal process was inhibited in the type II response by some neuronal mechanism when the temperature of the brain reached a certain level (Beckman *et al.*, 1976b).

In searching for this mechanism, they have recorded the electrical activity of single cells in the POAH area. They report that the activity of an anterior hypothalamic cell, as measured by firing rate per second, decreased as the hypothalamic temperature rose, and increased with the decline in temperature, suggesting that some cells in the POAH are activated at a certain temperature to inhibit the arousal process. Once the brain temperature exceeds the temperature band where the type II response can occur, they postulate that there is a rapid recruitment of neural activity and arousal goes to completion. Recording from a lateral hypothalamic neuron during this period showed that the cell was silent at 17°C, but increased its activity as the temperature rose to 23°C, and remained at this level of activity until it was "lost" at 32°C (Beckman, 1978).

Parenthetically it should be noted that single cell recording in acute preparations of hibernators and nonhibernators have been reported from other laboratories. Boulant and Bignall (1973) compared the number of thermosensitive cells in the hypothalami of rats and golden mantled ground squirrels and found no significant difference. Wünnenberg and associates (1978) found that the preoptic neurons in the guinea pig became inactive at temperatures of 28°–30°C, whereas single units in the Syrian hamster became inactive between 10° and 15°C. Most of the cells from the hamsters had a positive temperature coefficient over the temperature range of 10°–40°C, but a few cells showed a negative coefficient from about 30°C, so that the firing rate of the cell became progressively slower as hypothalamic temperature increased. It is not known whether this difference in temperature sensitivity obtains between all hibernators and nonhibernators, for a comparison between closely related species has not been made. Hamsters and guinea pigs are only distantly related, and are usually classified in separate suborders.

Beckman also showed that there were changes in the magnitude of the thermogenic response when ACh was infused into the midbrain reticular formation during various phases of the hibernating cycle. During entrance into hibernation, 200 μg of ACh produced only a slight rise in brain temperature and only a type I response when the brain temperature had reached about 6.2°C. After three quarters of the bout was completed, the same dose produced a type II response and in the fourth quarter the full arousal of the type III response occurred. Using lower doses, they showed that the minimum dose required to produce a response decreased as a bout of hibernation continued and they attributed this to a progressive weakening of inhibitory influences. At any period of the bout a larger dose was needed to produce a type II response than a type I response. They suggest that the midbrain reticular formation may act as a gating mechanism for controlling neural input to hypothalamic and hippocampal neurons that are themselves responsible for the arousal reaction. When the POAH area was directly stimulated by microinjections, it was found that responsiveness was low during the first part of a bout of hibernation. In contrast to the experiments with the MRF, responsiveness increased rapidly so that type III was the response that usually occurred during the second half of a bout of hibernation.

It was observed that the POAH was sensitive to much lower doses of ACh than was the MRF, and this may be correlated with the reports of virtually continuous bioelectric activity in the POAH area during hibernation, though an exception to this activity has been reported in the golden hamster where this area was reported to be electrically silent during the beginning of arousal from hibernation (Chatfield and Lyman, 1954).

The increase in responsiveness of the MRF and POAH as a bout of hibernation continued, as found by Beckman and Stanton (1976), may be correlated with the "progressive irritability" of Twente and Twente reported earlier in this chapter. The latter investigators found that hibernating golden mantled ground squirrels became more sensitive to stimuli as a bout of hibernation continued. Intraperitoneal injection of various substances including saline (Twente and Twente, 1968a,b) and epinephrine were used as stimuli. Concentrations of epinephrine that produced no result early in a bout of hibernation were apt to cause a partial warming or complete arousal later in the bout, and total arousals occurred more frequently as time in the bout progressed. As mentioned earlier, a puzzling aspect of this report was that norepinephrine was reported to be ineffective. This observation was later found to be incorrect, and norepinephrine was reported to be only slightly less effective than epinephrine in producing the arousal response (Twente and Twente, 1978).

The important methodological difference between the two observations is that the Twentes produced an effect by intraperitoneal injection of various stimulatory substances, whereas Beckman and associates infused drugs directly into the midbrain reticular formation or the preoptic–anterior hypothalamic area. There appears to be little question that the Beckman infusions were acting on the specific area of the brain that he was attempting to stimulate, for control infusions into other parts of the brain failed to elicit a response. Thus, it is reasonable to conclude that an increasing sensitivity to the infusion of ACh occurs in the MRF during a bout of hibernation and that a similar sensitivity occurs in the POAH area, but in this case the development of irritability is more rapid.

The site of action of intraperitoneally infused catecholamines is unknown. Although there is some evidence that small amounts of the catecholamines can cross the blood–brain barrier (Weil-Malherbe *et al.*, 1961), the general consensus is that crossing does not occur (Rapoport, 1976). Rothballer (1959) has suggested that a neurochemical need not cross the barrier to be effective, and points out that areas of the brain can "taste" the constituents of the blood, such as the hydrogen ion, and react accordingly. Thus, there is a possibility that the intraperitoneally injected amines are acting directly on the brain. According to Beckman's findings, the POAH region should be the area that is affected, since the MRF was not responsive to infusion of catecholamines. Thus, one might postulate that progressive irritability was taking place only in the POAH. This would agree with Beckman's concept that active inhibition of thermogenesis was occurring in the POAH area at the start of a bout of hibernation, and that this inhibition gradually lessened as the bout continued.

Another alternative is that peripheral injection of catecholamines and other substances, including isotonic saline (Twente and Twente, 1968a,b), give rise to a neural signal which is then carried to the brain via nervous channels. If the latter explanation is valid, progressive irritability could be taking place either in the periphery or in the brain, or in both areas at the same time. If the periphery alone became increasingly sensitive to catecholamines, a neural signal would become more easy to generate as a bout of hibernation continued and this would be reflected in the increased sensitivity of the thermogenic response. If progressive irritability took place only in the brain, then sensitivity to an unchanging peripheral neural signal would also increase with the passage of time. Obviously, if both brain and periphery became progressively more sensitive, the neural signal would be increasingly easy to generate at the periphery and it would be increasingly effective in the brain.

It is possible that long-term hibernation disrupts the normal progression of storage and metabolism of catecholamines, resulting in an increased sensitivity in both body and brain. In the normal euthermic mammal, the tissue level of catecholamines does not vary greatly, and it is the rate of turnover that indicates the level of sympathetic activity in a tissue. An increase in sympathetic activity results in the liberation of catecholamines from the synaptic vesicles of the sympathetic neurons by the process of exocytosis. This is accompanied by an increased synthesis of catecholamines. In the five-step synthesis of epinephrine from phenylalanine, the rate limiting step is the hydroxylation of tyrosine to dopa by tyrosine hydroxylase. The activity of tyrosine hydroxylase is inhibited by norepinephrine and is high in the absence of catecholamines. Some of the released catecholamines are metabolized by catechol O-methyltransferase and monoamine oxidase, but a greater amount is taken up by the neurons and stored again in the vesicles. The increased amount of catecholamine in the neurons inhibits the tyrosine hydroxylase and the synthesis of catecholamines is slowed. This process of checks and balances occurs both in the peripheral sympathetic neurons and in the brain itself (Axelrod and Weinshilboum, 1972). Even from this oversimplified description it seems clear that the maintenance of the balance is complex and that a perturbation such as the drastically low body temperature of deep hibernation might shift that balance from the one found in the euthermic steady state.

Biochemical assays of nonadrenal catecholamines in active and hibernating animals are inconsistent. Uuspää (1963) reported that the norepinephrine content of the whole brain without the cerebellum and of the brain stem alone were significantly reduced in hibernating hedgehogs when compared with the euthermic animal, but that epinephrine content was virtually the same. There was no difference in the concentration of epinephrine and norepinephrine in the hearts of hibernating hedgehogs when compared with active animals.

Draskóczy and Lyman (1967) measured the epinephrine and norepinephrine content of the brain, heart, brown fat, and adrenal gland of hibernating thirteen-lined ground squirrels and of euthermic animals killed in the winter and in the spring. There was no difference in the catecholamine content of the animals sacrificed during these two seasons. The content of catecholamine in tissues from hibernating animals was consistently lower than in active ground squirrels, but, due to the large variation, only the norepinephrine content of brown adipose tissue showed a significantly lower measurement ($p < 0.01$). Twente *et al.* (1970) measured the epinephrine and norepinephrine content of the hypothalamus, medulla oblongata, and cerebellum of golden mantled ground squirrels at progressive periods during a bout of hibernation and also when the animals were euthermic. They found no evidence that there was a change in the concentration of epinephrine or norepinephrine as a bout of hibernation progressed. The concentration of epinephrine in the medulla oblongata of the euthermic animals was significantly lower than in any of the hibernating groups, but otherwise there were no differences between the active and hibernating animals. Feist and Galster (1974) used the arctic ground squirrel to study the concentration of epinephrine, norepinephrine, and 5-hydroxytryptamine in the hypothalamus during euthermia, hibernation, and arousal. They found that norepinephrine was lower during a bout of hibernation than in the prehibernating state and that it rose during arousal. There was no evidence that it increased during a bout of hibernation. 5-Hydroxytryptamine was low during early arousal, but otherwise remained unchanged, and epinephrine increased during the latter part of the arousal process.

The measurement of turnover of catecholamines during hibernation appears to be limited to a single study from our laboratory (Draskóczy and Lyman, 1967). Using [³H]dopa, we found that turnover varied greatly in different tissues during the period of hibernation. In the heart and in brown fat, turnover continued, albeit at a greatly reduced rate when compared with the euthermic animal. Though the turnover rate in the whole brain in euthermic animals was high, our measurement showed that turnover stopped completely, not when the animals reached the low temperatures of deep hibernation, but actually at the time the animals started to enter the hibernating state. On the basis of this observation, it was postulated that the complete lack of activity of the adrenergic neurons of the brain might be the cause of the induction of natural hibernation. Presumably when the hibernating animal is disturbed, there is a burst of sympathetic activity,

turnover increases, thermogenesis ensues, and, if the animal is fully primed, full arousal from hibernation occurs. It is conceivable that the magnitude of the sympathetic discharge in response to a standard stimulus increases as a bout of hibernation proceeds, and this would be reflected in an increase in turnover of catecholamines. Unfortunately, there is no known way of comparing turnover rates accurately during the beginning of the arousal process.

The concept of increasing sensitivity of receptor mechanisms during a bout of hibernation, either at the periphery or in the POAH area, is illustrated in some of our unpublished observations on hibernation in the Turkish hamster. When some individuals of this species are mechanically and forcefully stimulated during the first day of hibernation at a T_b of 5°C, heart rate, metabolic rate, and body temperature increase transiently, but the pattern of arousal only continues for a few minutes and the animal sinks back into the hibernating state. No single stimulus that we have yet devised will result in arousal from hibernation. After several days in hibernation, the animal will arouse if stimulated and eventually arouses spontaneously. This obviously involves progressive irritability, but in an exaggerated form since the animal initially is not capable of arousing. The lack of arousal is in sharp contrast to the situation in the golden hamster, an animal of the same genus and of comparable size, which is extremely sensitive to mechanical stimuli during any period in its bout of hibernation and will arouse from hibernation in response to a very slight stimulus.

The effect of circulating catecholamines on Turkish hamsters in their first day of hibernation has been tested in a few experiments. Epinephrine or norepinephrine (0.12–0.24 mg/kg) in 1 ml saline was infused intraperitoneally over a 0.5-hr period after the animal had failed to arouse when stimulated violently. After infusion, some, but unfortunately not all, of the animals tested to date are more sensitive to any peripheral stimulus, as measured by increase in electromyographic activity and heart rate, and are capable of complete arousal. The same amount of epinephrine or norepinephrine injected into an animal which will not arouse when otherwise stimulated also may result in complete arousal, but clearly these experiments must be checked by infusion of catecholamines directly into the bloodstream.

These observations suggest that circulating catecholamines are indeed depleted at the beginning of a bout of hibernation and that the lack of sensitivity at that time is associated with this low titer. The reports by Twente et al. (1970) and Feist and Galster (1974) that the amount of norepinephrine in the hypothalamic area is unchanged during a bout of hibernation do not necessarily run counter to this view. With the Turkish hamster the infusion of norepinephrine must raise the concentration of the amine peripherally, and the observed result may well be due to peripheral stimulation and resulting action. If the POAH area is involved in the process of arousal, the peripheral signal may be strong enough to elicit the reaction in the POAH area irrespective of the concentration of norepinephrine in that area. Furthermore one may question whether measurements of concentration of norepinephrine in the POAH area, no matter how accurate, will give the proper indication of the physiological activity of that area. The POAH area is involved in a multitude of regulatory mechanisms and the critical area which may be involved in hibernation and arousal from that state may be very small, so small, indeed, that a change in its catecholamine content would be masked by the unchanged concentration of the rest of the POAH tissue surrounding it.

As far as peripheral concentration of catecholamines is concerned, the contrast between the hibernation of the Syrian and the Turkish hamster is worth considering. Experimenters have long been exasperated by the failure of Syrian hamsters to enter hibernation even when exposed to presumably ideal laboratory conditions. With the hamsters that eventually hibernate, the onset of hibernation is often delayed for months after being exposed to cold, and the shortest period for the onset of hibernation in over 1000 animals was 3 days (Lyman, 1954). Bouts of hibernation are also short, the longest period in a large series of animals being seven days (Lyman, 1948). In contrast, the great majority of Turkish hamsters hibernate when exposed to the cold. Some hibernate within the first 24 hr, and their bouts of hibernation can be three times as long as in the Syrian hamster. The difference in sensitivity has already been mentioned, but it must be emphasized that almost any physical disturbance, even gentle displacement of the hairs of the back, may result in arousal in the Syrian hamster.

A comparison between the sensitivity in hibernation of two widely different genera such as *Citellus* and *Mesocricetus* is like comparing apples and oranges, and little can be gained from it. It can be argued that hibernation as a physiological adaptation is polyphyletic and that two such genera came upon the same general result by two somewhat different evolutionary routes, and hence differences in physiological mechanisms are to be expected. On the other hand, the Turkish hamster (*Mesocricetus brandti*) and the Syrian hamster (*M. auratus*) are so closely related that they will copulate in the laboratory, though viable young have not been produced (Lyman and O'Brien, 1977). With such a comparison, it is reasonable to expect that a physiological variable such as different levels of sensitivity in hibernation is due to changes in a single mechanism. Is it too fanciful to suggest that this mechanism is the level and rate of turnover of one or more of the catecholamines?

From another point of view, some light can be shed on the overall cause of hibernation by examining in detail the process of entering into the hibernating state. As was mentioned in Chapter 3, the relationship between sleep and hibernation has been a favorite debating point, and, as early as 1881, Horvath wrote that Der Winterschlaf ist erstens kein Schlaf, und zweitens hat er gar nichts mit dem Winter zu tun'' (Horvath, 1881). The vehemence of the statement is such that it needs no translation, but from the accumulated knowledge of about 100 years it appears safe to say that Horvath was wrong on both counts, but particularly on the first one. Hibernation, or winter sleep, now appears to have certain attributes in common with the normal sleep which occurs in euthermic mammals.

The study of sleep and hibernation compounds the problems of studying the two phenomena individually. There appears to be no precise way of defining sleep without the use of electrical recording. With this method, slow wave sleep (SWS) can be separated from the awake state and from paradoxical or rapid eye movement sleep (REM), but this, of course, involves the use of encumbering electrodes as well as thermocouples or thermistors to measure T_b. Thus it is that only a few investigators have had the patience and skill to undertake a study of the relationship of sleep and wakefulness as the animal enters the hibernating state. Satinoff (1970) noted that muscle tonicity was present in golden mantled ground squirrels during entrance into hibernation at a T_b of 18°C. She did not report the EEG at T_b's between 37° and 29°C, and the transition from euthermia to the hibernating state must start somewhere near these body temperatures. She came to the tentative conclusion that animals entered hibernation either while they were awake or in the preliminary stages of SWS. South *et al.* (1969) concluded that the proportion of SWS and REM sleep was about the same in the euthermic marmot and in the marmot entering hibernation, but they did not report the total sleep time during euthermia and compare it with sleep time while entering hibernation.

Heller *et al.* (1978) have explored the relationship of sleep, temperature regulation, and hibernation in various species of rodents. Working with the kangaroo rat, *Dipodomys ingens,* which does not hibernate, they varied the temperature of the POAH area by means of chronically implanted thermodes, while holding the animal at various temperatures. They found that the rise in oxygen consumption was immediate when the POAH area was cooled at ambient temperatures between 10° and 25°C. However, at a T_a of 30°C, the delay in the response was sometimes as much as 10–15 min. They observed that the animals at this temperature spent much of the time asleep, and suggested that the loss of sensitivity might be related to the sleep state of the animal (Glotzbach and Heller, 1975). In further experiments, they measured the metabolic heat response to manipulations of the POAH at a T_a of 30°C when the animals were awake, in SWS, and in REM sleep. As had been shown in earlier experiments, the awake kangaroo rats increased their metabolic rate proportionately as the POAH temperature was lowered further from the temperature which first resulted in an increase in oxygen consumption. During SWS, the sensitivity to changes in hypothalamic temperature was still present, but greatly reduced, with a proportionality constant for metabolic rate production (α) at about one-half the value found in awake animals. Animals in REM sleep showed no reaction to lowering of the POAH temperature (Glotzbach and Heller, 1976).

The round tailed ground squirrel (*Citellus tereticaudus*) is not a deep hibernator, but enters a condition of torpor when denied food. During these periods of torpor, REM sleep was greatly reduced, and was absent when the brain temperature reached 26°C. At this temperature, SWS dominated the electroencephalogram, but periods of wakefulness were also recorded (Walker *et al.*, 1979). Studies with the deep hibernators *Citellus beldingi* and *C. lateralis* showed that the animals slept 66% of the 24-hr day during the summer, nonhibernating, period. Of the total sleep time, 19% was REM sleep and 81% was SWS. During the annual period for hibernation, the distribution of sleep states was different. Hibernation was entered through sleep, and when brain temperature was 35°–25°C the total sleep time was 88% of a 24-hr period but SWS occurred in 90% of this time and REM sleep was reduced to 10%. The EEG amplitude declined with the drop in brain temperature so that SWS and REM sleep could not be identified below 25°C (Walker *et al.*, 1977). Thus, there appears to be a continuum among the mammals tested, from animals that do not hibernate, but experience a decline in T_b during SWS, to animals that enter shallow torpor from SWS, to deep hibernators that also experience SWS until it can no longer be measured. Heller postulates that SWS is an evolutionary development that results in metabolic saving and that torpor and deep hibernation are extensions of SWS (Heller *et al.*, 1978). Parenthetically, Heller *et al.* (1978) have found that during the return to wakefulness, the round tailed ground squirrel, which is not a deep hibernator, spent 62% of the time in SWS. This did not occur in the deep hibernators that were studied and suggests that arousal from deep hibernation and exit from torpor may be very different physiologically.

These imaginative and interesting studies advance the knowledge of the relationship of hibernation to sleep, but the knotty problem of the cause of the change in threshold for temperature regulation remains. If the animal enters hibernation during SWS, why does hibernation occur during the winter months, but rarely in the summer in many species that hibernate? In this regard, Heller *et al.* (1978) have shown that there is an annual rhythm of total sleep time in the golden mantled ground squirrel. When kept at 22°C throughout the year, these animals sleep more during the winter months when they would normally be hibernating.

Dawe and his group approached this problem from a very different angle. In early March of 1968 they withdrew blood from the aorta of a hibernating thirteen-lined ground squirrel and injected 1 ml into each of three active animals maintained in a warm room. Two animals received the blood intravenously via the saphenous vein, whereas the other was injected intraperitoneally. When moved to the cold, the former two animals hibernated after 48 hr, but the latter animal did not. Blood was obtained from the two hibernating animals and transfused into three active ground squirrels kept in the warm room. When these animals were moved to the cold they also hibernated, and infusion of their blood into five other active, warm room animals resulted in hibernation when these animals were exposed to cold. Dawe and Spurrier (1969) named the blood-borne substance which presumably caused hibernation "trigger." Ground squirrels that had not hibernated all winter in the cold remained active when infused during March with blood from hibernating ground squirrels. Four animals that had been kept in the warm room were transfused at the same time and moved to the cold. All four of these animals hibernated within 22 days, and the time each animal spent in hibernation for the next 60 days correlated with the time that the donor had spent in hibernation during the previous 60 days. Further, infusion of blood from a donor that was in a prolonged continuous bout of hibernation resulted in the earliest occurrence of hibernation among the recipients. In this regard, it should be mentioned that, according to one figure, the length of the continuous bout was 36 days, which is extraordinary for a thirteen-lined ground squirrel (Dawe *et al.*, 1970).

In the same report, blood was drawn from hibernating ground squirrels and a single woodchuck in January and February and frozen in liquid nitrogen either as whole blood or after separating cells and serum. When these were tested on June 20 with animals just moved to the cold, hibernation occurred within 52 days in 20 of the 23 ground squirrels injected with ground squirrel blood, and in all three of the ground squirrels injected with woodchuck blood. Whole blood, washed cells, and serum were equally effective, and in this larger sample there was no correlation between the length of time the donor had been in a continuous bout of hibernation and the time of onset of hibernation in the recipient (Dawe *et al.*, 1970).

In attempting to isolate the "trigger," serum from hibernating ground squirrels was dialyzed using a membrane which would block molecules of over 5000 daltons. It was found that injection of the dialyzate caused hibernation in summer ground squirrels exposed to cold, whereas the residue did not. Further, a mixture of dialyzate and the residue did not result in hibernation. In another experiment, a series of blood samples was drawn from six ground squirrels and one woodchuck while the animals were either active, arousing from hibernation, or in the hibernating state. The serum from each sample was frozen and stored at $-15°C$. Forty-five recently trapped wild thirteen-lined ground squirrels were each infused with a thawed sample, or with a saline control, and transferred to a cold, dark room in June. Only the animals that had been infused with serum from the hibernating woodchuck or ground squirrels entered hibernation within an observation period of 16 days. Since the blood samples were taken in the winter, it was concluded that it was the state of the animal, rather than the season of the year, which determined whether the sample would "trigger" hibernation in the recipient ground squirrels. Again the duration of hibernation of donor animals is longer than that reported from other laboratories, for the woodchuck is listed as hibernating continuously for 6 weeks (Dawe and Spurrier, 1972).

Further studies were reported using pregnant and young thirteen-lined ground squirrels. Six pregnant ground squirrels were injected 4 days before parturition with the dialyzate of the blood of a hibernating woodchuck. These animals, with

26 other pregnant ground squirrels, were transferred from the warm room to the cold (7°C) at various times during the summer after the young were over 2 days old. After 6 weeks with their young, the females and the young were separated and put in individual cages. The females that had been injected with the wood-chuck dialyzate hibernated in July and August. The uninjected females, and a female injected with cold saline, did not hibernate until winter.

Remarkably, the 13 infants born of mothers injected with the woodchuck dialyzate all hibernated after being separated from their mothers. Twenty-six infant ground squirrels were injected intraperitoneally with 0.25, 0.5, or 0.75 ml of the serum dialyzate from a hibernating woodchuck when they were 2, 3, or 4 weeks of age. Of these animals, 25 out of 26 hibernated during the summer. Animals receiving a saline infusion, a dialyzate from an active or arousing woodchuck, or a dialyzate which had been heated to 37°C or above failed to hibernate, as did infants which were kept in a warm room and whose mothers had received the dialyzate or who had received the dialyzate directly. The tempera-ture of the warm and cold rooms and the method of injecting the dialyzate are unclear, as the table and the text do not agree (Dawe and Spurrier, 1974a).

From other experiments, it was concluded that the dialyzate contained a small molecule that could cause hibernation, and that the nondialyzable remainder contained a large molecule which inhibited the effect of the small molecule (Dawe and Spurrier, 1974b). Dawe (1978) has developed a theory involving the interaction of the large and small molecules over the annual season and the effect of this interaction on the occurrence of hibernation.

As might be expected, these experiments have resulted in considerable con-troversy, both in published articles and by word of mouth. From an operational point of view, several investigators have been unable to repeat these results using other species in spite of following the protocol as precisely as possible. It has also been pointed out that periods of uninterrupted hibernation of 36–42 days in ground squirrels and woodchucks (Dawe and Spurrier, 1972) have not been observed in other laboratories. Another criticism has been that the methods have not been consistent, and the inconsistencies have not been explained. For exam-ple, it has been emphasized (Dawe, 1978) that intravenous injection of the "trigger" substance caused hibernation, but intraperitoneal injection did not, yet the infant ground squirrels were injected intraperitoneally and these animals hibernated after injection (Dawe and Spurrier, 1974a). Evidence is presented to indicate that the trigger substance is denatured and is no longer active at tempera-tures above 20°C, yet the substance is injected into the blood stream of active animals and must reach the body temperature of 37°C very quickly. Dawe has recognized this contradiction and has suggested that the "trigger" substance must combine with a postulated "anti-trigger" in the blood which protects it from denaturation (Dawe, 1978).

Perhaps the main question that has been put forward concerns the possible mode of action of the "trigger" substance. Dawe and Spurrier (1975) have postulated that a preponderance of "trigger" substance in the body of the animal actually prepares the tissues for the hibernating state, and this is followed by hibernation. The amount of "trigger" injected into the experimental ground squirrel is not known, but it must be very small for it is the diluted dialyzate from blood of a hibernating animal. Usually 1 ml is injected into the animal. In some experiments, if the experimental animal hibernates within 60 days after the injection, it is assumed that "trigger" caused the hibernating state. Furthermore, the injected animals no longer exhibit an annual cycle of hibernation, but continue bouts of hibernation for the rest of their lives if they are kept in the cold (Dawe, 1973). It has been suggested that these are extraordinary results to be wrought by such a small amount of "trigger" material.

Questions have also arisen concerning the preparation of the "trigger" substance. The dialyzate that has proved effective is referred to as SM (small molecule) and believed to have a molecular weight of less than 5000. However, a recent report concludes that the "trigger" is bound to albumin or closely associated with it, and the molecular weight of albumin is over 70,000 (Oeltgen *et al.*, 1978). Such a large molecule would not dialyze across a membrane which blocks molecules of over 5000 in molecular weight. In this report the substance was extracted from the plasma, but in the previous work of Dawe's group the substance was obtained from the serum.

There have been reports from other laboratories concerning the "trigger" substance. Steiner and Folk (1978) injected thirteen-lined ground squirrels with "trigger" substance from hibernating hamsters during the summer. When exposed to cold, four out of six animals injected with "trigger" hibernated within 60 days, compared to four out of seven control animals injected with saline, and 11 out of 17 animals that were simply exposed to cold. The time between exposure to cold and hibernation was shorter in the four trigger-injected animals, and the difference between this and that of the four saline-injected controls was of borderline significance (t is reported as being 2.44, but it should be 2.447 for p to be 0.05). There was no significant difference in the elapsed time before hibernation if the "trigger"-injected animals were compared with the uninjected animals, or if the latter were compared with the saline-injected controls. The data were interpreted to show that the "trigger" substance was present in a nonseasonal hibernator and was effective in producing hibernation in a seasonal hibernator. However, Steiner and Folk (1978) emphasize that there was greater varation between the time of exposure and the time of hibernation in all the groups, and this has been noted in many other laboratories. Thus, the reported difference may well be fortuitous. Furthermore, one may question the validity of using the elapsed time before entering hibernation as a criterion when some animals of each group did not hibernate at all during the arbitrary observational period of 60 days, but might have hibernated at an undetermined later date.

In contrast to the results obtained by Steiner and Folk (1978), Minor *et al.* (1978) compared the effect of "trigger" substance withdrawn from hibernating golden hamsters on cold-exposed animals of the same species. In these experiments, there was no significant difference in hibernation between the three experimental animals and the four saline-injected controls, and it was concluded that the "trigger" substance did not occur in golden hamsters, and it was emphasized that hamsters were not a seasonally hibernating species.

In another experiment, Rosser and Bruce (1978) explored the concept that the "trigger" substance must be excreted to permit the animal to arouse from hibernation. They prepared a "trigger" extract from the plasma of hibernating thirteen-lined ground squirrels and collected urine from animals arousing from hibernation. When ground squirrels were exposed to cold in June, all four of the animals injected with "trigger" hibernated before August 1, whereas two out of the three animals injected with urine also hibernated and two saline-injected controls failed to hibernate. The authors reached no conclusion concerning the presence of "trigger" in the urine of arousing ground squirrels.

In the middle of the hibernating season, Galster (1978) stored blood from arctic ground squirrels (*Citellus undulatus*) cryogenically and injected the plasma into nine animals of the same species in the summer. He also withdrew blood from woodchucks (*Marmota monax*) in deep hibernation and injected the plasma into 10 active arctic ground squirrels. Neither group of experimental animals hibernated prematurely, and their basal and thermogenic metabolism did not differ from that of the controls.

Abbotts *et al.* (1979) prepared a dialyzate from hibernating Richardson's ground squirrels and injected it intravenously or by cardiac puncture into thin animals of the same species. Some of the experimental animals had been in captivity for more than 2 years and their annual hibernating cycles were out of phase with the actual time of year. Other experimental animals were wild-trapped young of the year. Various feeding regimes were employed. On the theory that the "trigger" substance might be a peptide, bacitracin was added to some of the dialyzate injection to retard degradation. In trials involving more than 50 cold-exposed animals, there was no evidence that the dialyzate had any effect on nest building, change in weight or occurrence of hibernation. The authors suggest that the effect of the "trigger" substance has been reported only with thirteen-lined ground squirrels, and that the reaction may be peculiar to them.

Recently Meeker *et al.* (1979) have reported that an albumin fraction isolated from hibernating woodchuck serum caused a decline in mean daily food intake when infused either intravenously or into the cerebral ventricles of macaque monkeys. This would indicate that the "trigger" substance may have other actions than inducing hibernation in animals that hibernate.

A puzzling array of data thus exists concerning the effect of "trigger" on inducing hibernation, and part of the puzzle must certainly be due to the difficulty of accurately testing the effect of the substance in question. It has apparently been assumed by many that seasonal hibernators will not hibernate during the summer months and, therefore, hibernation during that period must be due to some perturbation of the normal yearly cycle. Actually, experience in our laboratory and in many others has demonstrated that hibernation in thirteen-lined ground squirrels can occur at any time of the year, though it is less apt to occur in the summer, especially in newly caught specimens that have not been long exposed to the unchanging conditions of the laboratory. Furthermore, there is a great variation, both in seasonal and in nonseasonal hibernators, in the time between cold exposure and the first bout of hibernation. For example, in a group of 252 golden hamsters exposed to $5° \pm 2°C$ at all times of the year, the average time before attaining hibernation was 56.6 days, with the large standard deviation of 33.2 days. The shortest period before hibernation occurred was 3 days and the longest was 218 days. Furthermore, 121 animals died without entering hibernation, and the average time of death was 71.5 days after cold exposure, again with the large standard deviation of 67.2 days (Lyman, 1954). With variations such as these, it is virtually impossible to obtain unequivocal results with only a few experimental animals, and some of the reported data and conclusions must be viewed with great caution, even if the comparisons are statistically attractive.

Viewing the experiments on the "trigger" substance as a whole, it is clear that the large amount of positive information cannot be dispelled by categorically stating that the method of testing is not sufficiently precise. Some of the other criticisms have been listed here, but they do not completely explain away the results that have been reported. At the present time, the putative mode of action of the "trigger" substance has been explained only in theory (Dawe, 1978). Clearly more research is required to unravel these provocative observations.

REFERENCES

Abbotts, B., Wang, L. C. H., and Glass, J. D. (1979). Absence of evidence for a hibernation "trigger" in blood dialyzate of Richardson's ground squirrel. *Cryobiology* **16,** 179–183.

Axelrod, J., and Weinshilboum, R. (1972). Catecholamines. *N. Engl. J. Med.* **287,** 237–242.

Beckman, A. L. (1978). Hypothalamic and midbrain function during hibernation. *In* "Current Studies of Hypothalamic Function" (W. L. Veale and K. Lederis, eds.), Vol. 2, pp. 29–43. Karger, Basel.

Beckman, A. L., and Satinoff, E. (1972). Arousal from hibernation by intrahypothalamic injection of biogenic amines in ground squirrels. *Am. J. Physiol.* **222,** 875–879.

Beckman, A. L., and Stanton, T. L. (1976). Changes in CNS responsiveness during hibernation. *Am. J. Physiol.* **231,** 810–816.

Beckman, A. L., Satinoff, E., and Stanton, T. L. (1976a). Characterization of midbrain component of the trigger for arousal from hibernation. *Am. J. Physiol.* **230,** 368–375.

Beckman, A. L., Stanton, T. L., and Satinoff, E. (1976b). Inhibition of the CNS trigger process for arousal from hibernation. *Am. J. Physiol.* **230,** 1018–1025.

Boulant, J. A., and Bignall, K. E. (1973). Determinants of hypothalamic neuronal thermosensitivity in ground squirrels and rats. *Am. J.Physiol.* **225**, 306-310.

Chatfield, P. O., and Lyman, C. P. (1954). Subcortical electrical activity in the golden hamster during arousal from hibernation. *Electroencephalogr. Clin. Neurophysiol.* **6**, 403-408.

Dawe, A. R. (1973). Autopharmacology of hibernation. In "The Pharmacology of Thermoregulation" (E. S. Schönbaum and P. Lomax, eds.), pp. 359-363. Karger, Basel.

Dawe, A. R. (1978). Hibernation trigger research updated. In "Strategies in Cold: Natural Torpidity and Thermogenesis" (L. C. H. Wang and J. W. Hudson, eds.), pp. 541-563. Academic Press, New York.

Dawe, A. R., and Spurrier, W. A. (1969). Hibernation induced in ground squirrels by blood transfusion. *Science* **163**, 298-299.

Dawe, A. R., and Spurrier, W. A. (1972). The blood-borne "trigger" for natural mammalian hibernation in the 13-lined ground squirrel and the woodchuck. *Cryobiology* **9**, 163-172.

Dawe, A. R., and Spurrier, W. A. (1974a). Summer hibernation of infant (six week old) 13-lined ground squirrels, *Citellus tridecemlineatus. Cryobiology* **11**, 33-43.

Dawe, A. R., and Spurrier, W. A. (1974b). Evidences for blood-borne substances which trigger or impede natural mammalian hibernation. In "Circannual Clocks: Annual Biological Rhythms" (E. T. Pengelley, ed.), pp. 165-196. Academic Press, New York.

Dawe, A. R., and Spurrier, W. A. (1975). Effects on cardiac tissue of serum derivatives from hibernators. In "Temperature Regulation and Drug Action" (J. Lomax, E. Schönbaum, and J. Jacob, eds.), pp. 209-217. Karger, Basel.

Dawe, A. R., Spurrier, W. A., and Armour, J. A. (1970). Summer hibernation induced by cryogenically preserved blood "trigger." *Science* **168**, 497-498.

Draskóczy, P. R., and Lyman, C. P. (1967). Turnover of catecholamines in active and hibernating ground squirrels. *J. Pharmacol. Exp. Ther.* **155**, 101-111.

Feist, D. D., and Galster, W. A. (1974). Changes in hypothalamic catecholamines and serotonin during hibernation and arousal in the arctic ground squirrel. *Comp. Biochem. Physiol. A* **48A**, 653-662.

Galster, W. A. (1978). Failure to initiate hibernation with blood from the hibernating arctic ground squirrel, *Citellus undulatus*, and eastern woodchuck, *Marmota monax. J. Therm. Biol.* **3**, 93.

Glotzbach, S. F., and Heller, H. C. (1975). CNS regulation of metabolic rate in the kangaroo rat *Dipodomys ingens. Am. J. Physiol.* **228**, 1880-1886.

Glotzbach, S. F., and Heller, H. C. (1976). Central nervous regulation of body temperature during sleep. *Science* **194**, 537-539.

Heller, H. C., Walker, J. M., Florant, G. L., Glotzbach, S. F., and Berger, R. J. (1978). Sleep and hibernation: Electrophysiological and thermoregulatory homologies. In "Strategies in Cold: Natural Torpidity and Thermogenesis" (L. C. H. Wang and J. W. Hudson, eds.), pp. 225-265. Academic Press, New York.

Horvath, A. (1881). Einfluss verschiedener Temperaturen auf die Winterschläfer. *Verh. Phys.-Med. Ges.* **15**, 187-219.

Lyman, C. P. (1948). The oxygen consumption and temperature regulation of hibernating hamsters. *J. Exp. Zool.* **109**, 55-78.

Lyman, C. P. (1954). Activity, food consumption and hoarding in hibernators. *J. Mammal.* **35**, 545-552.

Lyman, C. P. (1958). Oxygen consumption, body temperature and heart rate of woodchucks entering hibernation. *Am. J. Physiol.* **194**, 83-91.

Lyman, C. P., and O'Brien, R. C. (1963). Autonomic control of circulation during the hibernating cycle in ground squirrels. *J. Physiol. (London)* **168**, 477-499.

Lyman, C. P., and O'Brien, R. C. (1977). A laboratory study of the Turkish hamster *Mesocricetus brandti. Breviora, Mus. Comp. Zool. Harv.* No. 442.

Meeker, R. B., Myers, R. D., McCaleb, M. L., Ruwe, W. D., and Oeltgen, P. R. (1979). Suppression of feeding in the monkey by intravenous or cerebroventricular infusion of woodchuck hibernation trigger. *Physiologist* **22**(4), 86.

Minor, J. G., Bishop, D. A., and Badger, C. R., Jr. (1978). The golden hamster and the blood-borne hibernation trigger. *Cryobiology* **15**, 557–562.

Oeltgen, P. R., Bergmann, L. C., Spurrier, W. A., and Jones, S. B. (1978). Isolation of a hibernation inducing trigger(s) from the plasma of hibernating woodchucks. *Prep. Biochem.* **8**, 171–188.

Rapoport, S. I. (1976). "Blood-Brain Barrier in Physiology and Medicine." Raven, New York.

Rosser, S. P., and Bruce, D. S. (1978). Induction of summer hibernation in the 13-lined ground squirrel. *Cryobiology* **15**, 113–116.

Rothballer, A. B. (1959). The effects of catecholamines on the central nervous system. *Pharmacol. Rev.* **11**, 494–547.

Satinoff, E. (1970). Hibernation and the central nervous system. *Prog. Physiol. Psychol.* **3**, 201–236.

South, F. E., Breazile, J. E., Dellmann, H. D., and Epperly, A. D. (1969). Sleep, hibernation and hypothermia in the yellow-bellied marmot (*M. flaviventris*). *In* "Depressed Metabolism" (X. J. Musacchia and J. F. Saunders, eds.), pp. 277–312. Am. Elsevier, New York.

Steiner, M., and Folk, G. E., Jr. (1978). Spontaneous and induced summer hibernation in 13-lined ground squirrels. *Cryobiology* **15**, 488–491.

Twente, J. W., and Twente, J. A. (1968a). Progressive irritability of hibernating *Citellus lateralis*. *Comp. Biochem. Physiol.* **25**, 467–474.

Twente, J. W., and Twente, J. A. (1968b). Effects of epinephrine upon progressive irritability of hibernating *Citellus lateralis*. *Comp. Biochem. Physiol.* **25**, 475–483.

Twente, J. W., and Twente, J. (1978). Autonomic regulation of hibernation by *Citellus* and *Eptesicus*. *In* "Strategies in Cold: Natural Torpidity and Thermogenesis" (L. C. H. Wang and J. W. Hudson, eds.), pp. 327–373. Academic Press, New York.

Twente, J. W., Cline, W. H., Jr., and Twente, J. A. (1970). Distribution of epinephrine and norepinephrine in the brain of *Citellus lateralis* during the hibernating cycle. *Comp. Gen. Pharmacol.* **1**, 47–53.

Uuspää, V. J. (1963). The 5-hydroxytryptamine content of the brain and some other organs of the hedgehog (*Erinaceus europaeus*) during activity and hibernation. *Experientia* **19**, 156–158.

Walker, J. M., Glotzbach, S. F., Berger, R. J., and Heller, H. C. (1977). Sleep and hibernation in ground squirrels (*Citellus* spp): electrophysiological observations. *Am. J. Physiol.* **233**, R213–R221.

Walker, J. M., Garber, A., Berger, R. J., and Heller, H. C. (1979). Sleep and estivation (shallow torpor): continuous processes of energy conservation. *Science* **204**, 1098–1100.

Weil-Malherbe, H., Whitby, L. G., and Axelrod, J. (1961). The blood-brain barrier for catecholamines in different regions of the brain. *Reg. Neurochem.; Reg. Chem., Physiol. Pharmacol. Nerv. Syst., Proc. Int. Neurochem. Symp., 4th, Varenna, Italy, 1960*, pp. 284–292.

Wünnenberg, W., Merker, G., and Speulda, E. (1978). Thermosensitivity of preoptic neurons and hypothalamic integrative function in hibernators and nonhibernators. *In* "Strategies in Cold: Natural Torpidity and Thermogenesis" (L. C. H. Wang and J. W. Hudson, eds.), pp. 267–297. Academic Press, New York.

Index

PHYSIOLOGICAL ECOLOGY

A Series of Monographs, Texts, and Treatises

EDITED BY

T. T. KOZLOWSKI

University of Wisconsin
Madison, Wisconsin